A Comparative Geography of China and the U.S.

GeoJournal Library

Volume 109

For further volumes:
http://www.springer.com/series/6007

Rudi Hartmann • Jing'ai Wang
Editors

Tao Ye
Associate Editor

A Comparative Geography of China and the U.S.

Springer

Editors
Rudi Hartmann
Department of Geography
and Environmental Sciences
University of Colorado Denver
Denver
Colorado
USA

Jing'ai Wang
School of Geography
Beijing Normal University
Haidian District
Beijing
China

Associate Editor
Tao Ye
State Key Laboratory of Earth Surface
Processes and Resource Ecology
Beijing Normal University
Haidian District
Beijing
China

ISSN 0924-5499 ISSN 2215-0072 (electronic)
ISBN 978-94-017-7805-3 ISBN 978-94-017-8792-5 (eBook)
DOI 10.1007/978-94-017-8792-5
Springer Dordrecht Heidelberg New York London

Printed on acid-free paper

Springer is part of Springer Science+Business Media (www.springer.com)

Preface

The People's Republic of China and the United States of America are two countries with outstanding qualities and features. Both countries, in a mid-latitude location, have a vast territory as well as large populations. The U.S. and China represent world powers which exert significant influence globally. As permanent members of the United Nations Security Council, they have important roles in international affairs. China is an ancient oriental civilization with tremendous historic depth, while the U.S. is a modern western society founded as an independent state just little over two hundred years ago. The U.S. and China are the two leading economies in the world. Despite rapid economic growth over the past three decades China has retained some characteristics of a developing country as expressed, for instance, in a moderately high GDP per capita by international comparison. The U.S. has been the economy with the highest total amount of produced goods and services, and the country's GDP per capita is also markedly high.

Exploring the geographical similarities and differences between the two countries, especially in terms of their populations, resources, cultures and economies, helps to provide more insights into the human-environment relations in both countries as well as to enhance and broaden the mutual understanding and common ground between them. Moreover, such comparative perspectives on the geographies of China and the U.S. could be valuable for addressing and reaching the goal of global sustainable development.

An indispensible component of regional geography is a focus on a home area, state or country with a mission to further national geographic education. Comparative research on the geographies of China and the U.S. contributes to a better understanding of the differences and similarities of both countries. A comparative study of the two countries has become—in the wider context of global change and globalization—an important teaching content and research method in regional geography.

The comparative approach dates back to the earliest stages of geography in the fourth and third centuries B.C. It re-emerged as a viable perspective in the foundation of modern geography in the 17th, 18th and 19th centuries. Organizing and compiling relevant teaching materials about the similarities and the differences in the geographies of China and the U.S. can be the basis for providing scientific and objective insights into the national conditions for the two countries' citizens, for promoting spatial and geographic thinking, for enhancing an understanding of the other country's cultural values and ultimately for a sound development of the human-environment relations in China and the U.S. Thus, this book is a new attempt in the field of comparative geography, with potentially great relevance in the field of geographic education.

The main thread of the book is the theme of human-environment interaction, with core contents on economic and regional sustainable development. The major approach employed throughout the book is that of a regional comparative study of China and the U.S. The book is organized in four parts. Part I includes three overview chapters: *Introduction*, *Physical Geography* and *Population & Ethnic Geographies*. Part II consists of four thematic chapters: *Agriculture and Food Production*, *Economic Geography*, *International Trade Issues and Status*, and *Mega-Regions of China and the U.S.* Part III focuses on regional comparisons in three chapters: a comparison of *Regional Urban Economic Clusters*, a comparison of four metropolitan areas in China and the U.S. which have seen rapid *Urbanization & Urban Sprawl*, and the *Main Agricultural Regions* of China and the U.S. in comparison. In Part IV, the concluding chapter, we finally pay particular attention to *Research Cooperation between Chinese and American Geographers* and its significance in the *Quest for Sustainable Development.*

A Comparative Geography of China and the U.S. is a book project that invited 20 geographers from American and Chinese Universities to collaborate on a Sino-U.S. regional geography textbook. In each chapter, American and Chinese authors joined in the task of a given theme in the regional comparative analysis of China and the U.S. The eleven chapters were completed in co-authorship by the following teams: Chapter 1: Rudi Hartmann and Jing'ai Wang; Chapter 2: Jing'ai Wang, Honglin Xiao, Rudi Hartmann and Yaojie Yue; Chapter 3: Lucius Hallet, Jing'ai Wang and Rudi Hartmann; Chapter 4: Mark Leipnik, Yun Su, Robert Lane and Xinyue Ye; Chapter 5: Hongmian Gong and Huasheng Zhu; Chapter 6: Gregory Veeck and Yuejing Ge; Chapter 7: Russell M. Smith, Yuejing Ge, Rudi Hartmann, Xiaping Dong and Yang Cheng; Chapter 8: Susan Walcott and Huasheng Zhu; Chapter 9: Tao Ye, Brian Muller and Peijun Shi; Chapter 10: Mark Leipnik, Yun Su and Xinyue Ye; Chapter 11: Peijun Shi, Clifton Pannell and Tao Ye. Rudi Hartmann and Jing'ai Wang designed the outline of the book and co-organized the Chinese-American team work in the completion of the book. While Rudi Hartmann had the main responsibility in the final editing of the texts, Jing'ai Wang took on the task of designing the layout of the maps and figures. Peijun Shi reviewed the book several times and provided useful comments and suggestions. Tao Ye was responsible for communication and coordination as well as for the formatting of the draft manuscript. Fang Lian reproduced most of the maps in GIS operations.

The outcome of the book is the result of an innovative team effort in establishing the foundations of a new comparative geography of China and the U.S. Due to the huge amount of information and data presented in this volume errors and mistakes may have occurred as well as imperfections in the layout of the book. The editors and chapter authors look forward to comments and suggestions from the readers.

October 2013

Jing'ai Wang
Rudi Hartmann

Acknowledgements

We would like to thank our universities, schools and departments, Beijing Normal University, School of Geography, and the University of Colorado Denver, Department of Geography and Environmental Sciences, for their encouragement and help in the completion of the book project.

In the process of editing this book, the Beijing Normal University Regional Geography National Teaching Team provided strong financial support. Other agencies and units, most notably the Chinese Universities "985" Project, the State Key Laboratory of Earth Surface Processes and Resource Ecology and the Key Laboratory of Regional Geography of Beijing Normal University gave additional financial support to the project and assisted the Beijing Normal University Team in many other ways.

Special thanks go to the students in the Key Laboratory of Regional Geographical Research at Beijing Normal University who served as secretaries of the book project: Jing Chen, Lu Hao, Jiansong Zhang, Yongdeng Lei, Yuanyuan Yin, Sheng Chang, Weixia Yin and Fang Lian.

The support received from the University of Colorado Denver, College of Liberal Arts and Sciences, for the initial development of the project, in form of a grant "Teaching the Geography of China" is greatly appreciated.

Thanks also go to several colleagues of the AAG China Geography Specialty Group, in particular to Gregory Veeck, Western Michigan University, Clifton Pannell, University of Georgia, and Susan Walcott, University of North Carolina at Greensboro, who were supportive in the early efforts of addressing the potential of a geographic education agenda for China on the Higher Education level in the U.S, China and other academic settings abroad. A first joint result was a panel session "Teaching the Geography of China" held at the AAG Meeting 2008 in Boston. Two special sessions focusing more directly on a Comparative Geography of China and the U.S. followed in 2010, at the occasion of the Washington, D.C. AAG Meeting.

We would also like to thank our spouses and families for their longtime support during the many years of work and travel relating to the book project.

Finally, our thanks go to Evelien Bakker and Bernadette Deelen-Mans, as well as Stefan Einarson and Mireille van Kan from the Dordrecht office of Springer Press who strongly and consistently supported the book project over the years.

Beijing and Denver
October 2013

Jing'ai Wang
Rudi Hartmann

Contents

About the Editors

Rudi Hartmann is an Associate Professor (C/T) at the University of Colorado Denver where he has taught geography and tourism planning since 1992. He received his Ph.D. from the Technical University Munich, Germany in 1983. His interests include geographic education, the geography of tourism and the history of geographic thought. His regional specialties are China, Europe and the American West. He has visited or taught at Clark University, Worcester, Massachusetts, Lund University, Sweden, University of California, Berkeley, Simon Fraser University, Burnaby, British Columbia, the University of Colorado, Boulder, and Beijing Normal University.

Jing'ai Wang is Professor at the School of Geography, Beijing Normal University, the winner of the national laureate for teaching, the leader of the Regional Geographical Education Team of China, the chair of the board of the "China Geography" Teaching Research Society, deputy director of the Geographical Education Steering Committee of the Ministry of Education of China, co-chair of the board of directors of the Society for Geography of Beijing, and the Director of the Key Laboratory of Regional Geographical Research of Beijing Normal University. For several decades, she has been actively engaged in research and teaching in the fields of geographic education, regional geography, natural areas assessment and natural disaster mapping. She was also named project leader of several national programs, National "863" Program, National Natural Science Foundation Project, National "973" Program, and National Key Technology R&D Program of China. She has published and edited many textbooks including "China Geography", "China Geographical Atlas" and "The Spatial-Temporal Patterns of Natural Disaster in China" as well as written over 200 research papers. Among the classes she teaches is "China Geography", a nationally recognized course. She is the winner of the First-level Award of National Teaching Achievements, the Second-level Award of National Teaching Achievements, the First Award of Science and Technology Progress of the Ministry of Education, China.

About the Authors

Yang Cheng is a Lecturer in the School of Geography at Beijing Normal University. Her research interest is in Health Geography, focusing on the application of qualitative and quantitative methods to the study of health, place, and health care.

Xiaoping Dong is Professor of Folkloristic Studies in the School of Chinese Language and Literature at Beijing Normal University. She is the Director of the Research Institute for Chinese Folk Culture. Dr. Dong' s research focuses on the study of Modern Folkloristic in a Globalizing World, Theory of Folkloristic, Folklore Fieldwork and Digital Folkloristic.

Yuejing Ge is Professor at Beijing Normal University. As Professor in the School of Geography, she mainly teaches World Regional Geography, Human Geography and Economic Geography.

Hongmian Gong is an Associate Professor in the Department of Geography at Hunter College of the City University of New York. Her research focuses on Urban Geography, GIS, and Transportation. The courses she teaches mainly include Urban Geographic Theory, Urban Applications of GIS, Geography of East Asia, Geography of China, and World Regional Geography.

Lucius Hallett IV is an Associate Professor in the Department of Geography at Western Michigan University. His teaching focuses on Human Geography courses and the Travel and Tourism Major requirements. His research interests lie in Agricultural Tourism in China, Japan, England and the state of Michigan. He is currently involved in numerous studies on the apple production and hard apple cider possibilities of south-western Michigan.

Robert A. Lane is a Professor in the Department of Agricultural and Industrial Sciences at Sam Houston State University. His research interests include forage crop quality and management, nitrogen fixation in legumes, persistence of forage legumes in pastures, natural and synthetic fertilizers for forage and turf application and cultivar evaluation and adaption.

Mark R. Leipnik is a Professor of Geography in the Department of Geography and Geology at Sam Houston State University in Huntsville, Texas. He earned his PhD in Geography in 1995 from the University of California at Santa Barbara. He has been a professor for 18 years and prior to that he was an environmental scientist with the United States Government Department of Interior in the Bureau of Reclamation. His expertise is in applications of geographic information systems particularly in marketing, law enforcement and natural resources. He is author of two books and over 20 book chapters and numerous journal articles on geospatial technologies.

Brian Muller is an Associate Professor in the Environmental Design program at the University of Colorado Boulder. His research interests concern land use and environmental policy, spatial analysis and urban growth dynamics.

Clifton W. Pannell is Emeritus Professor of Geography at the University of Georgia where he taught human geography, regional courses on East and Southeast Asia, and urbanization in developing countries. His research focuses on the changing economic and urban geography of China. He has taught and conducted research in Hong Kong, Taiwan, and mainland China. Dr. Pannell also served as Associate Dean of the Franklin College of Arts & Sciences, where he was responsible for international and interdisciplinary programs, and was formerly Director of the Center for Asian Studies. He serves on the editorial boards of Eurasian Geography & Economics and the Asian Geographer and is a member of the China Research Center in Atlanta, GA.

Peijun Shi is Professor at Academy of Disaster Reduction and Emergency Management, the Executive Vice-President of Beijing Normal University and the Director of the State Key Laboratory of Earth Surface Processes and Resource Ecology. His research focuses on natural disaster theories and risk governance. He teaches The Principle of Natural Disaster Theory, Introduction to Resources Science and many other courses in natural hazards and disasters.

Russell M. Smith is an Associate Professor of Geography at Winston-Salem State University with research interests in urban and political geography. His specific research interests include local government boundary change and a variety of urban planning issues including urban sustainability and megaregions. The main courses he teaches include Introduction to Geography, Geography of North America, Environmental Geography, Introduction to Geographic Information Systems, Urban Geography, and World Economic Geography.

Yun Su is Associate Professor in the School of Geography at Beijing Normal University with research interests in Risk Assessment & Response to Natural Disasters. The courses she teaches include Fundamental Topics of Regional Geography and Geography of China.

Gregory Veeck is Professor in the Department of Geography at Western Michigan University specializing in economic geography, agriculture, rural development, and rural environmental/ecological issues. His teaching interests are East Asia, Economic Geography, Ecology and Environment. His research focuses on Agriculture in China, Korea, and Japan, Rural Economic Development in East Asia and Rural Environmental Issues and Small-Scale Farmers in East Asia and the United States.

Susan Walcott is Professor in the Department of Geography at the University of North Carolina, Greensboro. She is an urban-economic geographer specializing in regional economic development, both in the U.S. and abroad. She has conducted extensive research on the Special Economic Zones and high-tech parks in China, publishing articles on Shanghai, Xi'an, Chengdu, and Chongqing. She has lived in Taiwan, traveled extensively in Asia, and speaks Mandarin Chinese.

Honglin Xiao is an Associate Professor in the Department of History and Geography at Elon University with research interests in applications of GIS and remote sensing in physical geography and environmental studies, human-environment interactions, environmental change, and quantitative methods.

Tao Ye is an Associate Professor in the State Key Laboratory of Earth Surface Processes and Resource Ecology at Beijing Normal University. His research focuses on Disaster Economics and Financial Management of Natural Disaster Risk. He teaches a bilingual course on Disaster Economics.

Xinyue Ye is Founding Director of the Computational Social Science Lab and an Assistant Professor of GIS and Economic Geography at the Department of Geography, Kent State University. His research addresses questions in economic inequality and its dynamics across multi scales and various dimensions (statistical, spatial, and temporal) through the design and implementation of new visualization and analytical methods.

Yaojie Yue is a Lecturer at Beijing Normal University. He teaches Introduction to GIS, Pedogeography and Introduction to Cadastral Sciences in the School of Geography. His research focuses on Land Use and Ecological Security, Integrated Risk Evaluation on Natural Disaster as well as Application of "3S" Technology.

Huasheng Zhu is an Associate Professor at the School of Geography, Beijing Normal University. He initially started his professional career as an economic geographer. The main courses he teaches include Foundation of Economics, Corporate Geography and Fieldwork of Human Geography. His academic interests are in the study of Regional Clusters and Industrial Upgrading, Entrepreneurship, Innovation and Local Development, Urban Regeneration and a New Industrial Geography.

Chapter 1
Introduction to A Comparative Geography of China and the U.S.

Rudi Hartmann and Jing'ai Wang

The People's Republic of China (see Fig. 1.1) and the United States of America (see Fig. 1.2) are two nations evoking a multitude of associations, both good and bad, for many people around the world. Both countries have fostered distinct national identities and pride within their own territories; both realms have, lightly and not so lightly, exerted powerful influences beyond their borders which have resulted in pivotal roles throughout the world. All these interactions have created complex forms of interdependence leading to extensive friendly cultural exchanges and mutually beneficial trade relations but have also resulted in fears among some citizens in neighboring countries of a continued Americanization and Sini-cization. In short, 'America' and 'China' represent cultures, societies, economies and geographical regions which matter in our world.

This text book offers a comparative assessment of both countries' diverse geographies. While not comprehensive, it explores important aspects and dimensions of the two countries' human and physical features from the perspective of professional geographers from both nations. A comparison of the two countries invites both cross-cultural and inter-cultural perspectives, with the goal of providing more complete, multi-dimensional perspectives for these two important nations. For instance, our chapter providing a comparative look at agriculture and food production in China and the U.S. not only looks at the types of foods produced in particular regions, but also examines the increasing exchanges of food products and food consumption practices between the two countries. For example, shrimp caught in the South China Sea or raised in ponds along the Yangtze River are shipped to Dim Sum restaurants in New Orleans (U.S.), while 'American' fast food items like hamburgers are served in McDonalds in Beijing.

R. Hartmann (✉)
Department of Geography and Environmental Sciences,
University of Colorado Denver, Denver, Colorado, USA
e-mail: Rudi.Hartmann@ucdenver.edu

J. Wang
School of Geography, Beijing Normal University, Haidian District, Beijing, China
e-mail: jwang@bnu.edu.cn

R. Hartmann, J. Wang (eds.), *A Comparative Geography of China and the U.S.*,
GeoJournal Library 109, DOI 10.1007/978-94-017-8792-5_1,

Fig. 1.1 The People's Republic of China

Prior to a broader summary of the project, two fundamental questions should be addressed, to dispel concerns regarding the usefulness or appropriateness of the comparative approach we have chosen for the text:

Why not focus on just one of two nations—given that both nations are so complex and full of intricacies that understanding even one of these nations present quite a challenge?

Why select only two geographies, those of China and of the U.S., from the many—given that globalization has essentially affected the whole world?

The first question is rooted in the ideographic tradition of regional geography: all regions are unique. As a region's 'distinct character' is carved out and/or assembled

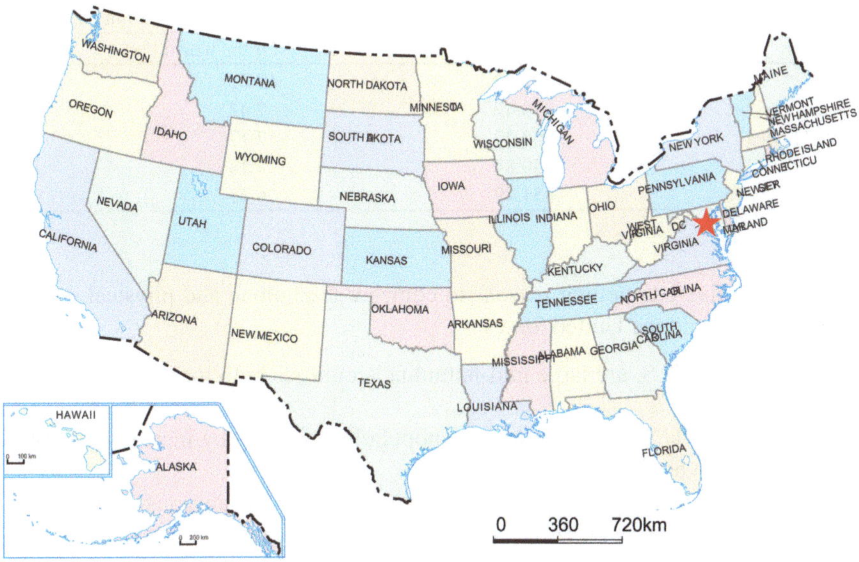

Fig. 1.2 The United States of America

to a 'regional portrait', scholars have not been encouraged to formulate generalities across neighboring or distant regions. This practice contradicts the reasonable interests of both the regional expert and the lay person who wish to see regions placed into broader contexts: "In which way is this region different from my home region? Why does this region have certain features or qualities I see in other regions represented as well?" The authors of this text believe that regional geographies can be successfully developed, while raising new types of questions, through the use of a comparative framework.

The second concern questions the common practice of regional geography in an era of globalization: Is it still valid to deal with the geographies of one or two countries to explain current changes, e.g. of resource uses in one country or region? Obviously, not only are the economies of China and the United States increasingly connected, rather all national economies worldwide have grown more and more "enmeshed" through the forces of globalization (Dicken 2007). Free market proponents of contemporary globalization ("hyper-globalists") go a step further; they argue that the power(s) of the state have been significantly eroded in the past decades and will (eventually) fade altogether. By contrast, the authors of this text book argue that the power(s) of the state continue to be central to almost all economies, and state actions continue to substantially shape the geographies of China and the United States. As both the economies and societies of these two great nations are among the more important agents of change in the world, a regional analysis of these two 'major players' may help promote a better under-standing of these processes throughout the world.

Table 1.1 Top five countries by land area size. (National Bureau of Statistics of China 2012)

Countries	Land area (km^2)	Of the world (%)
Russia	1709.82	12.72
Canada	998.47	7.43
United States	963.20	7.17
China	959.81	7.14
Brazil	851.49	6.33

There are five compelling reasons to compare the human and physical geographies of China and the United States:

1. China and the U.S. are large mid-latitudinal countries both located in the northern hemisphere
2. China and the U.S. are home to large populations marked by increasing cultural and ethnic diversity
3. China and the U.S. represent cultures with important contributions to world civilization (Eastern and Western traditions of world civilization)
4. China and the U.S. are leading economies in the Pacific realm and worldwide
5. China and the U.S. have significant political power in the world

1.1 Geographical Positions on Earth

The area size of the People's Republic of China and the United States of America are very comparable at about 9.6 million km^2 (or 3.6–3.7 million square miles). Depending on which claims one uses for both nations the two nations rank third and fourth after Russia and Canada (see Table 1.1). China and the U.S. each occupy approximately 6.4–6.5 % of the world's total land area (Fig. 1.3).

Located in East Asia, on the western shore of the Pacific Ocean, China has an extensive coastline home to much of the nation's population. From north to south, the national territory of the People's Republic of China extends some 5,500 km stretching from the center of the Heilongjiang River north of the town of Mohe (latitude 53°30' N) to the Zengmu Reef at the southernmost tip of the Nansha Islands (latitude 4°N). From west to east, the nation extends about 5,200 km from the Pamirs (longitude 73°40' E) to the confluence of the Heilongjiang River and Wusuli River (longitude 135°05' E). China share land borders with 14 nations extending over 22,800 km in length, in part due to the mountainous terrain that composes many of the borders. Moving from northeast to southeast, the nations include: Korea to the east; the People's Republic of Mongolia to the north; Russia to the northeast; Kazakhstan, Kirgizstan and Tajikistan to the northwest; Afghanistan, Pakistan, India, Nepal and Bhutan to the west and southwest; and Vietnam, Laos and Myanmar to the south. Across the seas to the east and southeast are the Republic of Korea, Japan, the Philippines, Brunei, Malaysia and Indonesia. As noted above, many of these borders emerged along natural barriers formed by deserts and high rugged

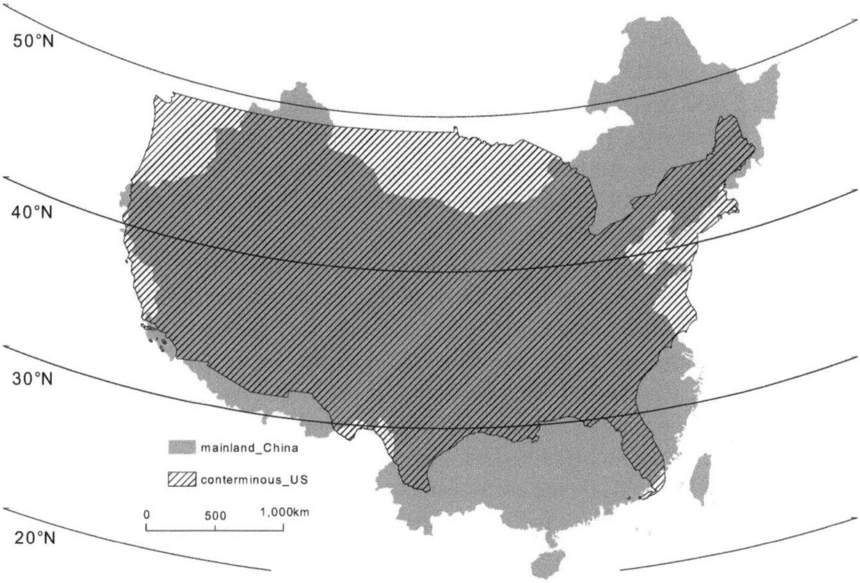

Fig. 1.3 Comparison of area size and latitudinal range of mainland China and the conterminous U.S

mountains. The North China Plain in eastern China is the largest area of lowland in the world. Tibet in western China is called "The Roof of the World", basically a plateau with a mean elevation of over 4,000 m (13,000 ft) surrounded by even higher mountains. To the south are the Himalayans, while to the northwest are the Kunlun Mountain, Tianshan Mountain and Altai Mountain. Eastern China has also mountains and hilly areas but much of the eastern regions are at a much lower altitudes than those of western China. The most important East-West running mountains in Eastern China are the Qinling Ranges commonly considered the natural boundary between North China and South China.

The United States of America are located in the western hemisphere with three major coastlines, with the Atlantic Ocean and the Gulf of Mexico to the east and southeast and the Pacific Ocean to the West. The U.S. shares borders with only two nations: Canada to the north and Mexico to the south. If all the widespread U.S. territories are taken into account including inhabited and uninhabited coastal areas in the Pacific and the Caribbean Oceans the United States is one of the world's most extensive cross-latitude countries of the world: from Point Udall, St. Croix in the U.S. Virgin Islands in the Caribbean as most Eastern point (64°34' W) to Orote Point, Guam (144°37' E) as most western point in the Pacific Ocean. The East-West extension of its fifty states (48 contiguous states of the U.S. plus Alaska and Hawaii) alone still makes up an overall remarkable length of close to 10,000 km (or 6,000 miles), from Peaked Island, the most western Aleutian Island, Alaska (172°26' E) to Sail Rock, just offshore of West Quoddy Head, Maine (66°56' W).

The South–North extension of the U.S. is—compared to China—less pronounced; it is largest between Alaska (most northern point: Point Barrow at 71°23') and Hawaii (most southern point: La Lae at 18°54') excluding the Rose Atoll, uninhabited but still a part of official U.S. territory. The North-South extension within the 48 contiguous United States is even smaller than that including Alaska and Hawaii: only about 2,900 km (or 1,800 miles) from 49°23' (Lake of the Woods, Minnesota) to 24°31' (Ballast Key, Florida). Most of the mountain ranges in the U.S. run north to south including the Appalachians, the Rocky Mountains, the Sierra Nevada and the Coastal Mountains/Cascade Range of California. There are no major mountain ranges running east to west, though.

The U.S. borders with its main neighbors Canada and Mexico are very different in nature; whereas the border with Canada is generally considered the longest peaceful borderline in the world (Mayda 2010) the U.S. border with Mexico has become one of the most contentious and violent border zones in recent history. China's official territorial boundaries are disputed in a number of places by other Asian countries although the current situation is peaceful and negotiations with these nations are ongoing.

The large land area incorporated within both countries as well as the implications of their general locations on the earth seems to suggest the benefits of a comparison of their physical geographies including climates:

1) China and the U.S. are located in the northern hemisphere, within a similar latitudinal range.
2) Mainland China and the continental U.S. are situated within a larger continent, China within Asia and the U.S. within North America. For both nations, the southeastern coasts adjacent to oceans have proven to be vital economic engines for the rest of the nation. A major difference is that the U.S. has two coasts, with the Atlantic Ocean in the east and the Pacific Ocean in the west; the western sections of China surrounded by the dry steppe and deserts of Central Asia.
3) Both countries have a wide East-West span of their respective territory, with a time difference of over four hours (or longitudinal difference resulting in a natural time difference).
4) The climates of the two countries are both heavily influenced by the presence of large mountain ranges and major river basins, including the Himalayans and the Rocky Mountains, the Yangtze River Basin and Yellow River Basin, the Mississippi-Missouri River Basin.
5) The interior sections of the two countries have similar continental climates, dry and cold in the winter and hot in the summer. In turn, the coastal areas share similar warm and humid weather conditions with abundant precipitation.

Many of the differences that do exist in climatic conditions at the national level are simply due to the larger tropical zone in South China—a zone only to be found in the U.S. in the most extreme southern portion of Florida and on the Hawaiian Islands. The northern and northwestern sections of Alaska close to and beyond the Arctic Circle with tundra vegetation and arctic climates are only found in China on the frigid Qinghai-Tibetan Plateau.

Both countries are marked by a wide range of distinct topographic regions. What is common to these individual topographical features for each country, though, is that both nations have extensive areas of all four major types of land-forms: plains, plateaus, hills and mountains.

The topography of China is high in the west and low in the east displaying a typical stair-step pattern moving to the eastern coastal lowlands. Within the eastern Eurasian continental slope, China is still a very mountainous country with 40% of the land classified as either hilly or mountainous. The Chinese territory descends in three great macro steps from west to east. The first step is the Qinghai-Tibetan Plateau which is the world's highest plateau at an average altitude of 4,500 m (above 14,000 ft). The altitude of the second step is 1,000–3,000 m; and the third step ranges from low mountains in Fujian from about 1,000 m to sea level. There are four major basins in China, namely the Tarim Basin, Junggar Basin, Qaidam Basin and the Sichuan Basin. China has also four major plateaus: the Qinghai-Tibetan Plateau, the Inner Mongolian Plateau, the Loess Plateau and the Yunnan-Guizhou Plateau. Last but not least, China has three great plains: the North China Plain, the Northeast Plain and the Lower Yangtze River Plain. This complex mosaic of mountains, basins, plateaus and plains contribute to distinct hydro-thermal conditions which in turn result in complex patterns of soils and vegetation. China's vast physical differences as discussed above result in complex and diverse geographical landscapes which have been further transformed by the actions of humans for more than 10,000 years.

The continental United States has plains, plateaus, hills and mountains as well ranging from west to east include: the Coastal Ranges of California, the Cascade Mountains, the high Sierra Mountains, the Rocky Mountains and in the eastern United States, diverse lower systems including the White Mountains, the Berkshires, the Alleghenies, the Blue Ridge Mountains and others on the Appalachian Plateau. Hilly areas are widespread in the central U.S., but again there are no large mountain systems running east to west, and few true mountains between the Alleghenies and the Front Range in Colorado. There are several distinct plateaus such as the Columbia Plateau, the Colorado Plateau and the Ozark Plateau. Finally, there are interior lowlands largely covering the Midwestern states of the U.S. as well as the expansive Gulf Atlantic Coastal Plain reaching from Cape Cod in Massachusetts and Long Island in the north to the Texan/Mexican Gulf Coast in the south central portion of the nation.

The similarities between the two countries' basic physical geographies are remarkable and result, in many instances, in similar patterns of population distribution and land use. However, several significant differences can be detected as well and these differences, when appropriate, will be examined more thoroughly in later chapters of the text.

1.2 Population Ranking in the World

China and the U.S. are both populous countries. Their populations rank first and third in the world (see Fig. 1.4): 1,343,239,923 and 313,847,465. With respect to population the second largest country is India, while the fourth and fifth largest populations are found in Indonesia and Brazil.

The populations of both China and the United States are unevenly distributed. In China, the highest population concentrations are found on the southeastern coast where 43% of the total population lives on 13.3% of the nation's total land area.

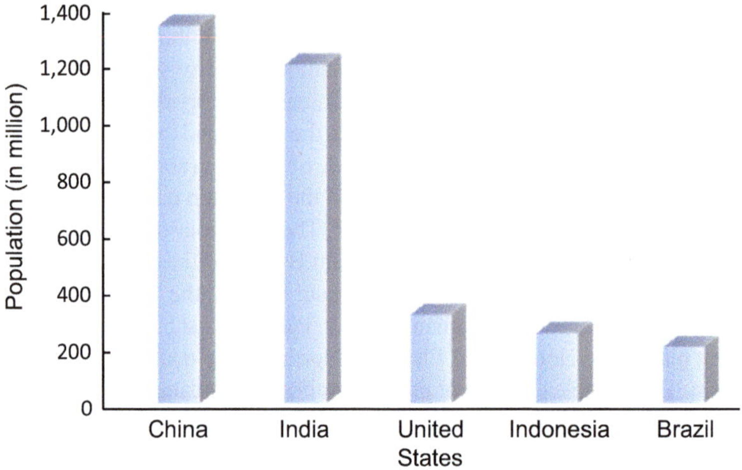

Fig. 1.4. Top five countries by population volume. (U.S. Census Bureau 2012)

If we divide China by the Huhuangyong line (running from Heihe in Heilongjiang Province to Tengchong in Yunnan Province) 95.4 % live east from the line, 4.6 % west of the line. There are great contrasts in population densities among those areas in China which are highly urbanized and economically more advanced such as the Pearl River Delta, the Yangtze River Delta and the Beijing-Tianjin-Hebei regional cluster and those areas that are more isolated and as a consequence less developed such as many rural areas within the Qinghai-Tibetan Plateau, the Loess Plateau and some of the interior basin regions.

In the U.S. population distribution and density is also highly variable. Whereas the original 13 states of the Union along the eastern coast and some of the earlier Midwestern states to the join the union contain higher population concentrations, the younger states in the Rocky Mountain West, for example, continue to have lower population densities despite recent growth spurs. Again, as in China, there are marked differences in the population densities among the heavily urbanized areas, such as the eastern 'Megalopolis' (Bosh-Wash city corridor along the eastern seaboard), the San Francisco Bay Area/Silicon Valley and the Southern California urban sprawl areas, and the rural areas of the Great Plains and the Mountain states, such as in Kansas/Nebraska, North/South Dakota and Idaho/Western Washington. The lowest population densities within the U.S. are found in Alaska, the largest U.S. state, but located in an extreme northern location.

Migration within the U.S. has contributed to significant population change in the past half-century, with rapidly growing areas since 1960 found particularly found in the West and South census regions of the country (California, Texas, Florida), at the expense of the older manufacturing or Rust Belt regions (Pennsylvania, Ohio, Michigan).

The ethnic composition of the populations of China is various, while the ethnic and racial compositions of the populations of the U.S. are both very complex, although, of course, there are also distinct differences between the two nations. Who are the 'Chinese' and who are the 'Americans'?

Officially, there are 56 recognized ethnic groups or nationalities in China. The Han people makes up China's (as well as the world's) largest ethnic group, with about 92 % of the population of the People's Republic of China (or about 1.2 billion persons). The 55 remaining ethnic minorities include the following with more than 1 million persons: the Zhuang, Manchu, Hui, Miao, Uygur, Yi, Tujia, Tibet-an, Mongolian, Bouyei, Dong, Yao, Korean, Bai, Hani, Li, Kazah and Dai. The largest group is the Zhuang nationality with more than 15 million persons. While considerably different in the past, at the present time ethnic minorities live mostly in areas that are less densely populated (such as in Inner Mongolia), areas at higher elevations (such as the Tibetans in Tibet, Qinghai, Sichuan) or in border regions (such as the Uygurs in Xinjiang). In times of war, migration and during other major political and economic events various ethnic minorities mingled and some nationalities are now widely scattered all over the country like the Manchu nationality—once concentrated only in Northeast China. Subsequently, some of the 55 ethnic minority groups have assimilated—more or less—to what might be called Han culture although there have been also efforts made to revive their cultural traditions. It is evident and widely recognized that every ethnic group has rich traditions which often go back many centuries; the folklore if many of the minority groups continues to be celebrated in the People's Republic of China in many ways. In five cases, the status of selected nationalities has led to the foundation of autonomous regions: the Guangxi Zhuang Autonomous Region, the Inner Mongolia Autonomous Region, the Ningxia Hui Autonomous Region, the Xinjiang Uygur Autonomous Region and the Xizang Autonomous Region (Tibet). The largest non-Han proportion of the population is found in Guangxi (about 17 million or 38 % of the total population). In two of the autonomous regions the minority populations surpass the Han populations: the Tibetans in Xizang and the Uygurs in Xinjiang. What unites all the various ethnic populations of China is the fact that more than 99 % are people indigenous people to the territory of the People's Republic of China—whereas in the United States this proportion is reversed: 1 % of the population are "indigenous" to U.S. territory, while 99 % claim ancestry from people who lived originally outside the territory of the U.S.

The United States of America has seen tremendous changes regarding its population's ethnic and cultural composition during the country's comparatively short history. While the dominant culture group in the newly founded United States of America (1776, with a first census 1790) has often been described as WASP (White Anglo Saxon Protestants), current trends in the structure of the U.S. population point to an evolving multi-cultural society. An expression of these ongoing complex changes is the presidency of Barack Obama, himself the son of a father from Africa and a mother with European ancestry, with leading members of his administration comprising most of the ethnic and culture groups in the United States.

In the last few decades, the percentage of non-Hispanic whites in the U.S. population has been gradually and consistently decreasing, from about 80% in 1980 to about 69% of the population (2000 Census figure, see Perez and Hirschman 2009). There are many regions of the U.S. including the L.A./Southern California Metropolitan Area where non-Hispanic whites no longer constitute the majority of the population but are simply another minority. The most rapidly growing population groups in the United States include Asian-Americans and Hispanics. In overall proportions, Hispanics have surpassed Americans in African descent (as of 2003) as the largest minority group. Hispanics' proportion the U.S. population are currently estimated at about 15%, while the proportion of African-Americans has remained at 13%. Americans with Asian descent including those of Chinese descent make up about 4% of the U.S. population. The proportion of Native Americans (or Indigenous Americans, or simply 'Indians') has remained unchanged over the past decade at 0.9% (or at 1.3%, when multiple ancestry is included) of the U.S. population. Many Native Americans continue to live on reservation lands since they were relocated in the eighteenth and nineteenth centuries.

America has been and remains an immigration country. The various immigration waves over the past 200 years reflect a very complex ethnic and cultural population structure. Whereas immigrants from Northern and Western Europe, from countries such as England, Scotland, Ireland, the Netherlands, France, Germany, Austria, Switzerland, Sweden, Norway and Denmark, made up the large majority of new arrivals to the country during the first 100 years (1790–1890), more and more immigrants to the U.S. 1890–1920 had ancestry from Southern and Eastern Europe. During this time period a large number of Greek, Italian, Polish and Russian immigrants made the U.S. their home. After a period with low immigration rates 1930–1950 (Great Depression, WWII) an increasing number of people started to come from Latin America, the Middle East and Asia. Since the 1970s these groups have been the main source regions for immigrants to the U.S. After 1990 many immigrants from Africa arrived in the U.S. as well. Whereas the main Latin American immigrant group was from neighboring Mexico, there are several leading immigrant groups from Asia among them from China (including Taiwan), the Philippines, India, Korea, Vietnam and Japan. The cultural complexity of immigrants from Asia is enormous and comprises small but distinct ethnic groups like the Hmong culture group (from Laos, Thailand and China), the Mongolian and the Nepalese.

There have also been significant changes as to the various religious groups making the U.S. their home. The U.S. census does not include information about religion, but other non-governmental associations provide useful information on this topic. While most people in colonial America and in nineteenth century U.S. belonged to Protestant church denominations such as the Episcopalian, Methodist or Lutheran Church over time, there was a slow and steady increase in the number of Roman Catholics. In addition to older largely Catholic immigration groups such as the Irish, Polish, Italian and German settlers that constituted much of the pre-WWI urban industrial work force, the large number of newly arriving Hispanics have made Roman Catholicism the largest denomination, if Protestant groups are not combined. Other rapidly growing U.S. religious groups include the Church of Latter Day Saints ('Mormons') and a number of Islamic denominations (Sunni, Shiite, etc.). The complexity of religion and belief systems in the U.S. is a striking feature of American culture and incorporates a multitude of religious groups. Jews, for instance, who account for about 2–3 % of the U.S. population are organized in very different ways. Jewish community life in the U.S. has developed both in distinctly fundamentalist ways such as among Orthodox and Ultra-Orthodox Jews as well as among Reform Jews who have made deliberate adaptations to mainstream American life. Last but not least, new forms of religion in the U.S. such as a form of Zen Buddhism and New Age philosophies have become acceptable for many honoring the principle of religious freedom in the U.S.

While the 'melting pot' model—as a pathway for immigrants to assimilate to social and cultural life in the U.S.—was the dominant paradigm for most of U.S. history, different forms of bi-cultural or multi-cultural models have emerged more recently. The number of Spanish only speakers has increased considerably, mostly among first generation Hispanic immigrants which has caused many communities, local and regional agencies as well as business groups to communicate with the new residents in their language (and not in English only). There are several large culture groups where non-English languages are predominantly spoken at home including first and second generation Chinese immigrant families in the U.S.

It is important to keep in mind that we have increasingly seen a 'blending' of the various racial, ethnic and culture groups, and that the boundaries between the Census groups are no longer clear cut. More and more Americans belong now to two, three or more racial, ethnic and/or ancestry groups. This more complex demographic situation, with new "emerging American identities" (Perez and Hirschman 2009), has been confirmed in recent analyses of the Census 2000.

Finally, it should be emphasized that population census categories or nomenclatures for the various ethnic, cultural and racial groups have frequently changed. An example of the changing practices in this respect is the chosen names for the descendents of the former slaves from Africa: from 'negroes' and 'coloreds' (in use up to the 1950/1960s) to 'Black Americans' (during and after the civil rights movement) and 'African-Americans', currently used as the politically correct term.

A multi-cultural America is no longer defined along previously hardened racial, ethnic and cultural lines but on a consensus that diversity contributes to the strength and vitality of the nation. In present-day society the many facets of a culturally diverse population have been widely acknowledged—though not in all cases and instances.

Both China and the U.S. have seen conflicts and tensions over racial, ethnic and cultural issues in the distant and recent past. Some minority groups in both countries have had a hard time to be fully recognized, and their status in society has been lacking. The continued practices of their language and religion have been occasionally met with suspicion and disregard—both in 'mainstream' America and China.

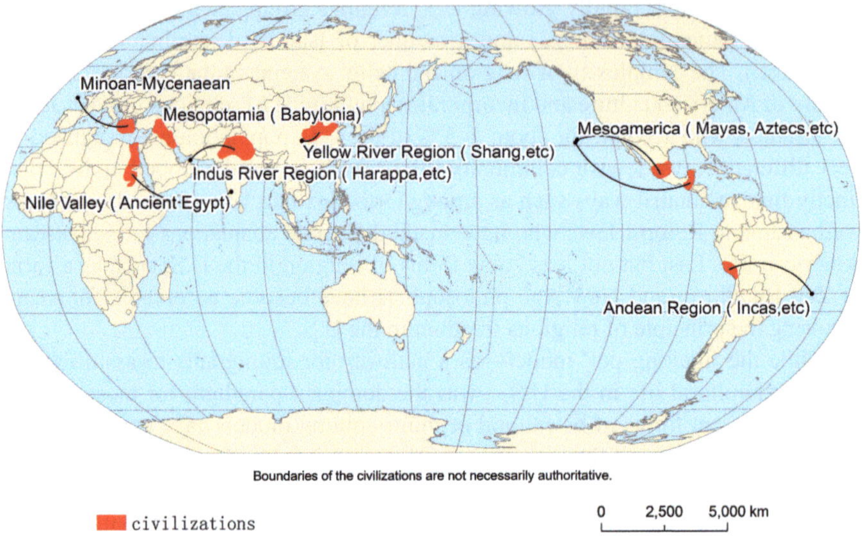

Fig. 1.5 Civilizations in the world

1.3 Cultural Contributions to World Civilization

The United States is a country in the Western Hemisphere, while China is a country in the Eastern Hemisphere. Both nations represent important cultures which have contributed significantly to world civilization—though contemporary American culture and Chinese civilization are grounded in very different traditions.

China is the site and origin of one of the greatest civilizations of the world. Ancient Chinese civilization extends backwards in time for more than 4,000 years with comparable cultural advancements made in early human history only in Egypt, Babylon and India during the height of their civilization (see Fig. 1.5). China's four great ancient inventions, namely the compass, gunpowder, papermaking and printing, are major contributions to the modernization of the world's economy and to humanity in a wider sense.

The early formation and development of Chinese civilization has been characterized more aptly in the following way:

The first light of Chinese civilization revealed itself 7,000–8,000 years ago, as indicated by the ruins of the Daxi Culture in Sichuan Province and Hubei Province, the Majiapang Culture in Jiangsu Province and Zhejiang Province, the Hemudu Culture in eastern Zhejiang and the Yangshou Culture along the middle reaches of the Yellow River and its main tributaries.

According to legend, the primitive tribes that inhabited the middle and upper reaches of the Yellow River were unified into two powerful tribes under the Yellow Emperor and Fiery Emperor, and began their push southward 5,000 years ago. After years of warfare, they conquered the Sanmiao and Jiuli tribes active in south China under the leadership of Chi You. Part of the defeated tribe was incorporated into the tribes under the Yellow and Fiery emperors to become a component part of the Han people, which marked the beginning of the Chinese nation. This history has also given rise to the term "descendants of the Yellow and Fiery emperors" that Chinese often use to refer to themselves.

Archaeological studies have revealed that around 5,000 years ago the Chinese entered the stage of patriarchal society. Not only did villages begin to appear but also the initial forms of cities began to become evident. Extensive communities indicated that the population at the time had already reached a fairly large size and agriculture had made great headway. The earliest discoveries took place during this period. ShenNong tried and tasted various kinds of wild plants to select crops appropriate to be cultivated for food and herbal medicine to cure disease. The Yellow Emperor invented the compass, which helped him defeat Chi You. More importantly, the appearance of chariots greatly reduced labor intensity. Lei Su, wife of the Yellow Emperor, discovered silk making by raising silkworms, and produced the first garments, which allowed the ancient people to bid goodbye to the period when they wore animal skins and tree leaves. The tribe under Chi You in the south learned how to make weapons with copper, creating the conditions for making bronze vessels, metallurgy and alchemy of later times.

During the Xia Dynasty, 4,000 years ago, China entered the period of slave society. The Shang Dynasty (sixteenth to eleventh centuries BC), which replaced the Xia, saw the height of bronze culture, when superb smelting and casting techniques brought forth beautiful wares made of bronze. Pottery making also developed very rapidly with the appearance of primitive pottery wares. Sericulture and silk weaving reached maturity at this time.

From 475 BC to the end of the nineteenth century, China went through a long feudal period. Before the fifteenth century, China was one of the most powerful countries of the world, occupying a leading position in the development of productivity and technology. Ancient China enjoyed a developed agriculture and advanced irrigation system, an independent tradition of medicine and advanced botanical knowledge. China's four great inventions, namely, the compass, gunpowder, movable type printing and papermaking, not only changed the world but also accelerated the evolution of world history. Besides, China was rich in ceramics and silk textiles which were inventions that exerted a great impact worldwide. China also kept the world's most detailed and earliest astronomical records. The first people to take note of such astronomical phenomena as comets, sunspots and new stars were all Chinese. It was also the Chinese who produced the most advanced astronomical observatory apparatus of the time. In metallurgy, China long held a leading position. When Europeans still could not turn out a single piece of cast iron in the fourteenth century, Chinese people had already produced cast iron on an industrial scale four centuries earlier.

In the field of thought, Confucius, founder of Confucianism, not only had far-reaching significance for China, but for the whole of East and Southeast Asia. The warfare strategies introduced by the noted military strategist Sun Zi are still studied and referred to today. Taoism was an important school of thought, and is known for its simple dialectical elements. Its position of "quietude and inaction" has many identical views with the thoughts of modern man. Taoism, based on the Taoist doctrines, is an independent religion established in China.

When commenting on the relationship between China's civilization and that of the rest of the world, the late Joseph Needham, historian of China's science and technology and professor at Cambridge University, once said that people must remember that in early times and into the Middle Ages China was way ahead of the West in almost every discipline of science and technology, from chart making to gunpowder. Western civilization, he went on to say, did not begin until the era of Columbus, and China had left the Europeans far behind in science and technology before that time.

Unfortunately, the country's feudal bureaucratic system held back science and inventions from making further progress, and prevented Chinese society from developing modern science, resulting in China staying long in the experimental stage in science and technology.

Modern China is experiencing a completely new era in which respect for science and inventions and encourage creativity have become the guiding principles of society. Looking back at the contributions China's civilization has made to the world, we have reason to believe that a more prosperous and stronger China will surely make new contributions to the civilization of mankind.

The First Americans, the first known inhabitants of modern-day United States territory, are believed to have arrived in the Americas over a period of several thousand years beginning 14,000–16,000 years ago. Migration to the North American continent was possible by crossing over the Bering Land Bridge (a land bridge which was in existence during extended time periods of the Wisconsin glaciation). It is now widely assumed—and confirmed by both fossil records and genetic research results—that Native American people are descendants of early big game hunters migrating from Siberia to present-day Alaska. Proponents of an alternative 'long chronology' migration model argue that there is substantial evidence for earlier migration waves to the Americas as well.

In the following millennia several Meso-American civilizations evolved such as the Maya and the Aztec civilizations. Several core areas of early Native American cultures existed in what is now the United States. Socially complex, agricultural societies flourished in, for instance, the lower Illinois River valley or in the Southwest 1,000–3,800 years ago. Little has been left of what may have been early indigenous forms of North American civilizations. Native American languages and religions did most often not survive European colonization and if resistant enough were eventually suppressed in nineteenth and twentieth century American society. Surprisingly, some of the indigenous religions and tribal languages have made it into the modern era, such as a widely spoken Navajo Nation language and continued religious practices among the Hopi Indians, the descendants of the ancient Pueblo cultures. In general, though, Native American cultural traits have largely vanished and/or are found only in residual form in the broad spectrum of current American popular culture.

Geographers have tried to identify what are essential or enduring values of American Culture. Cultural geographer Wilbur Zelinsky came up with the following four themes which best characterizes the American World View (Zelinsky 1973; Stoddard et al. 1986):

1. An intense, almost anarchistic individualism which de-emphasizes obligations to other members of a family or social group (rooted in the American frontier experience)
2. Mobility and change: no other society is so characterized by frequent migrations, restless mobility, and a passion for speed and innovation
3. A mechanistic world vision: the belief that technology can solve all problems, even the environmental ones that it causes
4. Messianic perfectionism or an assumed superiority of the American way of life and the desire to project it on the rest of the world

Whereas the terms American Culture and American Popular Culture continue to be widely used in and outside the U.S. the concept of an American Civilization was never firmly established in the literature, with few exceptions. It has been argued that American (popular) culture is rooted in and/or is largely an expression of Western Civilization, a broader concept aptly applied. While the influences of western traditions of thought, from Classical Antiquity to the Renaissance and Enlightenment in Europe, are widely acknowledged for the formation of an early U.S. political culture and legal system, there is also a noticeable and widening rift between contemporary American (popular) culture and traditional European expressions of Western Civilization.

Political scientist Samuel Huntington identified Europe, North America and Australia/New Zealand as core areas of a 'Western Civilization' dominated region of the world—currently facing an ongoing or forthcoming "clash of civilizations" (1993). While most of his attention is given to a highly perceptive cultural fault line between 'Islamic' and 'Western' civilizations, several 'Confucian-Asian' cultures are included in his treatise about future geopolitical battle-lines as well. China, Korea and Vietnam make up the 'Confucian Civilization' distinctly separate from Asian 'Buddhist', 'Hindu' and 'Japanese' civilizations. Political geographers have rejected this simplistic view of the world where 'civilizations' merely replace territorial states as agents which have become more and more closely connected in a world system, last but not least by the current forces of globalization (Agnew 2003).

There is some fascination about the U.S.'s perceived role of sole superpower since WWII—which is a questionable notion considering major setbacks of the U.S. political-military power during the Vietnam War, the 2001, September 11th terrorist attacks as well as the Golf & Afghanistan-Iraq Wars. The processes leading to hegemony or changes in hegemony status have been examined in world system theories (see, for instance, Wallerstein 1974, 1980, 1989, 2003 and 2004). Here, the main explanation of a world power's persisting supremacy or declining hegemony is based on an analysis of three forms of economic dominance (productivity, trade and financial dominance). As we look at the currently more complex, multi-core world economy—propelled increasingly by forces of globalization—it is conducive

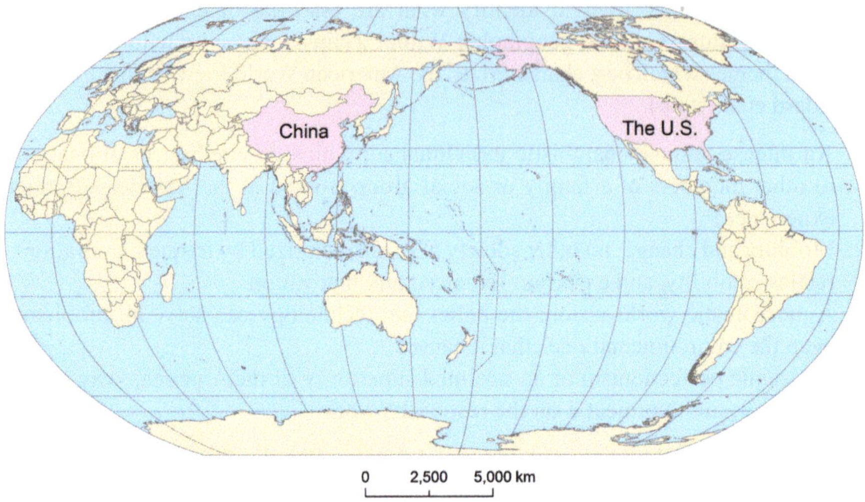

Fig. 1.6 Locations of China and the U.S

to assume that we have been living in a transitional period since the 1970s: from a world with the U.S. as single hegemonic power to a core group of competing countries including the European Union, the Soviet Union/Russia, Japan and China.

1.4 Economic Ranking in the World

China and the United States are both located at the circum-pacific belt which has become a leading trade region of the world (see Fig. 1.6). The economies of both China and the U.S. are part of this 'Pacific Rim' trade region. The Pacific Rim—originally a geographical term denoting the zone of crustal instability or 'ring of fire' around the Pacific Plate—includes most prominently Japan, South Korea, Taiwan, Hong Kong in Pacific Asia and Canada in North America—just to name of few of the leading economies which were part of a rapidly rising 'Pacific Rim' region in world trade during the 1970s and 1980s. Meanwhile the regional concept of Pacific Asia in world trade has been widely accepted. China and the U.S. are both members of the Asia-Pacific Economic Cooperation (APEC) which is a forum for 21 'Pacific Rim' countries. APEC members account for approximately 54 % of world GDP and about 44 % of world trade.

The rise of Pacific Asia in world trade is increasingly marked by the re-emergence of China as a leading economic power in Asia. The national economy of the People's Republic of China has gone through crucial periods of restructuring and modernization since new economic policies and the first special economic zones were established and implemented in the late 1970s and early 1980s.

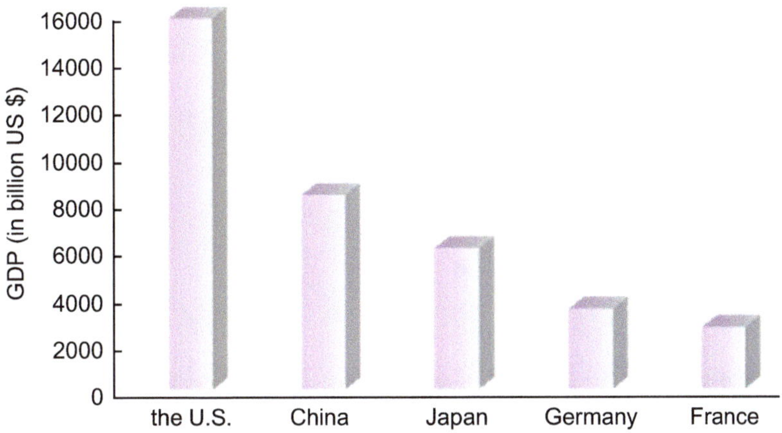

Fig. 1.7 Top five countries by GDP (nominal) in 2012. (International Monetary Fund 2013)

Consequently, the rank of China's GDP among the major economies worldwide improved dramatically: from rankings between numbers 20–30 in the 1950s and numbers 10–20 in the 1980s to currently a number two ranking (2012). Only the United States surpassed—according to statistics published by the IMF and the World Bank—China's GDP (see Fig. 1.7). China has had the fastest growing economy among the major countries for the past quarter of a century, with an average annual GDP growth rate of about 9%, even reaching14.2% in 2007 (Fig. 1.8). In recent years, the growth has slowed. Still, the country's per capita income ranks in the lower middle field by world standards, at about $ 4,940 (nominal, 113th of 213 countries/economies), and $ 8,430 (PPP, 94th of 213 countries/economies) in 2011, according again to statistics published by the IMF. While China's per capita income may lag behind other economic growth figures and rankings, it is worthwhile mentioning that poverty in the People's Republic of China has been significantly reduced. Since 1978 hundreds of millions of Chinese have been lifted out of poverty bringing the country's poverty rate down from 53% in 1981 to 8% in 2001 and 2.5% in 2005.

The economy of the United States is the largest national economy in the world. Its GDP was estimated at $ 15.68 trillion in 2010. The U.S. economy maintains a high level of output per person (GDP per capita (PPP), $ 49,922 in 2012, and ranked number six in the world according to IMF). Until the most recent economic down-turn, the U.S. economy has maintained a stable overall GDP growth rate, a low unemployment rate and high levels of research and capital investment funded by both national and, because of decreasing saving rates, increasingly by foreign investors. In 2008, 72% of the economic activity in the U.S. came from consumers.

Trade relations between the United States of America and the People's Republic of China have seen great improvements and are expressions of a greater connectedness of the economies of both nations. The total trade volume (imports and exports combined) between China and the U.S. has risen from $ 85.4 billion in 1998 to

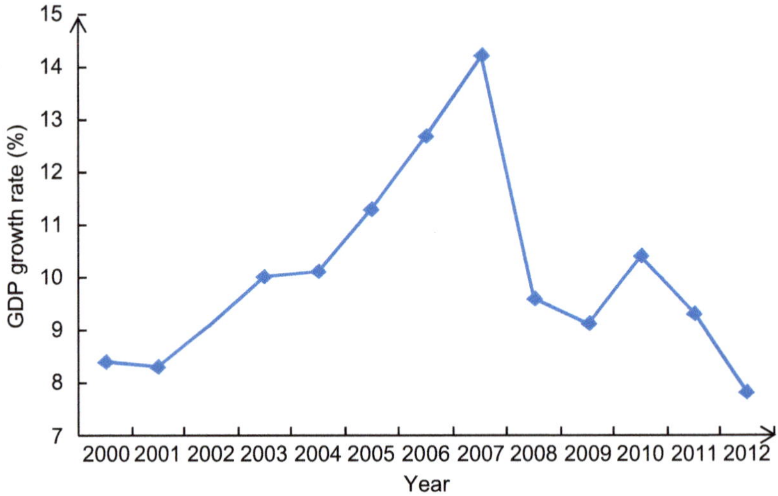

Fig. 1.8 GDP growth rate over the past 12 years in China. (http://data.worldbank.org/indicator/NY.GDP.MKTP.KD.ZG)

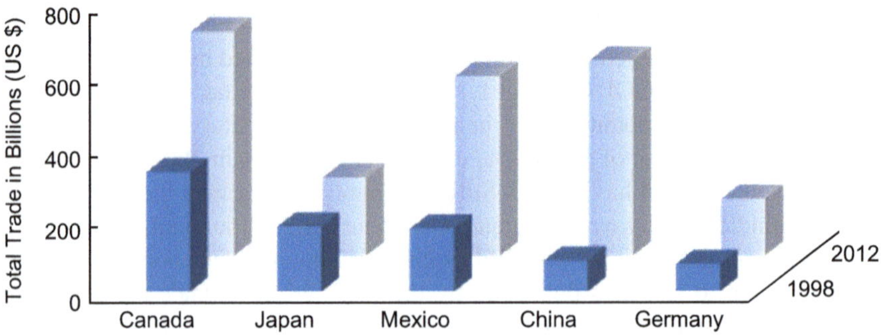

Fig. 1.9 Top five countries with which the U.S. trades. (Country trade report http://countryreport.mofcom.gov.cn/record/view.asp?news_id=23023; U.S. International Trade Data http://www.census.gov/foreign-trade/top/dst/2012/12/balance.html)

$ 536.23 billion in 2012 (see Fig. 1.9). China is now the second most important trade partner of the U.S. after Canada. In 2003, China surpassed Japan as the leading trade nation for the U.S. in Asia. Vice versa, the U.S. has also become more salient in China's trade relations worldwide. The U.S. is now—with the European Union and Japan—among the three leading trade nations of China. The U.S. remains China's biggest importer; Hong Kong, a Special Administrative Zone of the People's Republic of China since 1997 yet separately listed in trade statistics, comes in second. The trade deficit between the U.S. and China—with rapidly rising Chinese exports to the U.S.—has consistently increased over the past 5 years 2003–2008. Worldwide, The People's Republic of China is now ranked either number 3 or 4 in total international trade volume after the European Union, the United States and

Table 1.2 List of countries by carbon dioxide emission (millions of tons). (EDGAR: CO_2 time series 1990–2011 per region/country)

	2009	2010	2011
China	8,270,000	8,900,000	9,700,000
United States	5,330,000	5,530,000	5,420,000
India	1,750,000	1,860,000	1,970,000
Russia	1,740,000	1,780,000	1,830,000
Japan	1,180,000	1,260,000	1,240,000

Japan—with slightly diverging total trade figures for the two leading trade nations in Asia in the past few years.

While the U.S. and China represent the two leading economies of the world with far reaching ramifications for the health of the global economy, there are equally demands on both countries to take a lead on a greater responsibility for the ecological well-being of Planet Earth. One of the main concerns is the high energy consumption of both countries to sustain their economic growth. The consequences include the continued high and/or an increased output of greenhouse gases which is considered one of the main causes for climate change. China and the U.S. are the leading emitters of carbon dioxide and other pollutants that have negatively affected the atmosphere. In terms of carbon footprint both countries rank number one and two (see Table 1.2). China and the U.S. jointly contribute to more than 40 % of the carbon dioxide emissions worldwide.

There are, in particular, two critical trends as to carbon dioxide emissions. First, the U.S. shows a continued high output of greenhouse gases per capita. Very little progress has been made in the reduction of carbon dioxide emissions which has held steady at about 17.3 per capita which is the highest number among the industrialized countries in this respect. While the per capita figure is significantly lower for China, at an estimated 7.2 per capita for 2011, the total output of emissions has seen a dramatic increase from 2001 to 2009/2010/2011, from below 3,000,000 to above 8,000,000 million t of carbon dioxide emissions (estimates for 2010/2011).

Another way to measure a country's overall economic and social achievements beyond the mere use of the GDP per capita indicator is the Human Development Index (HDI)[1]. The annually released HDI index figures are a composite of three indicators: life expectancy at birth, an education index and a standard of living (GNI per capita) index. A HDI of 0.8 or more is considered to represent "high development". The 2013 report based on the 2012 figures saw 16 countries above 0.9 ("very high human development"). The top ten includes the U.S. (Number 3) and Japan (Number 10) with the remaining nations from the top ten list coming mostly from European countries. For several years Norway has had the highest recorded HDI. China and India, the two most populous countries of the world, are found in the mid field of these annual lists. China's current HDI is at 0.6999, which is close

[1] HDI is an index which is calculated and published by the United Nations Development Programme (UNDP); it gives a comparative assessment of the development of 185 nations represented within the UN.

to the "high human development" category. It is an indication that China while having gained enormously ground in recent years can still be considered a "developing country", though it is certainly in a rapid development phase.

1.5 Political Power in the World

Along with its impressive economic growth, China has attracted a lot of attention because of the increasing political power the country exerts in the world. In recent years, China has attained a status in Asia and worldwide that rivals that of the United States. While it is not the purpose of the book to examine the competition for world hegemony and to speculate on the changes that might have occurred in this respect, it is certain to say that U.S. and China are two countries with significant political power in the world that cannot be ignored. Both countries have an important impact on the globe and the world.

The political power of China and the U.S. is of concern in the context of the book as both countries have a rising responsibility for the wellbeing of humanity and its habitat. Our world faces the challenges of traditional risks such as natural disasters, economic recession, regional conflict, resource shortages and environmental degradation but also a bundle of new risks such as climate change, terrorism, genetic modification and other unknowns. Sustainable development, millennium goals and the future earth require close collaboration among the more powerful countries. China and the U.S. have to wisely use their political power to fulfill their joint responsibility for the future of our planet earth.

1.6 The Structure of the Book

The present text consists of 11 chapters which are organized in four parts:

1) General Overview Chapters: Introduction (Chap. 1) and the general presentation of the physical geography and population—ethnic geographies (Chap. 2 and 3) and a discussion of the main themes
2) Thematic Chapters focusing on agriculture and food production, economic geography, international trade, and mega-regions. (Chap. 4, 5, 6 and 7)
3) Regional Comparisons: Regional urban economic clusters, urban growth and sprawl, and main agricultural regions. (Chap. 8, 9 and 10)
4) Conclusions: The concluding chapter examines an increasing collaboration of Chinese and American scholars in international institutions, in particular, on global issues. The chapter stresses a need for continued and joint stewardship for our planet, with a list of issues and questions of sustainability for the future of China and the U.S. (Chap. 11)

References

Agnew, J. (2003). *Geopolitics: Re-visioning world politics* (2nd ed.). New York: Routledge.

Dicken, P. (2007). *Global shift: Mapping the changing contours of the world economy* (5th ed.). New York: The Guilford Press.

Huntington, S. P. (1993). The clash of civilizations. *Foreign Affairs, 81,* 44–60.

Mayda, C. (2010). *The regional geography of the United States and Canada*. Boulder: Rowman & Littlefield. (forthcoming).

Perez, A., & Hirschman, C. (2009). The changing racial and ethnic composition of the US population: Emerging American identities. *Population and Development Review*, *35*(1), 1–51.

Stoddard, R., Wisehart, D.J., & Blouet, B.W. (1986). *Human geography: People, places and cultures*. Upper Saddle River, New Jersey: Prentice Hall.

Wallerstein, I. (1974). *The modern world system, volume I: Capitalist agriculture and the origins of the European world-economy in the sixteenth century*. New York: Academic Press.

Wallerstein, I. (1980). *The modern world system, volume II: Mercantilism and the consolidation of the European world-economy* (pp. 1600–1750). New York: Academic Press.

Wallerstein, I. (1989). *The modern world system, volume III: The second great expansion of the capitalist world-economy* (pp. 1730–1840). San Diego: Academic Press.

Wallerstein, I. (2003). *Decline of American power: The U.S. in a chaotic world*. New York: New Press.

Wallerstein, I. (2004). *Alternatives: The U.S. confronts the world.* Boulder: Paradigm Press.

Zelinsky, W. (1973). *The cultural geography of the United States*. Englewood Cliffs: Prentice-Hall.

Chapter 2
Physical Geography of China and the U.S.

Jing'ai Wang, Honglin Xiao, Rudi Hartmann and Yaojie Yue

China and the U.S. receive currently a lot of attention in the mass media as well as in the academic literature. A comparative physical geography of both countries is an important step to better understand their complex relationships. Landforms, climates, river basins and vegetation patterns of China and the U.S. are respectively introduced, as well as their natural regionalization which is based on the above mentioned factors. The following section focuses on natural disasters including earthquakes, hurricanes/typhoons, floods and droughts. Further, environmental issues and problems are discussed. Finally, all the physical geography features for both countries are compared to show the similarities and differences.

J. Wang (✉)
School of Geography, Beijing Normal University, Haidian District, Beijing, China
e-mail: jwang@bnu.edu.cn

H. Xiao
Department of History and Geography, Elon University, Elon, U.S
e-mail: hxiao@elon.edu

R. Hartmann
Department of Geography and Environmental Sciences,
University of Colorado Denver, Denver, Colorado, USA
e-mail: Rudi.Hartmann@ucdenver.edu

Y. Yue
School of Geography, Beijing Normal University, Beijing, China
e-mail: yjyue@bnu.edu.cn

R. Hartmann, J. Wang (eds.), *A Comparative Geography of China and the U.S.*,
GeoJournal Library 109, DOI 10.1007/978-94-017-8792-5_2,

2.1 Physical Geography

2.1.1 Landforms

2.1.1.1 Topography and landforms of China

The most obvious topographic feature of China is the higher elevation in the west and the lower elevation in the east, which could be visualized in form of steps of a stairway (Fig. 2.1). On top of the stairway—the first topographic step—is the Qinghai-Tibet Plateau in the southwest of China, a mountainous plateau with an average elevation of more than 4,500 m. Here are found a series of east-west or northwest- southeast trending mountain ranges, whose elevation are more than 5,000–6,000 m, including mainly the Hoh Xil Mountain, the Bayan Har Mountain, the Tanggula Mountain, the Gangdise Mountain and the Nyainqentanglha Mountain. The Qaidam Basin, known as a "treasure basin", is embedded inside these mountains. The plateau is surrounded by numerous towering mountains. In the south rises the highest mountain range in the world, the Himalayas Mountains, with an average elevation of more than 6,000 m. The Kunlun Mountains, Altun Mountains, Qilian Mountains in the north, and the Longmen Mountains, Minshan Mountains, Hengduan Mountains in the east form the boundary between the first and the second topographic step (Wang, 2007).

The second topographic step consisting of mainly plateaus and basins, is between the outer margins of the Qinghai-Tibet Plateau and the Daxing'an Ranges, Taihang Mountains, Wushan Mountains, and Xuefeng Mountains. This region contains a series of high altitude mountains from 1,500 to 2,500 m, such as Yinshan Mountains, and Qinling Ranges etc., as well as plateaus from 1,000 to 2,000 m including the Inner Mongolia Plateau, the Ordos Plateau, the Loess Plateau, and the Yunnan-Guizhou Plateau from north to south. The largest basin (Tarim Basin), the second largest basin (the Junggar Basin) and the basin with the lowest elevation (the Sichuan Basin) in China are located here.

The third topographic step consists of plains and peneplains, which extend along the boundary between the Daxing'an Ranges and the Xuefeng Mountains. From north to south, the Northeast Plain, the North China Plain, and the Middle-lower Yangtze River Plain are located with elevations mostly below 200 m. The vast area to the south of the Yangtze River consists of mainly hills with elevations of less than 500 m. To the east of these plains and hills, narrow and long mountains which are north-east trending are found. These mountains include the Changbai Mountains, Central-Shandong Mountains, Xianxia Mountains, Wuyi Mountains and so on, ranging from 500 to 1,500 m. To east of the coastline is a broad continental shelf. Many Islands are distributed here; and the most famous are the Hainan Island and the Taiwan Island.

China contains five kinds of landform types, including mountain, plateau, hill, basin, and plain. Among the five landform types, mountains and plateaus are most

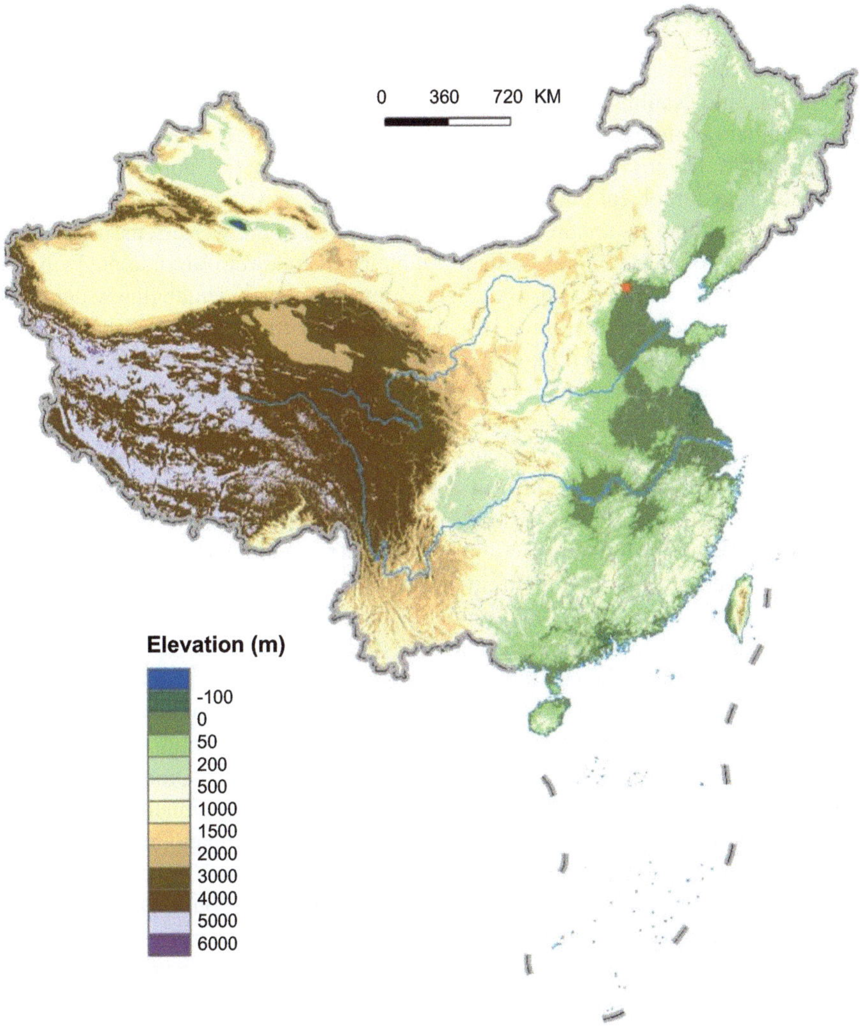

Fig. 2.1 The topography/landforms of China

extensive, accounting for 59 % of the total China land area followed by basins (19 %), plains (12 %), and hills (10 %) (Chinese Academy of Sciences, 1985).

China is a mountainous country, where the mountainous regions account for about 65 % of its total land area. The crisscrossing mountains constitute the backbone of the landscapes of China and have an impact on the spatial distributions of other landscape types in China. (1) The south-north trending mountain ranges include the Helan, Liupan, and Hengduan (transverse) mountain ranges. These north-south running mountains divide China into an eastern and a western half. In the West, most of the mountains which are northwest or northwest-west trending are above 3,500 m high, while most of the northeast-north trending mountains in the East are below 3,500 m. (2) There are three major east-west trending mountains, including Yinshan Mountains-Tianshan Mountains, Kunlun—Qinling—Huaiyang Mountains and Nanling Mountains from north to south. The latitudinal distance between them is almost equally 8°. The Qinling Mountain is not only the watershed of the Yellow River, Yangtze River, and Huai River systems, but also an important natural geographical boundary between northern and southern China. (3) The north-west trending mountains are mainly distributed in the western part of China, including the Altay Mountains, Qilian Mountains, Karakorum Mountains, Hoh Xil Mountain, Tanggula Mountain, Gangdise Mountains, Nyainqentanglha Mountains and so on. The western part of the Himalayas in the north of Qinghai-Tibet Plateau is also north-west orientated; however, its eastern portion gradually shifts into east -west orientation, forming an arcuate mountain belt bulging southward. Most of these mountains are high and steep with a frigid climate and generally are covered by glaciers. (4) The north-east trending mountains are mainly distributed in eastern China. They are arranged in form of a West row, an East row, and an outer row respectively from west to east. The West row includes Daxing'an Range, Taihang Mountains, Wushan Mountains, Wuling Hill, Xuefeng Mountains etc. The East row starts from the Changbai Mountains in the north, through Qianshan and the low hills in Luzhong, and then reaches to the Wuyi Mountains in the south. The Outer row consists of the Taiwan Mountains on the island of Taiwan.

The Qinghai-Tibet Plateau, the Inner Mongolia Plateau, the Loess Plateau, and the Yunnan-Guizhou Plateau are the four biggest plateaus in China. The Qinghai-Tibet Plateau lies to the north of the Himalayas Mountains, to the south of the Kunlun, Altun, and Qilian Mountains, and to the west of the Minshan—Qionglai—Jinping Mountains. It is the largest and highest plateau with a frigid climate and glaciers. The Inner Mongolia Plateau, the Loess Plateau, and the Yunnan-Guzhou Plateau, are located in the second topographic step and divided by the Yinshan Mountains, the Qinling Range, and the Dalou Mountains from north to south. Due to the differences in the composition of the materials and the exogenic processes, there are obvious differences among the morphotypes of the landforms on the plateaus. The Inner Mongolia Plateau located in the northern interior, where the climate is dry with little rainfall, is less impacted by fluvial forces but dominated by eolian forces. The plateau surface is relatively well preserved. Covered by soils of loose texture and affected by a strong fluvial process, the surface of the Loess Plateau was severely cut into pieces full of gullies and ridges everywhere. The Yunnan-Guizhou Plateau, also known as the "karst plateau", has a full set of karst landforms with a subtropical humid climate, widespread carbonate rocks, and very well developed karst processes.

The Tarim Basin, the Junggar Basin, the Qaidam Basin and the Sichuan Basin all located in tectonic fault zones, are the four biggest basins in China. The Tarim Basin that contains the largest desert in the country, the Taklimakan Desert, is the biggest basin in China. It has obvious Aeolian denudations and erosions with a closed terrain, an extremely drought prone climate, and a sparse vegetation cover. The second largest basin in China, the Junggar Basin, is a semi-closed basin with a slightly increased precipitation and a relatively denser vegetation cover. It has vast areas of grasslands and a very developed livestock sector. Due to the abundant water from the melting snow and ice, agriculture thrives in the oases at the edges of the Tarim Basin and the Junggar Basin. In addition, there are many oil and natural gas fields in southern and northern Xinjiang Province. China's third largest and highest basin, the Qaidam Basin, has a dry climate, long hours of sunshine, abundant solar energy resources, and it is rich in salt, metal ores, oil and natural gas resources. The smallest basin in China, the Sichuan Basin, also known as the "land of abundance", is surrounded by many mountains. It has a warm and humid climate, numerous rivers systems, fertile soils, rich natural resources, a dense population, and a developed economy.

China's three largest plains, the Northeast Plain, the North China Plain, and the Middle-Lower Yangtze River Plain, are all concentrated in the third topographic step, among the east-west or north-east trending mountain ranges. With the vast land areas, low and flat terrains, easily accessible by transportation, dense population, and developed cities and towns, they are the nation's major agricultural bases and densely urbanized areas. The Northeast Plain is the biggest plain in China. It is formed by the alluvial deposits of such rivers, as the Heilong River, Nenjiang River, Songhua River, and Liaohe River, and is characterized by the large area of black soils and the widely distributed marshes. The North China Plain is the second largest plain, mainly formed by the alluviums from the Yellow River, the Huaihe River, and the Haihe River. The lands are low lying and flat with gentle slopes where many river beds are above the plains on both sides of the rivers. The phase distribution of the above-ground rivers and the depressions is a unique feature of the North China Plain. The third biggest plain, the Middle-Lower Yangtze River Plain, includes the Dongting Lake Plain, the Poyang Lake Plain, the Jiangsu and Anhui plains along the rivers, and the Yangtze River Delta, distributed like a string of beads from east to west. The Yangtze River Basin, a famous "Land of Abundances" in China, is characterized by a low lying and flat terrain, dense lakes and canals, and large areas of connected rice paddy fields.

China's hills, generally called the "Southeast Hills", are mainly distributed in the third topographic step, and are particularly concentrated in the vast areas to the east of the Xuefeng Mountain and to the south of the Yangtze River. Among them, the hills to the south of the Yangtze River and to the north of Nanling Mountain are called the "South Hills"; whereas the hills to the east of the Wuyi Mountain and within Zhejiang and Fujian provinces are called the "Zhejiang-Fujian Hills". To the north of the Yangtze River, there are not many hills except the Shandong Hills and the Liaodong Hills. The Southeast Hills are largely distributed on the two sides of a series of north-east trending mountains with middle to low altitudes; inside the mountains are many different sizes of scattered red rock basins. Because of the differences in lithological features, the Jiangnan Hills are covered with thick, red sandstones or conglomerates. Zhejiang-Fujian contains extensive granite and rhyolite. The Shandong Hills and Liaodong Hills consist of metamorphic rocks and granite which have been severely cut with a cursive coastline, and numerous bays and islands.

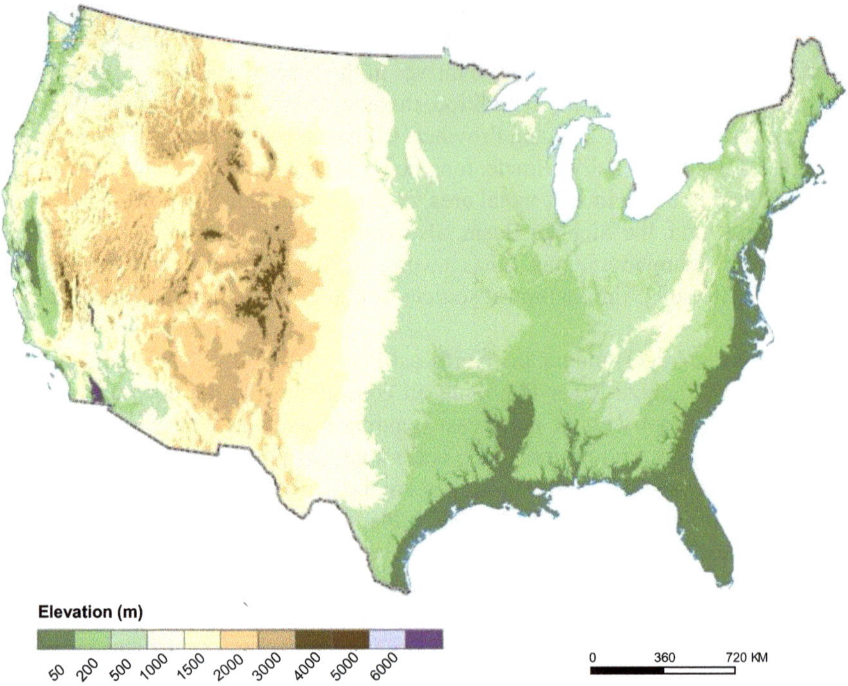

Fig. 2.2 The topography/landforms of the conterminous U.S.

2.1.1.2 The Topography and Landforms of the United States

The most obvious topographic feature of the conterminous U.S. is the higher elevations on the east and west sides and the lower elevations in the vast central plains. In addition, all of its major mountain systems have a north-south alignment. Its terrain can be roughly divided into three zones (Fig. 2.2):

The Western Cordillera System The Western Cordillera starts from Alaska in the north and extends all the way to the American Isthmus in the south, stretching the entire north-south extent of the western conterminous U.S. It is composed of a series of high and wide (usually 2,000–3,000 m high, 800–1,600 km wide) mountain ranges, plateaus, intermountain basins, and valleys that cover approximately one-third of the U.S. land area. Its components are arranged in a series of three large north-south trending bands, with the Rocky Mountains on the east separated from the pacific coastal mountains and valleys on the west by a series of high, heavily dissected intermountain basins plateaus.

The Rocky Mountains in the east, generally high and rugged, stretch more than 4,830 km (3,000 miles) from the Yukon plateau in the western Canada to the Rio Grande River in New Mexico in the southwestern United States. This bulky north-south trending mountains are about 4,830 km (3,000 miles) long; they run through almost the entire north-south extent of the U.S. This range serves as an important continental divide of the drainage systems. All major rivers in the US, including the Columbia, Colorado, Missouri, and Rio Grande Rivers, originate from the Rocky Mountains and eventually drain into three of the world's Oceans: the Atlantic Ocean, the Pacific Ocean, and the Hudson Bay and eventually Arctic Ocean. The water supply feeding into the rivers and lakes from the runoff and snowmelt off the peaks of the Rocky Mountains accounts for one-quarter of the entire freshwater resources in the United States. The zone of the Intermountain Basins and Plateaus includes, from north to south, the Columbia Plateau, the Central Great Basin, and the Colorado Plateau. This region contains a series of high altitude plateaus ranging from 1,219 to 2,134 m (4,000–7,000 feet) trenched by canyons or hundreds of linear ranges separated by basins of varying size, making it the most complex area in the western United States in terms of its geological structures. The western Pacific Mountains and Valleys are composed of two mountain ranges parallel to the Pacific coast, the Coastal Mountains along the eastern coast of the Pacific and the western Cascade—Sierra Nevada Mountains and the narrow lowlands in between.

The Appalachian Mountains System Running almost parallel to the coastline, the Appalachian Mountains stretch from Alabama in the United States in the south all the way to the Newfoundland and Labrador area of Canada in the north, about 2,600 km in total length and usually 1,000–1,500 m above sea level. Mountains in this region are heavily folded and dissected by the down cutting rivers, forming many parallel ridges, plateaus, and valley topography. In the south, it consists of several parallel mountain ranges that vary gradually southward from 100 to 500 km in width. The system has five unique topographic units, the Piedmont, the Blue Ridge and Great Smokey Mountains, the Ridge and Valley area, the Appalachian Plateau, and the New England section. The Blue Ridge—Smoky Mountains is the system's "backbone", with the highest peak of the entire system at Mount Mitchell at an elevation of 2,037 m. Between the Blue Ridge—Smoky Mountains and the Atlantic Coast is the narrow Piedmont, with general widths between 50–350 km and heights between 50–300 m. Due to the big altitudinal differences between the mountains and the plains, when the rivers flow from the steep slopes of the higher inlands down to the plains, many waterfalls and rapids are created along the edges of the piedmont, which is commonly known as the fall line. The Atlantic Coastal Plain contains many beaches, lagoons, swamps, and mud flats. It has a north-south extent of about 600 km, west-east width of less than 200 km, and local relief of mostly between 30–90 m. The nation's largest peninsula, the Florida Peninsula is located here. Glacial moraines are widely distributed in the Ohio and New York areas, due to the invasion of the glaciations during the Quaternary. The Atlantic Coastal Plain is America's most developed industrial area.

The Great Plains of the Central Area Between the Rocky Mountains and the Appalachian Mountains lie America's vast interior plains, which extend from the border between the U.S and Canada to the Gulf of Mexico. Its north-south extent covers the entire central US, east-west spans 5,000 km, and accounts for one half of the total US territory. It is composed mainly of three physiographic regions: the

Eastern Plain, the Western Plain, and the Southeastern Coastal Plain. The boundary between the Eastern Plain and Western Plain is roughly along the 100 °W meridian. Located north of the Missouri River and the Ohio River, the Eastern Plain contains mainly undulating terrains with lakes and basins of various sizes that were significantly shaped by the past glaciers. Its altitudes are normally below 500 m, with a slightly higher elevation in the southwest. Near the Great Lakes is the Lawrence Plain that is characterized by the low rolling hills and moraines left behind by the retreating glacier. It is the drainage divide between the Gulf of Mexico and the Arctic Ocean. Rivers north of the watershed flow into the Hudson Bay; rivers south of it drain into the Mississippi River and then eventually into the Gulf of Mexico. The Western Plain, also known as the Great Plain, is located between east of the Rocky Mountains and west of the Eastern Plain, with its north-south extent reaching the Canadian and Mexican borders. Its terrain rises gradually from east to west, with elevations at around 500 m near the 100 °W to 1,500 m at the foothills of the Rocky Mountains. It is a high elevation plateau. Although being cut deeply by the east-west trending canyons, its surface remains very flat with very little relief changes, except for the gullies formed by the river erosion. The glacial moraines and tills are widely distributed in the north part of the plain. With flat terrains, fertile soils, and lush grasses, the Great Plains are America's most important agricultural area, also known as the Prairie. The Atlantic Coastal Plains include two parts, the Atlantic Coastal Plain and the Mexico Coastal Plain. This zone is mainly composed of low rolling alluvial deposits from the Mississippi River. Its southern margins are the Mexico Coastal lowlands (below 200 m), including the Mississippi Delta, the world's largest delta with black oily and fertile soils. Many swamps are distributed in the Mississippi River mouth area. Since located in the rising area of the Gulf of Mexico coast, a bird's foot delta was formed at the mouth of the Mississippi River, extending about 100 meters into the ocean every year.
The US contains various kinds of landforms, including mountains, valleys, hills, plains, and plateaus etc. There are many classification systems of land-forms, but the system from the United States Geological Survey (USGS) is the most commonly used one. It divides the landscapes in the US into eight different types, with each of them being further divided into sub types.

The Laurentian Uplands Located near Wisconsin and Minnesota, the Laurentian Uplands is the largest outcrop of the oldest core in North America continent, the Canadian Shield, in the United States. It consists mainly of low altitude hills and mountains that are made of Precambrian igneous rocks or metamorphic rocks. These highly metamorphosed rocks are important sources of iron, copper and other important industrial minerals in the United States.

The Atlantic Coastal Plain The Atlantic Coastal Plain stretches over 3,500 km from Cape Cod in the northeastern U.S to the border with Mexico, and is one of America's flattes areas. The coastal plains from New Jersey to Texas are composed of the Late Cretaceous to Holocene sedimentary rocks that were mainly deposited in the marine environment. A large portion of the deposit is still sands or clays that have not been hardened into shale or sandstone yet. After several rounds of uplifts, those rocks formed a series of terraces tilting toward the ocean.

The Appalachian Highland The Appalachian Highland, stretching southwest from southeastern Canada to central Alabama in the U.S, is a highland of 2,400 km length and160–480 km width. It consists of a series of mountain belts at an average height of about 900 m. The highest peak of the system, also the highest point in the United States east of the Mississippi River, is Mt. Mitchell (2,037 m) in North Carolina. The landforms include low foothills, hillsides in the Blue Ridge Mountains formed by the metamorphic and granite rocks from the Precambrian to Paleozoic periods, valleys and ridges formed by the folded Paleozoic sediments, the St. Lawrence River valley covered by the glaciers and lake sediments, the gently rolling Appalachian plateau, and the low hills formed by the Paleozoic Cambrian igneous and metamorphic rocks in the New England region.

The Inland Great Plain The Great Plain is an immense inland area that is over 1,600 km long in its east-west extent from the Appalachian Mountains to the Rocky Mountains and north-south extent stretching from the U.S-Canada border to the coastal plain along the Gulf of Mexico. It was once a shallow inland sea. However, over millions of years, it has been gradually covered by the glacial deposits from the Canadian Shield in the north, and the sediments eroded away from the Rocky Mountains in the west and the Appalachian and the Ozark/ Ouachita Mountains in the east. Most of the fluvial sediments came from the marine and rivers during the Mesozoic and Cenozoic periods.

The Inland Highland The Inland Highland is a mountainous area with rugged terrains. It encompasses a large area including eastern Oklahoma, western and northern Arkansas, southern Missouri, and the southeastern corner of Kansas. It is the only main plateau between the Rocky Mountains and the Appalachians and is composed of the Ozark Plateau and the Ouachita Plateau.

The Ozark Plateau is covered by Paleozoic Cambrian igneous and metamorphic rocks, and by the Paleozoic sedimentary limestone or dolomite rocks. In the middle of the plateau, the mountain ridges are mainly about 300–400 m wide and 30–100 m high, gradually increasing toward the south. The Ouachita Plateau is composed of folded sedimentary layers formed in the Paleozoic period. The ridges extend parallel from the east to the west, with elevations generally about 600 m.

The Rocky Mountains The Rocky Mountains, a product with many faults and folds from the orogenies from the Precambrian to the Cenozoic periods, are the major mountains in the western United States. They extend more than 4,830 km from the border with Canada to New Mexico in the southern United States. Its highest peak with an elevation of 4,400 m is Mt. Elbert in Colorado. The eastern edge of the Rocky Mountains rises above the central plains; its western edge includes ranges such as the Wasatch near Salt Lake City and the Bitterroots along the Idaho-Montana border. The Great Basin and Columbia Plateau separate them from the other mountain ranges further west. The Rocky Mountains are the drainage divide of the conterminous United States.

The Intermountain Plateaus The intermountain Plateaus are located west of the southern Rocky Mountains. The linearly structured landforms are a combined result of the great thickness of nearly horizontal rock layers from the Paleozoic, Mesozoic, and Tertiary periods and the dry climate. Due to the continued uplifts and rifts of the crust during the geological history, the plateaus were cut into large rugged horsts and grabens along the fault lines. They are composed of the Colombia Plateau, the Colorado Plateau, and the Basins and Ranges. The plateaus are characterized by a flat surface and many canyons formed from deeply down cutting rivers.

The Pacific Mountain System The Pacific Mountain System is composed of a series of mountain ranges that extend from the US-Canada border along the west coast all the way to the US-Mexico border. It is the youngest and the most active structure in the geological history of the United States. Those folded mountain ranges with steep slopes reflect the ongoing orogenic activities. The major ranges include the volcanically active Cascade Mountains, the young and steep Pacific Borders, and the Sierra Nevada Mountains which are mainly made of granites. The Cascade Mountains form curved north-south belts that extend parallel to the Pacific coastline.

2.1.2 Climates

2.1.2.1 The Climate of China

China is located between the 3 ° 52'N and 53 ° 31'N bordering the largest continent (the Eurasia Continent) in the west and facing the world's largest ocean, the Pacific Ocean., in the east. The average annual precipitation is about 650 mm which decreases from the southeast to the northwest. The seasonal distribution of the rainfall is uneven, with most of the precipitation concentrated in the summer and very little in the winter. Due to the impact of the mountain surfaces, the up-lifts of the Tibetan Plateau, and the coastal currents, the climate of China has the following characteristics:

First, the monsoon climate is significant. Under the influence of the seasonal variation of the global wind belts and the land-sea heating sources, the world's most famous monsoon region is formed in the southeast of China. Compared with the east region of the North America at the same latitude in the same Northern hemisphere, the monsoonal effect here is much more noticeable. The boundary of the monsoonal influence is along the Daxinganling—Yinshan Mountain—Helan Mountain—Wushaoling—Bayan Har—Tanggula—eastern Gangdise Mountains. Due to the control of the cold high pressure system in the land surface during the winter, this region is dominated by the northerly wind and the cold and dry climate. On the other hand, during the summer, this region is controlled by the warm low pressure system. Consequently, the southerly wind prevails and brings warm and moist weather conditions to this area. The monsoons influencing China in the summer are classified as the southeastern and southwestern monsoons. The southeastern monsoon originates from the North Pacific Subtropical High pressure system, and mainly affects the eastern region of China. The southwestern monsoon has two sources, one comes from the southwestern monsoon in the Indian Ocean, which forms a low altitude jet stream in the Arabian Sea, and then affects southwestern and southern China through the Indian Peninsula; the other is from the trade winds in northern Australia, which cross over the equator and then continue to flow northward to affect southern and central China.

Second, the continentality of the climate is strong, which is formed by the large thermal variations of the landmass and the absences of moisture. Its main features in China are large annual temperature variations; the summer solstice and winter solstice are followed immediately by the hottest and coldest months respectively. The temperature in spring is higher than in autumn; the precipitation is concentrated in summer. This climatic continentality in China increases from southeast to northwest. The boundary between the continental climate and the oceanic climate is the line along the Huaihe River-Qinling-West Sichuan Mountains-the Himalayas Mountains.

Third, rain and heat closely correlate in the same period. In the summer, the monsoon from the low latitude area of the Pacific Ocean and the Indian Ocean is warm and humid with the result that the climate is hot and humid. In the winter, the monsoon wind coming off from the Eurasia continent brings cold and dry climate conditions. Consequently, nearly all areas in China receive most of their annual rainfall in the hot summer months. This provides a good climatic environment for the development of agriculture and livestock. However, the instability of the monsoonal precipitation sometimes exacerbates the frequency and magnitude of the damages from the hazardous weather conditions such as drought and floods.

Fourth, due to the vast land size and various landforms, the types of climates are complex and diverse in China, dominated by sub-tropical and temperate climates. Based on such temperature indicators as the accumulated temperature (the average daily temperature >= 100 °C) and the aridity index, Huang Bingwei divided China into 7 temperature zones, 21 natural regions, and 45 sub-regions, indicating the complexity of the climates in China. However, in terms of the classification or zonation of the climates, most regions of China belong to the subtropical and temperate climate zones. Only the islands of Hainan, southern Taiwan, Qiongzhou Peninsula, Leizhou Peninsula, southern Yunnan, and a few other areas are in the tropical climate zone.

China can be divided into six climate zones including the boreal, the temperate, the warm temperate, the subtropical, the tropical, and the highland climate (Zhou, 1993). The boundary between the warm temperate and the subtropical is roughly along the line between the Huai River—the Qinling Mountains, which is also the divide between North China and South China. The line separating the subtropical and the tropical zones roughly passes through the southern part of Taiwan Island, the northern Lezhou Peninsula, and southern Yunnan. Because of the large land extent and being one of the world's leading agricultural areas, the subtropical zone is further divided into the northern, the central, and the southern subtropics. Taking the crop overwintering conditions into account, the tropical zone can be further divided into the marginal tropic, the central, and the equatorial tropics. The climate on the Qinghai-Tibet

Plateau is unique. Unlike the eastern areas in the same latitude, it is not influenced by the cold air masses in the winter, and it has a low temperature in the summer.

Based on such an index as the number of days whose average temperature is ≥100 °C, which is the temperature required for the blossom and the fruit yielding of thermophilic crops, and the average temperature of the warmest month, the Qinghai-Tibet Plateau can be divided into the cold, the cold-temperate, and the highland temperate zones (Fig. 2.3).

2.1.2.2 The Climate of the United States

The conterminous United States is located between 25 °N to 49 °N Parallels. If Alaska and Hawaii are included the latitudinal range reaches from the Arctic Circle to the Tropic of Cancer. The U.S. borders the Atlantic Ocean, the Pacific Ocean, Mexico and the Caribbean Sea. In addition, the north-south alignment of all major mountain ranges on the east and west sides and the vast flat central Great Plains allow the air masses from north or south to sweep freely across the continent's vast interior at high speeds without encountering any natural barriers. Different areas are influenced by different air masses at different times of the year. The southeastern part and the coastal area along the Gulf of Mexico are influenced by the warm currents from the Gulf of Mexico; the west coastal area is impacted by elevations of the landforms and currents of the ocean; the climate in the central area is shaped by the polar cold air masses and the warm currents from the Gulf of Mexico. Consequently, the climates in different areas are very different. However, most of the area has a temperate and subtropical climate (Fig. 2.4).

The climate of the United States has the following characteristics:

Firstly, the types of climate are complex and varied significantly in different regions. The United States contains nearly every major type of climate on earth, varying from polar to subtropical from north to south. The climate in the conterminous United States can be divided into five climate zones. To the east of 100 °W Meridian, from north to south, the climate changes from humid continental, subtropical humid, and then humid tropical climate. The humid continental climate is distributed approximately to the north of the 40 °N Parallel, characterized by the dry-cold winter and mild-rainy summer. The subtropical humid climate is roughly located southeast of the 40 °N Parallel, including the Gulf Coast and most of the Florida Peninsula. The basic characteristics of this climate are its warm summer, mild winter, relatively evenly distributed precipitation throughout the year, concentrated summer precipitation, and no obvious dry season. Only the southern tip of Miami on the Florida Peninsula has the humid tropical climate type. Its main characteristics include hot and humid conditions year round and evenly distributed annual precipitation which is in summer and fall slightly higher, though. To the west of the 100 °W Meridian, the climate is divided into semi-arid, arid, subtropical semi-arid and arid climates from east to west. The climates in those areas have a pronounced continentality, with hot summer, cold winter, little precipitation, and high temperature fluctuation. The northern Great Plains and the Columbia Plateau are governed by a semi-arid climate. The arid climate is limited to the northern

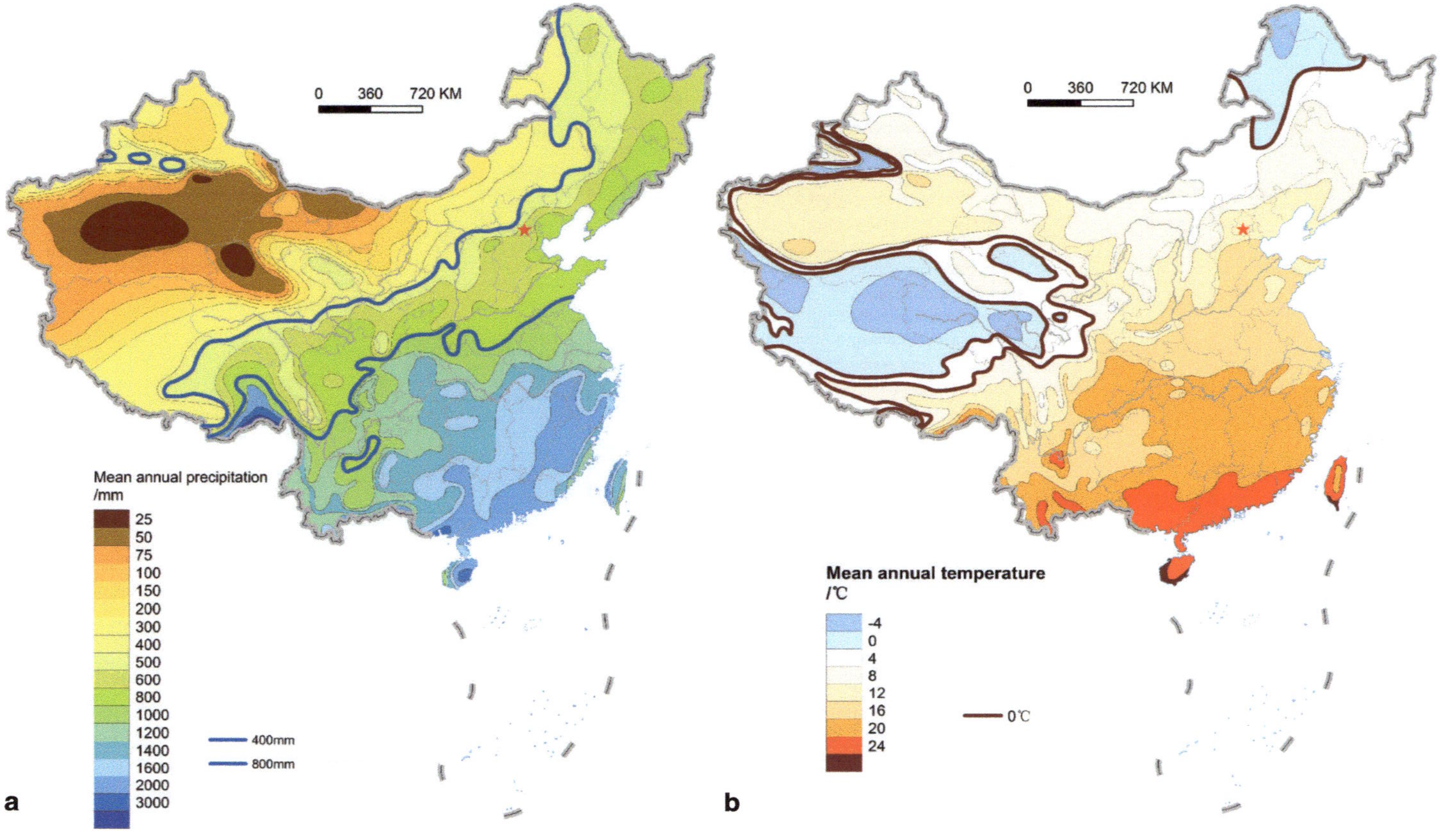

Fig. 2.3 **a** Mean annual precipitation of China. **b** Mean annual temperature of China. **c** Wet and dry zone of China. **d** Climate zone of China

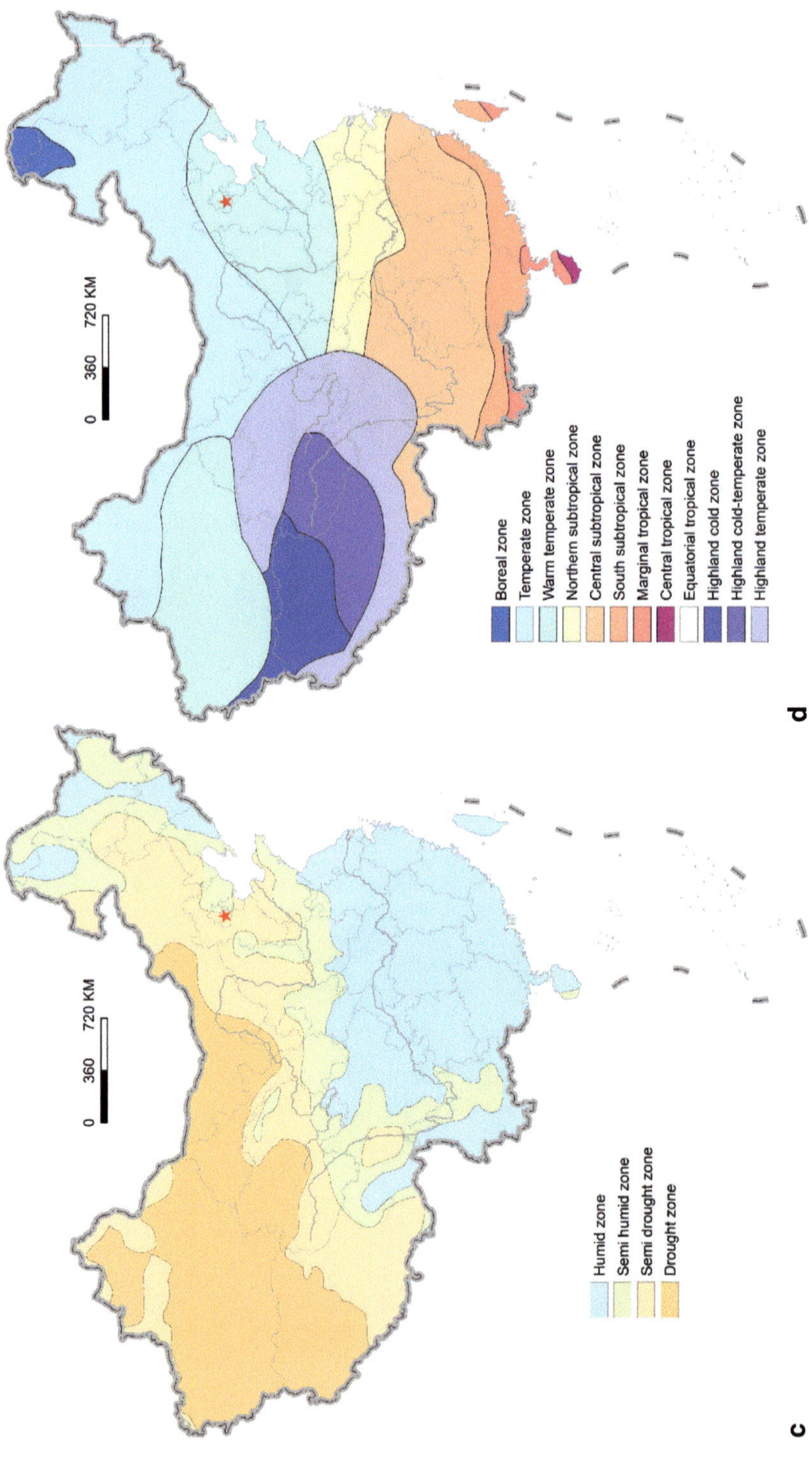

Fig. 2.3 (continued)

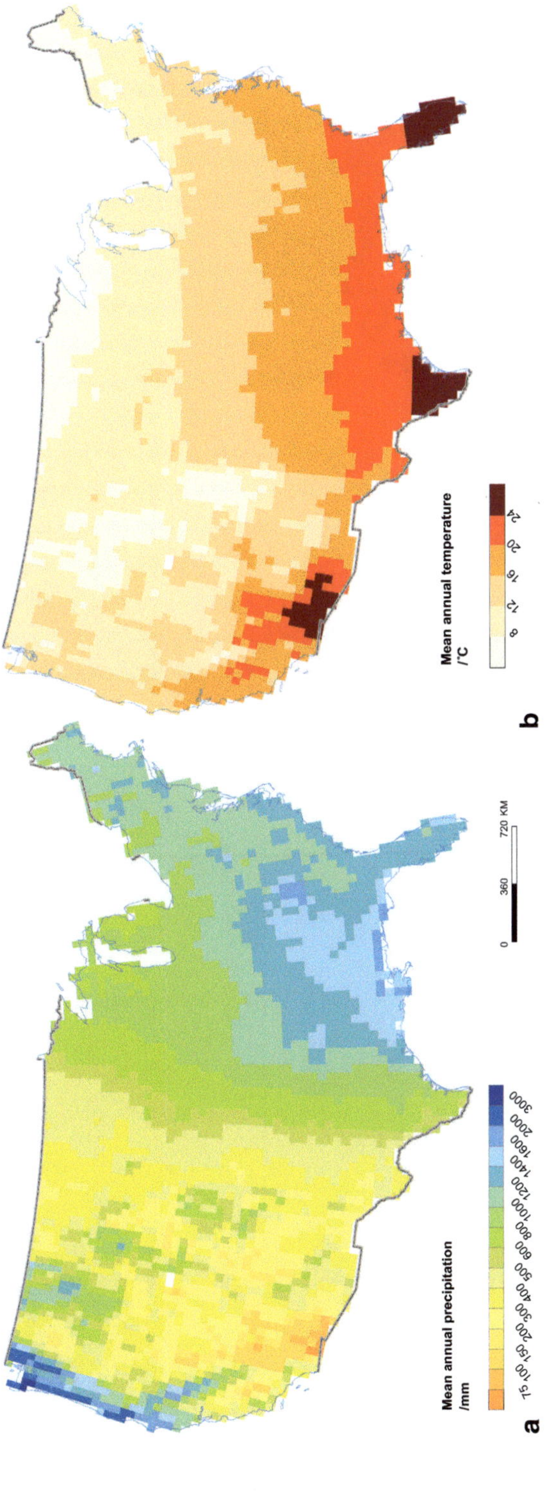

Fig. 2.4 **a** Mean annual precipitation of the conterminous U.S. **b** Mean annual temperature of the conterminous U.S. **c** Relative humidity of the conterminous U.S. **d** Climate zone of the conterminous U.S.

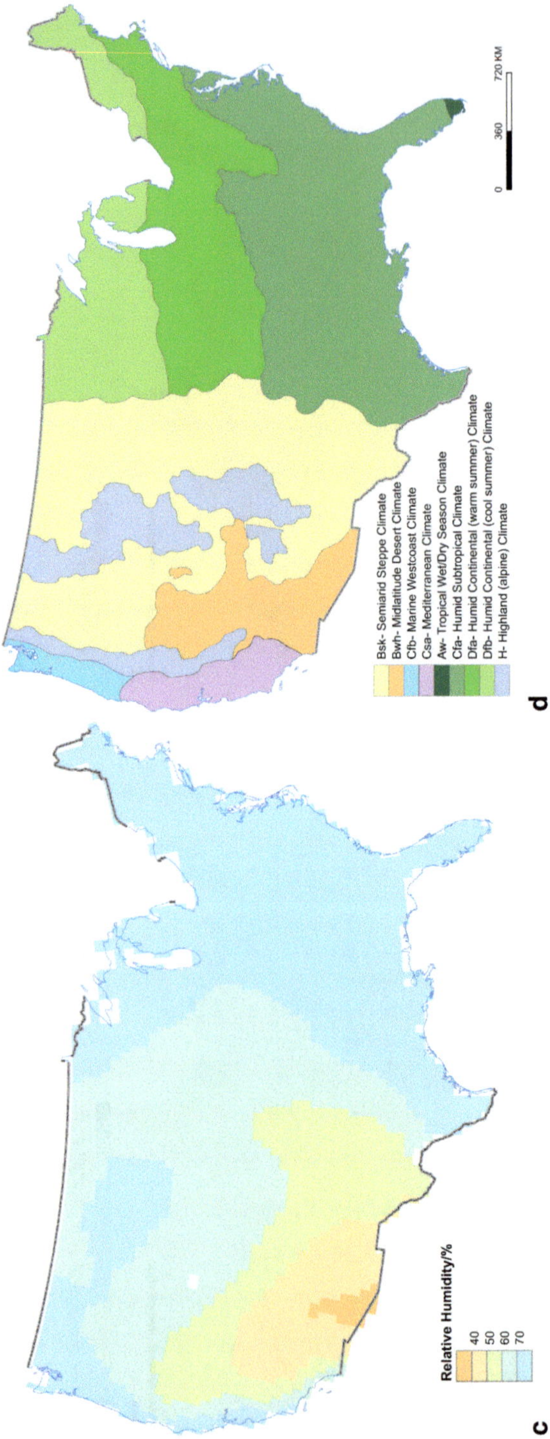

Fig. 2.4 (continued)

Great Basin and Range area, the semi-arid climate to the Great Basin, and the arid type to the southern Great Basin and Range and parts of the Colorado Plateau.

The Rocky Mountains and the Cascade—Sierra Nevada Mountains to the north of the 40 °N Parallel belong to the highland climate zone where the climate is extremely unstable compared to those of the surrounding areas. In addition, it is also characterized by low and rapidly changing temperature, large diurnal temperature range, and small annual temperature variation. The Pacific Coastal areas to the north of the 40 °N Parallel have a temperate maritime climate, with a cool summer, a warm winter, a small annual temperature range, and an evenly distributed precipitation although with autumn and winter accounting for more. Due to the impact of the California cold currents, the regions between 30 °N–40 °N meridians have a Mediterranean climate with a cool summer, a small annual temperature variation, and a decreasing precipitation from north to south.

Secondly, the temperate continental climate dominates. The lofty north-south aligned Cordillera System acts as a massive barrier to the flow of the maritime air masses. Thus, the dominating climate in the US is the temperate and the subtropical climate types, especially the temperate continental climate type. Here the general features of the climate are that it has a cold winter and hot summer with the coldest month in January and the warmest month in July, a large annual temperature range, a moderate annual precipitation, and a summer dominated rainy season. However, when compared with the climates from the Central Asian areas which are also marked by a strong continentality, its degree of continentality is not as extreme. The average annual temperature difference is smaller than in those areas at the same latitudes in Asia. In addition, the degree of concentrated summer rainfall and winter drought is not as obvious as those in Asia.

Thirdly, the distribution of the climate types is unique. All of the climate types in the US have unique distributions and arrangements. There is a contrast between the west and the east in terms of the climate types and alignments. Approximately to the east of the 100 °W Meridian, the climate changes from the humid temperate to the subtropical humid climate from north to south, reflecting the characteristics of a zonal structure. But to the west of the 100 °W Meridian, including the Great Plains and the Cordillera Mountain regions, the dryness increases from east to west, showing the characteristics of a non-zonal structure. In the western coastal areas of the continent the climate type changes sequentially from temperate maritime, to subtropical summer dry, and then to tropical arid and semi-arid climate. Although the climate types vary from north to south regularly, they also extend in west-east direction. Therefore, the climate has an integral embodiment of zonal and non-zonal structures.

2.1.3 Rivers and lakes

2.1.3.1 The River Basins of China

China has numerous rivers, of which more than 1,500 have a drainage area of more than 1,500 km^2. Most rivers flow into the Pacific Ocean, but some drain into the Indian Ocean as well. From north to south, the seven major rivers running into the Pacific

Ocean are the Songhua River, Liaohe River, Haihe River, Huanghe (Yellow) River, Huaihe River, Changjiang (Yangtze) River, and Zhujiang (Pearl) River. The drainage areas of those rivers accounting for 45 % of the nation's total territory, are within the Eastern Monsoon climatic zone. Due to the influence of the topography and climate, the distributions of the rivers are not even. In the areas affected by the summer monsoon, the river networks are developed with considerable amount of runoffs. Most of the rivers drain directly into the oceanbecoming China's out-flowing rivers and accounting for 65 % of the total land area. For the areas that are not reached by the monsoon, the river networks are sparse with little amount of runoffs. Most of the rivers are not able to flow into the ocean, becoming China's inland rivers and accounting for 35 % of the total land area. The watershed boundary between the out-flowing river basins and the inland river basins coincides with the 400 mm isohyet (Fig. 2.5).

The Yangtze River is the longest river in China and the third longest river in the world. It originates from the Glasgow Dandong Snow Mountain, the peak of the Tanggula Mountains, and then passes through 11 provinces, with a length of 6,300 km and a drainage area of about 1,800,000 km^2. In the upstream areas, due to the steep relief of the land and the severely damaged vegetation covers, there are frequent disasters such as serious soil erosions, landslides, and mudflows. They cause frequent droughts, floods, and sedimentations of the soils and sands in the middle-lower streams of the river. Therefore, the protection and construction of the ecosystem in those areas is a top priority. In the lower reaches of the river, the channel is jammed with mud and sand sedimentations. As a result, there are high frequency of droughts and floods, and severe water pollution. Because of the high concentrations of the population and economic activities in these areas, the disaster prevention and mitigation are essential.

The Yellow River, known as China's "mother river", is the second longest river in China. It starts from the Bayan Har Mountain in Qinghai, and then drains through 9 provinces or autonomous regions, before it finally empties into the Bohai Sea in Kenli County of Shandong Province, with a length of about 5,400 km. The Yellow River is famous in the world for its high sediment concentrations and frequent floods. The sediments mainly come from the Loess Plateau in its middle reaches. Once arriving at the lower reaches, the heavy loess soils deposits form the phenomenon of river beds being 3–5 m above the surrounding areas. In the mouth area of the river, the channels have been blocked, extended, and constantly changed, thus becoming the major factor for the formation of the Huanghuaihai Plain.

2.1.3.2 River Basins of the United States

The most distinctive hydrographic feature of the United States is its well-developed drainage networks. According to the statistical data, the U.S has more than 250,000 rivers of all sizes, with more than 20 of them being over 1,000 km long. In addition, there are thousands of big or small lakes in the U.S. The drainage networks formed by these rivers and lakes have become the important pillars for the water recourse, transportation, hydroelectricity, irrigation, inland river transportation, food, and

Fig. 2.5 River Basins of China (I: Eerqisi River Basin; II: Amur River Basin; III: Liaohe River Basin; IV: Haihe Basin; V: Yellow River Basin; VI: Huai River basin; VII: Yangtze River Basin; VIII: Southeast Coastal Rivers Basin; IX: Pearl River Basin; X: Brahmaputra River Basin; XI: Xinjiang Inland River Basin; XII: Inner Mongolia Inland River Basin; XIII: Qinghai Inland River Basin; XIV: Tibet Inland River Basin)

entertainment. Generally, the rivers and lakes can be divided into three major drainage systems (Fig. 2.6).

The Atlantic Drainage System All of the rivers that are located east of the Rocky Mountains and drain into the Atlantic Ocean belong to the Atlantic drainage system. America's largest river, the Mississippi River, occupies a drainage area that spans west to the eastern slope of the Rock Mountains, and east to the western

Fig. 2.6 Major rivers and lakes of the conterminous U.S.

slope of the Appalachian Mountains. It crosses over 31 states and covers the entire north-south extent of the conterminous United States, with a length of 3,734 km. Together with its tributaries, including the Missouri River, Ohio River, Tennessean River, and Arkansas River, it is the fourth longest river system in the world, after the Nile River, the Amazon River, and the Yangtze River. With the drainage area of about 3.1 million km^2, it accounts for 40% of the total continental area in the United States. After the Amazon River and Congo River, it is the world's third largest drainage basin.

The Pacific Drainage System All the rivers that drain into the Pacific Ocean belong to the Pacific drainage system. Its main rivers include the Colorado River and the Columbia River. Those rivers all run through the plateaus with fast moving water and rich water resources.

The Colorado River, the 7th longest river in the US, has its origin on the western slope of the Rocky Mountains. It is 2,333 km in length and covers a drainage area of 703,000 km^2. Up to 98.6% of its total length is inside the United States, and only the remaining 1.4% is in Mexico. It passes through Utah, Arizona, Nevada, and the arid and semi-arid regions of California, before it finally empties into the California Bay through Mexico.

The Columbia River, with a length of 2,001 km, also originates from the Rocky Mountains. It enters the U.S from Canada through the state of Washington. It flows backwards and then to the south for a short distance before finally turning westward. It acts as the border between Washington State and Oregon State. The area of the basin is 668,000 km^2, nearly equivalent to the area of France. Of all the major rivers in the U.S, it has the second largest amount runoff, second only to the Mississippi River. About 90 % of its total runoff is dumped into the Pacific Ocean, and only 10 % of its water is used for agricultural irrigation.

The Great Lake System The most famous lakes of the United States, the Great Lakes including Lake Superior, Lake Erie, Lake Huron, Lake Ontario, and Lake Michigan, are located between the northeastern United States and Canada. Together, they constitute the largest group of fresh water lakes in the world and cover a total area of 245,000 km^2, with about 2/3 of it being in the U.S and the rest in Canada. Of the five Great Lakes, Lake Superior is the largest that occupies an area of over 84,200 km^2. It is also the largest lake in the world. The five lakes are connected with each other through waterways. Their runoffs drain into the Atlantic Ocean through the Lawrence River. Between Lake Erie and Lake Ontario are the world-famous Niagara Falls.

2.1.4 Vegetations and Soils

2.1.4.1 The Vegetation Distribution of China

Accordingly, vegetation is the most sensitive indicator of the landscape and is often studied as a representative of the land cover. Under the combined influence of various factors, such as climate, topography, and soil, the geographic distribution of vegetation in China has an apparent zonal feature.

Latitudinal Zonation China's north-south extension is over 50 degree of latitude, resulting in solar radiation differences in different areas. Under similar atmospheric moisture conditions, the differences in the atmospheric energy (temperature) lead to the formation of different vegetation zones at different latitudes from north to south. The latitudinal zonal feature is more evident in the humid and semi-humid areas in eastern China. From south to north zones include the tropical seasonal rainforest or rainforest zone, the southern subtropical broadleaf evergreen forest zone, the broadleaf evergreen forest zone, the northern evergreen or broadleaf ever-green forest zone, the warm temperature deciduous broadleaf evergreen forest zone, the temperature mixed forest zone, and the cold temperature deciduous forest zone. The subtropical broadleaf evergreen forest zone is the main area for some of the fruit production, such as citrus, loquat, and bayberry. The northern subtropical evergreen forest zone and the deciduous broadleaf forest zone are most northern for the distribution of the subtropical economic trees in China. The warm temperate broadleaf forest zone is the main growing areas for the deciduous fruit trees such as apples, pears, dates and so on.

The Zonation of Dryness and Wetness China's east-west extension is about 5,000 km long, spanning about 62° of longitude. Under similar thermal conditions, the decreasing atmospheric moisture from east to west caused by the land-sea distribution leads to the zonal variations in vegetation types. The direction the various types of vegetation extend are generally perpendicular to the direction of the summer monsoon. The vegetation can be divided into two parts by the line of the Kunlun Mountains-Qinhuai River. Vegetation to the north of this boundary line belongs to the temperature, warm temperature zone; vegetation to the south of this boundary line belongs to the subtropical and Qinghai-Tibet highland zone. The zone in the northern half covers a longer west-east extension and has apparently a more varying vegetation feature. From east to west, the vegetation types include mixed forest, semi-humid forest and steppe, semi-arid steppe, semi-arid semi-desert or desert, extreme arid desert and so on.

The Altitudinal Zonation This altitudinal zonation of vegetation refers to the changes in vegetation types due to changes of such factors as air temperature, precipitation, and soil with the increasing elevations of the mountains. On the one hand, the altitudinal vegetation zones are affected by height, mountain trending, relief, rock, soil and other features of the mountains themselves. On the other hand, they are also affected by the location of the mountains. China's vast mountainous areas and the differences in the above mentioned factors lead to various vertical zones of the vegetation. On a large scale, the altitudinal spectrum of the vegetation can be divided into two types, "the humid" and "the dry" types.

In summary, the altitudinal spectrum of the vegetations in the mountainous areas of China has the following characteristics: (1) the altitudinal spectrum is based on the horizontal level of the local area; (2) from south to north, the structure of the eastern humid zones changes from complex to simple, the number of zones decreases and the elevation range where the same zone is prevalent also becomes lower; (3) from the eastern humid regions to the western semi-arid regions, the elevation range where the same vegetation zones are prevalent gradually increase with the increasing degrees of the dryness, thus leading to a simpler the zonation.

2.1.4.2 The Vegetation Distribution of the U.S.

The vegetation distribution in the U.S is closely associated with rainfall, air temperature, soil, and topography. Under the combined influence of various factors, the geographical distribution of the vegetation has several considerable features of zonation.

The Longitudinal Zonation In the U.S, a large area of deciduous broadleaf forests is distributed between the Atlantic Coast in the east and eastern Texas and western Minnesota in the west. West of it is America's vast area of grassland in the central plains. This vast grassland stretches from the state of Indiana all the way to the Rocky Mountains. Approximately at the 98 °W to 100 °W meridian is the boundary between the short grass and tall grass prairie. West of the 100 °W Meridian, because of the blocks from the Cordillera System, the precipitation decreases gradually from east to west. Except for the northwestern Pacific Coastal area, the entire western region receives less than 500 mm of annual rainfall. Consequently, most areas are governed by semi-arid or arid climates. The dominating vegetation is the temperate steppe and prairie. Shrub, thorn bush, and savanna vegetation are distributed in the southern Pacific coast and the southern Rocky Mountain areas.

The Latitudinal Zonation The Eastern United States borders the Atlantic Ocean in the east, the Great Lakes in the north, and the Gulf of Mexico in the south. Although the area contains the Appalachian Mountains, they are not high enough to constitute a climatic barrier. As a result, the East has a very good moisture and heat condition. However, it diversifies gradually form north to south. The dominating vegetation are forests. From north to south, it is classified into zones of temperate coniferous and deciduous mixed forest, temperate deciduous broadleaf forest, and subtropical evergreen forest. In the west, the narrow Pacific Coast area contains the temperature maritime, subtropical summer dry, and tropical arid and semi-arid climate types from north to south. Correspondingly, the vegetation varies from coniferous forest, subtropical evergreen, shrub to tropical desert from north to south.

The Altitudinal Zonation In the Rocky Mountains and western Cascade—Sierra Nevada Mountain areas of the American West, the lofty mountain heights have caused an altitudinal zonation of the climate. Correspondingly, the vegetation on the mountains has also developed an altitudinal zonal distribution. The vertical distribution is related to the latitude, mountain height, and side of the mountain, and so on. For example, in the Rocky Mountains near the south of the 40 °N Parallel, the vegetation varies in the order of deciduous forest, evergreen forest, coniferous forest, mountain steppe, mountain tundra, and permanent snow and ice from the bottom to the top. Each of the three forest zones has a vertical space of more than 650 m. However, in the Rocky Mountains north of the 40 °N Parallel, due to the reduced heights, the zones of the forests disappear. Only the mountain steppe, mountain tundra, and the permanent ice and snow zones are presented. Due to the impact of the Pacific moisture, in the Cascade—Sierra Nevada Mountains, the altitudinal zonation is different from that of the Rocky Mountains. The elevations of the altitudinal zones of the vegetation are lower than those of the Rocky Mountains. In the north, the alpine tundra is distributed at about 2,000 m; in the south, it is distributed at about 3,500 m. Below it, there are three altitudinal zones of the conifer forest. Further down is the woodland and shrub species of the foothills.

The altitudinal distribution is also associated with the aspect and degree of the slope. In the Rocky Mountains, zoning mainly takes place in the eastern or northern slopes. In the Cascade—Sierra Nevada area, the altitudinal zones have been compressed into very narrow bands. In the areas that are drier and steeper, the zoning becomes blurred.

2.2 Natural Disasters

2.2.1 Earthquakes

2.2.1.1 Earthquakes in China

Located between two of the major active zones for earthquakes in the world, the "Circum-Pacific Belt" and the "Mediterranean—Himalayan Belt", China is one of the most active countries for earthquakes. Historically, most of the provinces, autonomous regions, and the municipalities have had many destructive earthquakes (Fig. 2.7).

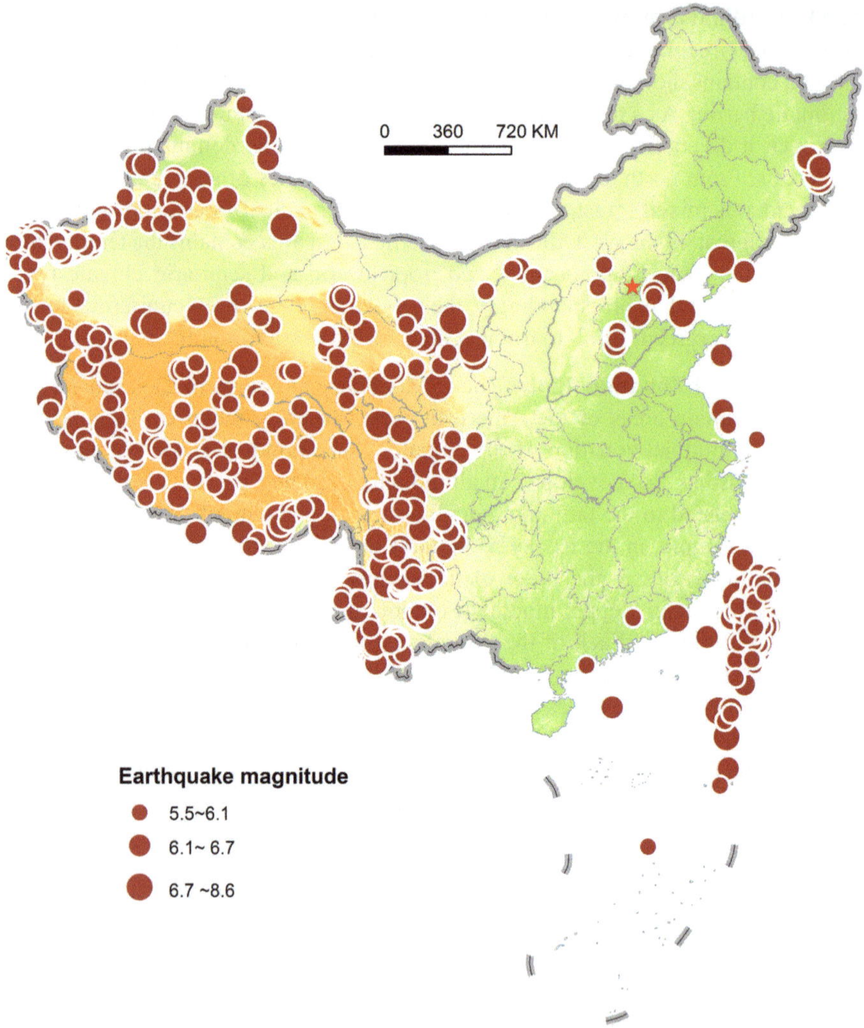

Fig. 2.7 Distribution of earthquake epicenters in China (1990–2012)

The earthquakes in China had high frequencies, large magnitudes, wide areas of distributions, and shallow hypocenters. Statistics indicate that approximately 35 % of all of the earthquakes above magnitude 7 in the world took place in China. Almost one third (1/3) of the entire country, one half (1/2) of the cities and megacities with populations of more than 2–3 millions are located in the zones with potential of having strong earthquakes of larger than 7 magnitude. Of all the 1,200,000 losses of lives in the earthquakes worldwide in the twentieth century, China accounted for 590,000. The earthquake in Wenchuan County of Sichuan Province on May 12, 2008 was the most destructive since the foundation of the People's Republic of

China in 1949. It caused direct economic losses of 845.1 billion Yuan and more than 87,000 fatalities or missing persons.

The distribution of earthquakes in China is uneven. In the western region earthquakes are more active and frequent. In the east, strong earthquakes are mainly found in northern China and southeastern coastal areas. They are stronger but less frequent. According to the existing historical records, taking the 107 °E Meridian as the boundary, the West had 91 earthquakes that were over magnitude 7, while the East only had 27 (excluding the earthquakes with deep hypocenters in Taiwan and Northeast China). However, the earthquakes in the East caused more damages than those in the West since the East has a denser population and a more developed economy. The major reason contributing to the stronger earthquakes in the West and weaker ones in the East is that the major forces driving the deformations of the continent in China come from the collisions between the Indian Plate and the Qinghai-Tibet Plateau.

The distribution of the earthquake hazards in China has an apparent zonal feature. They were mostly concentrated as linear bands in Taiwan, Xinjiang, southwestern and northern China. Different regions have different tectonic causes. The tectonic zone for earthquake in Taiwan includes the province of Taiwan and its adjacent oceans, which is the most frequent and active area in China. The occurrences of the earthquakes in this area are associated with the arc-shape structure of the Pacific Ocean and the tectonic movements of the Taiwan Island and its surrounding areas. They are mostly concentrated in the north-northeast (NNE) trending left-lateral reversal strike-slipping faults, the northeast-east (NEE) right-lateral strike-slipping faults, and the active folds on the thrust faults at the western foots of the Central Range.

The Southwest Tectonic Zone includes the entire Qinghai-Tibet Plateau and the western part of the Sichuan-Yunnan Plateau. It is one of the most active areas for earthquakes in the world. This area has been uplifted significantly since the Cenozoic Era, constituting the most magnificent plateau in the World and the area with the thickest crust in China. The earthquakes here are mainly concentrated in the northwest—east-west—north-northwest curved faults and the northeast and north-south faults. The earthquake in Wenchuan in 2008 was the result of the movement of two local faults in this area. The Xinjiang Earthquake Zone, one of the most active areas in China, is characterized mainly by its reverse faults and fault basins. The occurrences of the earthquakes were associated with the huge Cenozoic fault basins and the orogenic movements at those times. The earthquakes mainly take place in the boundary areas between the mountains and the plains.

The tectonic zone in northern China is the best studied area in China for its seismic structures. The most prominent feature of this area is the tension tectonic activities. Since the Neogene, those tectonic movements have been very active. The impacts of the tectonic movements deep underneath the Earth surface are that the crust has been stretched thin, and the upper mantle has been pushed up. In some areas, layers of low velocity are also found. In the shallow parts, rifts and graben-type basins were formed during the active tectonic movements of the faults.

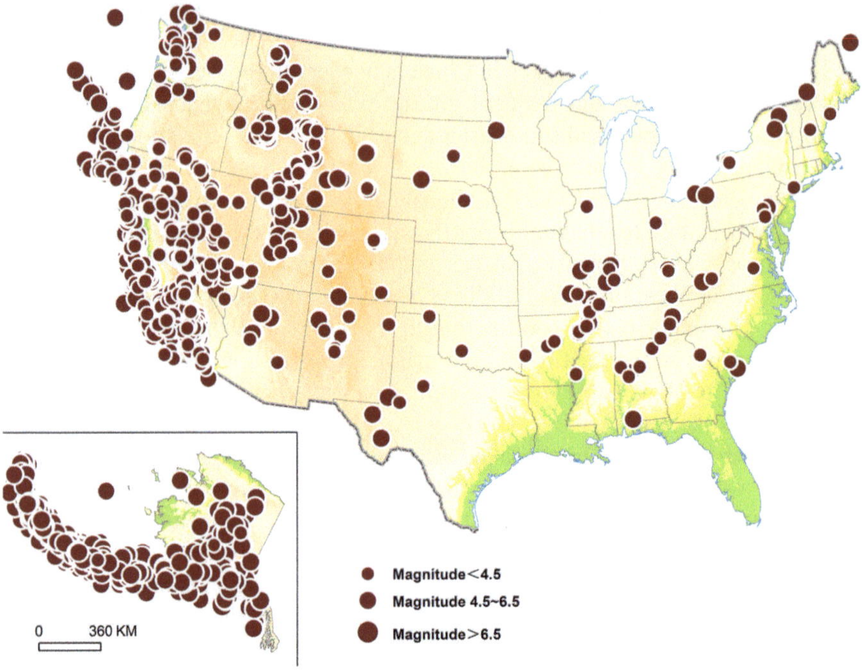

Fig. 2.8 Historical earthquake epicenters of the U.S.

2.2.1.2 Earthquakes in the U.S.

The United States is located on the most active area on earth for earthquakes, the "Circum-Pacific Seismic Belt"; consequently, it is one of the countries with most frequently occurring earthquakes (Fig. 2.8). Historically, earthquakes have taken place in many areas in the US, including many states in the central portions of the continent, and on the East Coast. Nationwide, all 48 conterminous states are vulnerable to earthquakes. At least 39 states have a moderate to very high risk for earthquakes. However, the Pacific Northwest is the most likely place to have earthquakes in the US. There, the Juan de Fuca oceanic plate around the states of Washington and Oregon is gradually subducting 25 km underneath the westward moving North American continental plate. It has the capability of causing magnitude 8 or greater earthquakes (it does not get greater than 9 or so!) that would pose great threats to Seattle, Portland, and other cities.

California is located on the junction of the Pacific Plate and the North American Plate. Due to the interaction between the two big plates, it is an earthquake prone zone. The San Andreas Fault is the major culprit for the earthquakes in California. It starts from the Gulf of California in the south, and then extends north to past Los Angeles and the San Francisco Peninsula. Historically, it has caused many catastrophic damages. The magnitude 7.8 earthquake in San Francisco in 1906 was caused by the San Andreas Fault. It turned most of San Francisco into piles of debris and caused nearly 3,000 deaths. The magnitude 6.6 earthquake in northwest of Los Angeles in the January 1994 was also on the San Andreas Fault. It caused the collapse of more than 11,000 buildings, collapse and destruction of highways and high rise buildings, breaks of gas or water pipes, distributions of communication systems, and widespread fires. In addition, it directly or indirectly killed 62, injured more than 9,000, and displaced over 25,000 people. The total property losses reached $30 billion.

Because earthquakes occur not very frequent in the Eastern U.S, this has led to a misconception of this area being immune from earthquakes among many peoples. On the contrary, studies indicate that earthquakes in the East could possibly cause more devastating losses because the severe shaking from a similar magnitude earthquake would affect a larger area than in the Western U.S. In addition, unlike the Western U.S, most structures in the Eastern United States are not designed to resist earthquakes. Furthermore, the population density in the Eastern United States is higher than that of the West. Of the largest earthquakes recorded in North America's history, three of them originated from the New Madrid Fault system in a two-month span from 1811 to 1812. The powers from these magnitude 7 earthquakes were enough to force the Mississippi River to temporarily flow backward. With recent tectonic activities in the area, scientists believe that the ancient Ramapo Fault running near New York City might be building up energy to reactivate itself again in the near future.

2.2.2 *Hurricanes/Typhoons*

2.2.2.1 Typhoons in China

Located on the Northwestern Pacific Coast, China has the highest number of tropical cyclones (typhoons) that reached on its territory and it is one of the countries in the world with the most severe damages from typhoons. On average, about 7.4 typhoons made landfall on China's soil each year, which caused economic losses of about $24.6 billion and 570 deaths. Nearly 20 provinces, municipalities, and autonomous regions in the East have been impacted by typhoons; however, they are more concentrated in the coastal provinces and cities (Fig. 2.9). According to the statistical data, nearly 89 % of typhoons made landfall on the southern coastal areas (Guangdong, Guangxi, Fujian, Hainan and Taiwan), while only 11 % made landfall on Zhejiang Province or the coastal provinces north of it. The number of typhoons which made landfall on Guangdong Province is highest, accounting for nearly 1/3 of the total landfall. It is followed by Taiwan (1/5).

The activity of typhoons in China has a clear seasonality. Typhoons are most likely developing during the period from late summer to early fall. The number of occurrences is the least in February, gradually increases to the highest number in August, and then decreases slowly from September onward. Typhoons take place mainly in July to October, accounting for 68.6 % of all occurrences.

The seasonality of the typhoons' landfall has the following features: (1) in May and June, they are located at the coasts of the southern provinces; (2) in July and

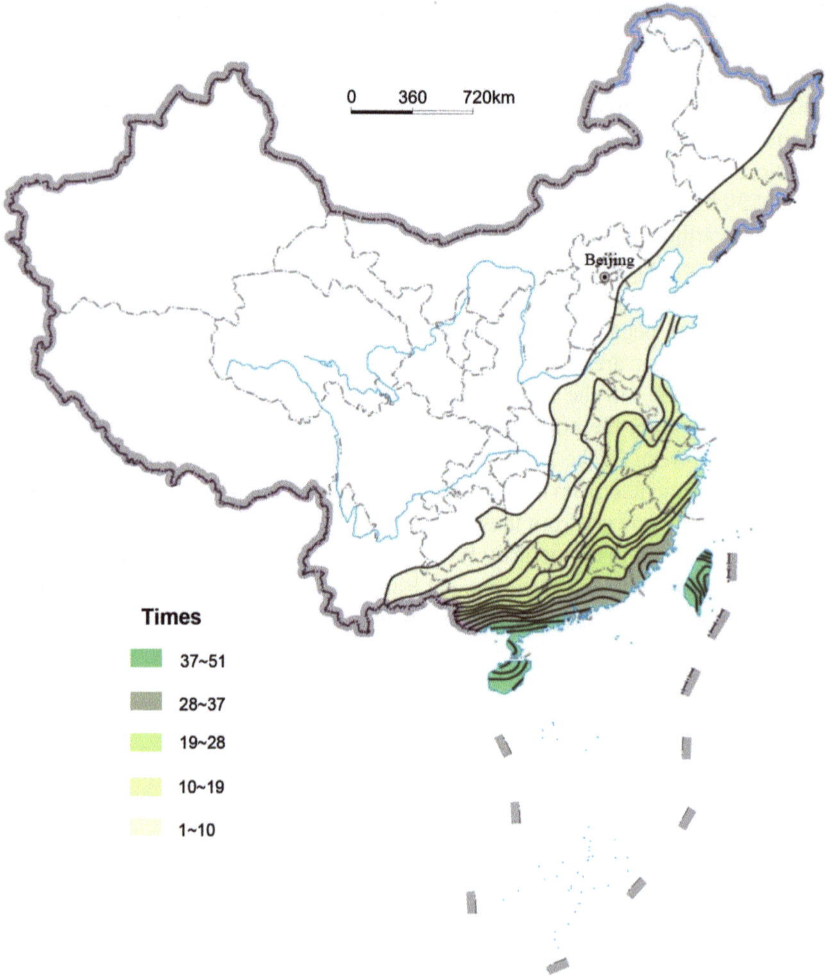

Fig. 2.9 Distribution of typhoons in eastern China (1949–2000 years). (Shi 2003)

August, they are in the areas between Guangdong, Guangxi, Hainan in the south, and the vast coastal areas of Liaoning in the north; (3) in September and October, they are active in the area along the coasts of the Yangtze River Estuary; (4) in November, the active areas are confined to the coasts of Guangdong, Hainan and Taiwan; (5) in December, they make landfall occasionally only on Guangdong.

In addition, the paths of typhoons also change with the seasons. From November to May, it travels northward in the ocean from east of the 130 °E Meridian. When it reaches the south of the 16 °N Parallel, it turns westward to the southern area of the South Sea or makes landfall from southern Vietnam. During June to October, typhoons head northward from the ocean east of the 125 °E Meridian, slightly shifting westward. During the hot summer season, its west moving path is located even further north.

2.2.2.2 Hurricanes/Tornados in the U.S

In the Atlantic Ocean, hurricanes mainly form over the warm tropical and sub-tropical ocean water of at least 26 °C that extends to a depth of 46 m below the surface between 5° and 20 °N. Most of the hurricanes that made landfall in the U.S. originated from a hot and humid area in the Atlantic Ocean, surrounded by west of North Africa, the east coast of the Central America and the Gulf of Mexico coast, known as "Hurricane Alley". Typically, when the low-pressure systems move westward from the west coast of Africa, they will encounter the warm waters of the tropical Atlantic. The warm waters will provide sufficient energy to fuel the development of hurricanes and turn them from the initial stage off the west coast of Africa into monsters to attack the U.S coast.

The strong winds, high waves, storm surges, floods, tornadoes, landslides, and coastal erosions brought forth by hurricanes can cause tremendous property damages and the losses of lives. Not only those in the hurricane-prone areas could suffer from the devastating destructions of the hurricanes, but the residents in the interior areas are vulnerable to the destructive winds, tornadoes and floods from hurricanes as well (Fig. 2.10). More than half of the total U.S. population lives within 80 km of the coast. As population concentrations and development have accelerated along the coastal areas, the damages caused by hurricanes have risen exponentially over time. Historically, six of the 10 costliest disasters in the U.S were caused by the tropical weather systems.

Among them, the Galveston hurricane in 1900 was the deadliest natural disaster in U.S. history, which claimed about 8,000 lives. The New England hurricane in 1938 was the costliest natural disaster of its time, causing about $3 billion in damages. Hurricane Hugo in 1989 caused losses of almost $11 billion for the Carolinas. Hurricane Andrew in 1992 accounted for more than $20 billion of property damages. Hurricane Katrina in 2002 killed 1600 people and caused over $100 billion in damage. Losses from the recent Atlantic hurricane seasons have been keeping rising, with losses of $42 billion in 2004 and hundreds of billions in 2005.

2.2.3 Floods

2.2.3.1 Floods in China

China has always been one of the countries in the world with frequent floods and most destructive damages with regards to its vast territory, complex terrains, and significant monsoonal influence. Floods are the most serious natural disaster in China. Since the twentieth century, the seven biggest rivers alone had 31 large or extremely large floods, 55 large floods, 127 regular floods (Wang, 2006) (2.13 times yearly on average). In recent years, floods have struck nearly 9,670,000 acres of lands with 4,729 fatalities and 2,025,300 destroyed houses each year. They have caused direct economic losses of up to 113,668,000,000 Yuan, accounting approximately for 1–3 % of the national GDP.

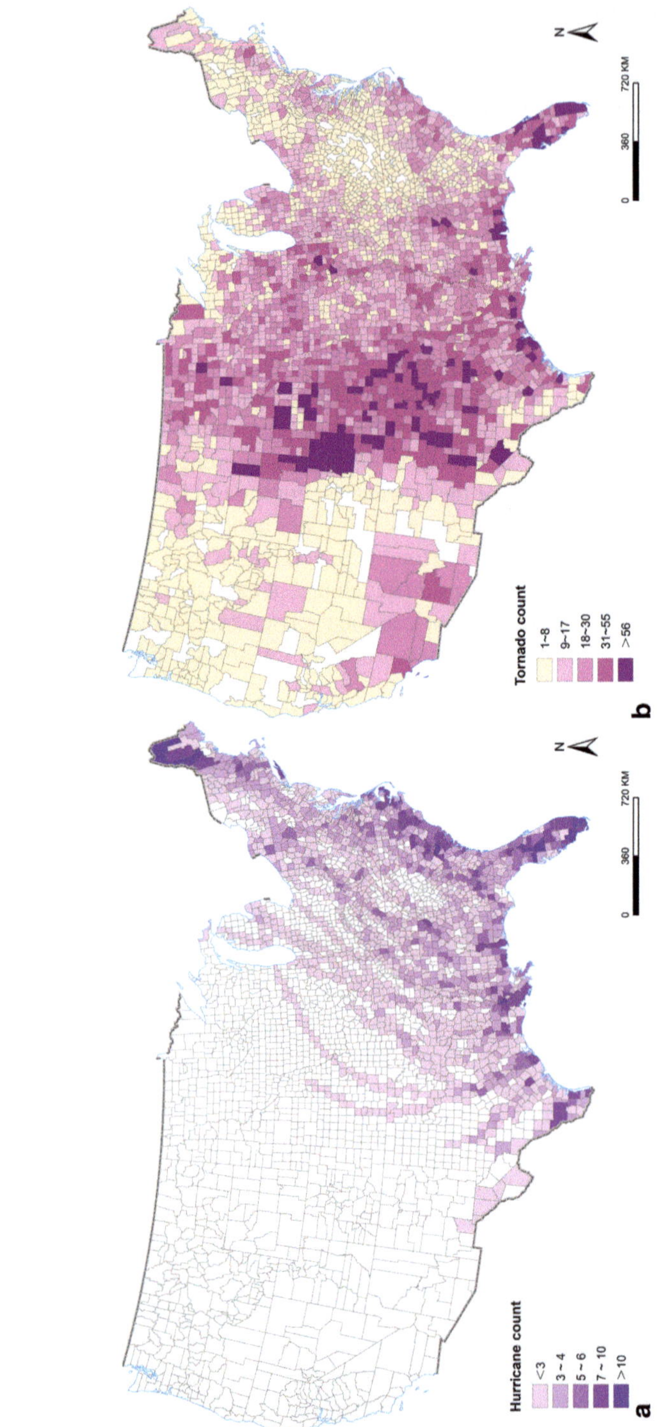

Fig. 2.10 **a** The hurricane distribution in the conterminous U.S. **b** The tornado distribution in the conterminous U.S.

In China, floods occurred in many areas. Historically, every province, autonomous region or municipality has had various kinds of floods (Wang, 2006). The flood patterns in 1949–1965 period and in 1978–2000 period (Fig. 2.11) provide us with some information about their spatial distributions in the past. During the 1949–1965 period, the distribution of the floods in China had an obvious north-east difference. Floods were mainly distributed in the places east of the Hu Huanyong line. Moreover, these eastern areas in the second topographic step suffered from more damages. During the 1978–2000 period, floods declined from the northeast to the southwest and from the southeast to the northwest. Flood frequency declined from the areas east of the Hu Huanyong line to the sub-arid area and from Xinjiang Province to the tropical area. Comparing the patterns in the two periods reveal that floods during the period from 1978–2000 affected larger areas and were more frequent. The center of the extreme floods shifted south-ward, northward, and southwestward from the North Plain (taking Henan as the center). The changes in the flood pattern were mainly the result of influences of changing land uses. On the one hand, humans extend their activities from the plains to the wetlands, particularly with the creation of farmlands from the low wetlands in northern China and from the lakes in the Middle-Lower Yangtze River Plain. On the other hand, the over exploitation on hilly lands and forest lands causes the deterioration of the ecosystem, and severe soil erosion.

2.2.3.2 Floods in the U.S.

Floods are the most costly and frequent natural disaster in the United States. Basically, any normal natural processes, such as summer rains, heavy thunder-storms, or winter snow melts can cause floods, let alone extreme weathers such as hurricanes and heavy thunderstorms. Sometimes, the failure of levees and dams and blocked drainage system in the urban areas can also result in flooding. Floods in the Eastern U.S and the Gulf Coast were mainly caused by hurricanes and storms, while snow-melt and rainstorms were the main causes in the Western U.S. Rainfall intensity and duration are the two most important factors that contribute to flood. When a river receives more water than it can drain, flood occurs. Floods can be slow or fast rising. Within small drainage basins, a brief intense rainfall in the local areas can cause fast-moving but short-lasting flash flood. However, within large drainage basins, floods result from widespread rainfall that last for hours or days.

In the U.S., on average, floods causes more than $6 billion of property damage and kill more people than tornadoes, hurricanes or lightning combined each year. Flooding has claimed more than 10,000 lives since 1900. Nearly half of these deaths were the result of the drivers' underestimations of the depths and power of the moving water. When they drove their cars into the flooded roads, their automobiles were swept downstream. Thanks to progress made in the development of warning systems, floods have claimed fewer lives than they used to during the past several decades. However, they have resulted in more economic losses due to the increased development in the coastal area. For example, the flooding brought by Hurricane Katrina in the late summer of 2005 accounted for a large proportion of the more than

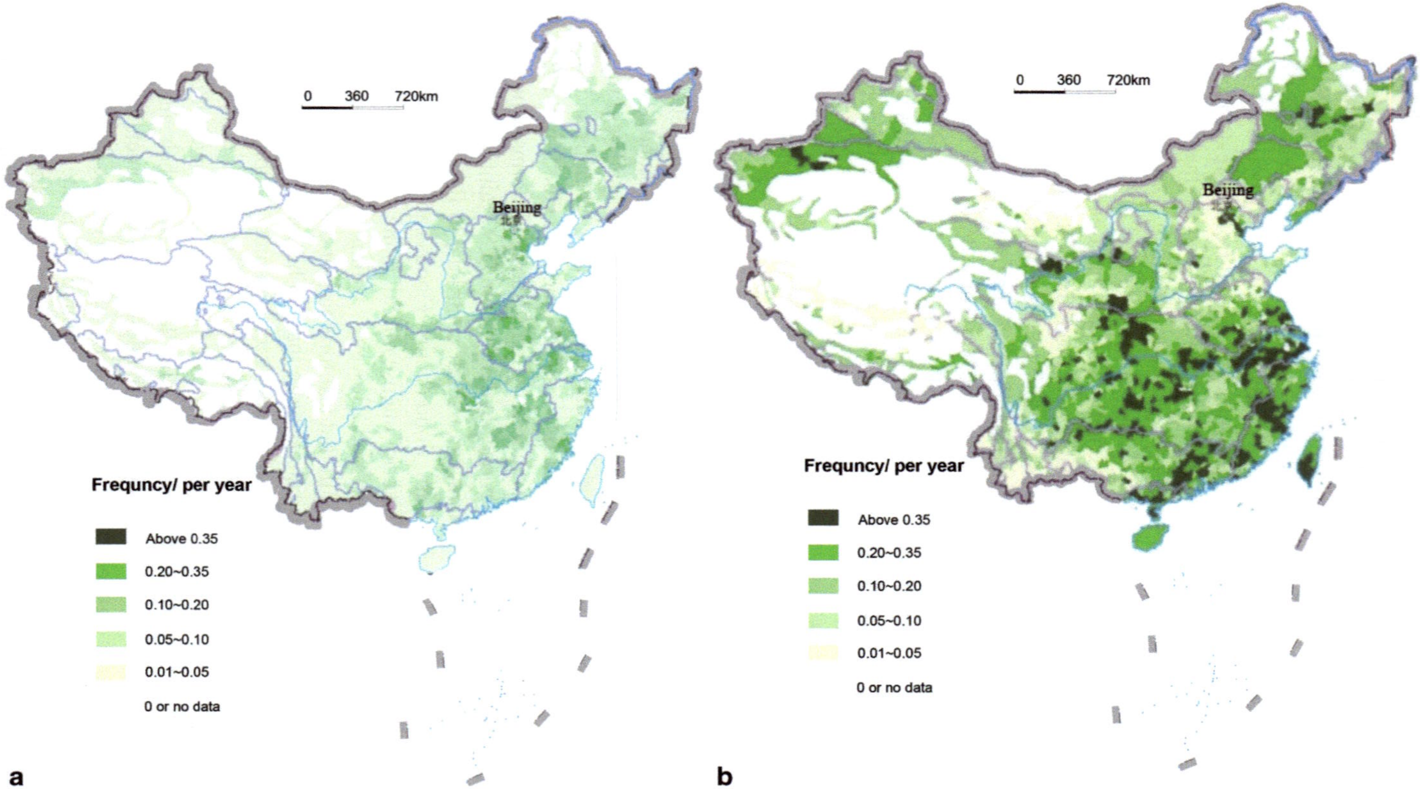

Fig. 2.11 **a** Distribution of flood frequency in China (1949–1965). (Shi 2003). **b** Distribution of flood frequency in China (1978–2000). (Shi 2003)

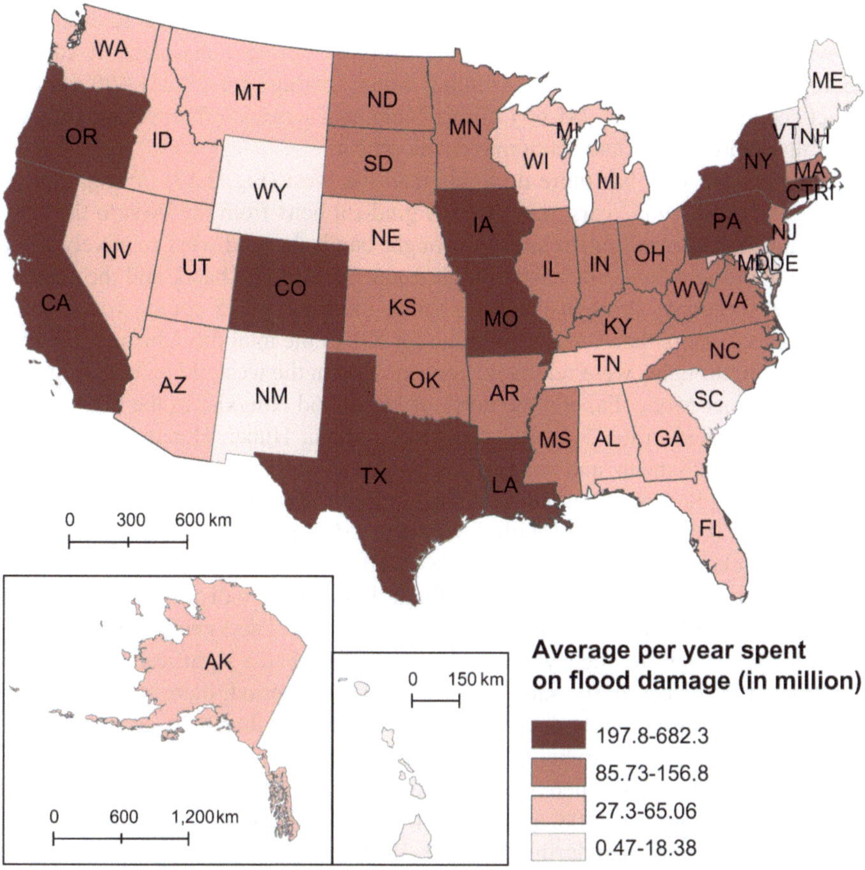

Fig. 2.12 Average per year spent on flood damage by state in millions US$. Quoted from http://www.bigrede.com/floods/

$200 billion of losses, the costliest natural disaster in U.S. history. The costliest flood in U.S history, the 1993 Midwest flood, affected 1/3 of the U.S states, and caused 50 deaths, the breaks of hundreds of levees along the Mississippi-Missouri River, nearly 10,000 home losses, the submergence of hundreds of towns, and the inundation of 15 million acres of farm-land, causing more than 20 billion of economic losses.

2.2.4 *Droughts*

2.2.4.1 Droughts in China

Droughts are frequent in China. According to the current statistics, China had 1056 droughts during the past 2155 years from 206 B.C. to 1949. With climate change, droughts have affected larger areas and have had higher frequencies in recent years.

Each year, on average, about 21.598 million acres of lands are subject to droughts; of which, 9,496,130 acres would suffer from the drought disasters. It caused 15,405,000 tons of grain losses, drinking water shortage for 29,231,400 people, direct economic losses of 231.664 billion Yuan, and serious influences on China's agricultural production and economic development.

In China, the East has more droughts than the West (Fig. 2.13) (Wang, 2006). During the 1949–1965 years, there were 4 gradient belts from the West to the East including the western low frequency drought band, the mid-west relatively high frequency drought band, the eastern high frequency drought band, and the eastern coast low frequency drought band. Of them, there were 115 counties with high frequencies of more than 0.20, accounting for 5 % of the total droughts nationwide. In the North, droughts were relatively concentrated in the west of Heilongjiang, the middle of Inner Mongolia, and the north of Hebei and Ningxia; in the South, they mainly distributed in the 5 central provinces (Anhui, Hubei, Hunan, Jiangxi and Henan), eastern Sichuan, and northern Guizhou and Yunnan. 480 counties, with approximately 22 % of the total number of droughts, had drought disaster frequencies of more than 0.08. Compared with the 1949–1965 period, the spatial distribution of droughts changed little during the 1978–2000 period, with only slightly reduced number of droughts in the North China Plain. However, the distribution of the severe droughts had a relatively large change in the South. The 5 central provinces all had a lower drought frequency. By contrast, Guizhou had a significant increase in the drought frequency in the same period. During the period, there were 64 counties nationwide with drought disaster frequencies of more than 0.20, accounting for about 2.8 % of the total number of national droughts and half of what was the case in the 1949–1965 period. There were 693 counties nationwide with drought disaster frequencies of more than 0.08, accounting for 34 % of the total of national droughts which are 12 % more than during the 1949–1965 period's level.

A comparison of the drought patterns during the two time periods reveals three main characteristics:

First, the spatial patterns of droughts were different in the East and West. In the later period, the drought frequency in the North was higher than that in the South. The frequencies of droughts had a zonal effect along certain directions. This was possibly the result of the step-like topography, the spatial variation of the precipitation, and the different material responses to the drought disasters. The distributions of the droughts with high values in the North were related to the Southeast monsoon and the ecological degradation of the environment from farming.

Second, during the second period, the extend of the drought area spread westward. Meanwhile, the area with severe drought had been expanded northeastward and southwestward, a reflection of human activities especially regarding exploitation of more dry agricultural lands. During the first period, the total number of counties with drought accounted for 83 % of the national total. However, it increased to 94 % during the second period.

Third, from the viewpoint of land use types, the center of droughts with high values was more stable in the livestock areas, generally in the vicinity of Central to Eastern Inner Mongolia. However, the center in the agricultural areas changed from south to north and from single center to multiple centers (Fig. 2.13).

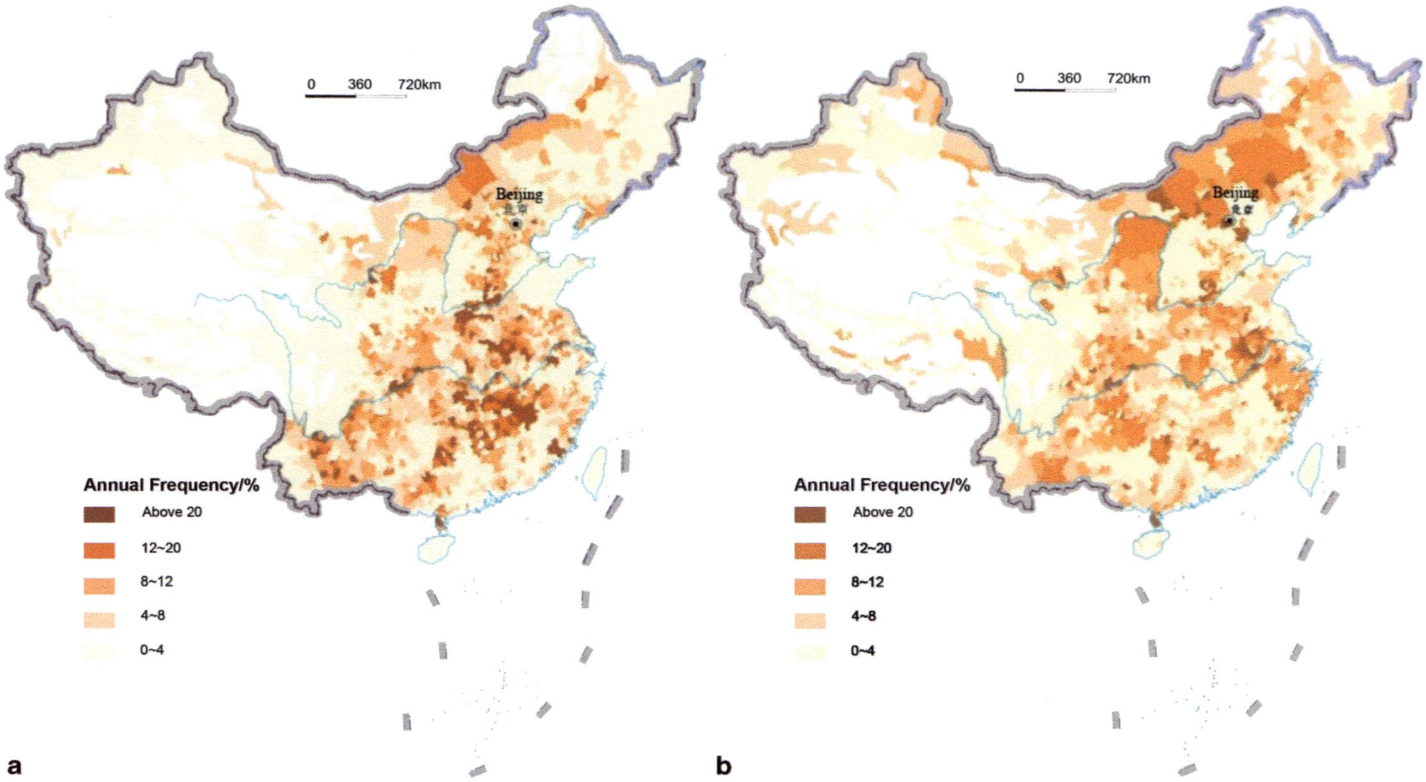

Fig. 2.13 **a** Annual frequency of drought in China (1949–1965). (Wang 2002). **b** Annual frequency of drought in China (1978–2000). (Wang 2002)

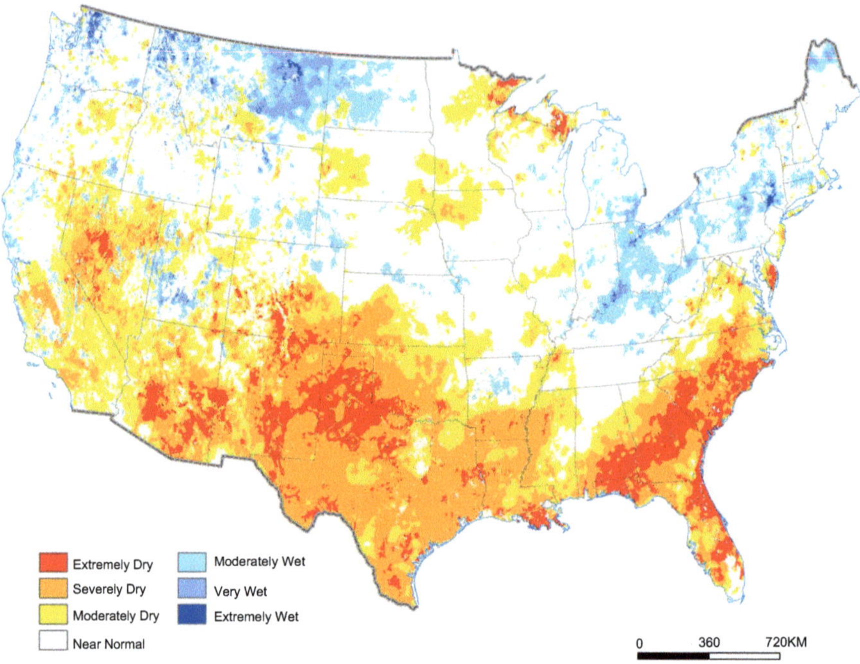

Fig. 2.14 Drought severity index of the conterminous U.S. (August 2012)

2.2.4.2 Droughts in the U.S.

Drought has affected more people in the United States than any other natural hazard. In fact, almost every year, a region somewhere in the U.S. experiences drought. Evidence from the instrumental record indicates that drought can affect any region in the U.S. While the central High Plains seems to have the most persistent droughts, the southern Plains and parts of Texas have the highest frequency of drought. Although much of the natural disasters are brief and short-lived, drought is a more gradual phenomenon, slowly spreading to more areas over time. In extreme cases, drought could last for many years, and could have destructive influences on agriculture and water supplies. The two major droughts of the twentieth century, the 1930s Dust Bowl drought and the 1950s drought, both lasted five to seven years and covered large areas of the conterminous U.S. The drought that affected the largest area during the past 100 years (the period of instrumental record) was the one in July 1934, which threw nearly 80 % of the continental U.S. into moderate to extreme drought.

Historically, drought has caused tremendous damages to the United States, averaging $6–8 billion each year. However, more losses have taken place in recent decades. For example, every year from 1987 to 1989, the losses caused from droughts reached as high as $39 billion, becoming one of the most costly natural disasters documented in U.S. history. Nationwide losses from the drought of 1988 exceeded $40 billion, more than the losses caused by Hurricane Andrew of 1992, the Mississippi River floods of 1993, and the San Francisco earthquake of 1989.

Drought has had a great impact on American societies. For instance, the Dust Bowl in the 1930s caused a massive migration from the Southern Plains to California, changed the agricultural policy on the Plains, and synchronized with the Great Depression, exacerbated the sufferings of millions. Even today, the hundreds of drought related records set up in the 1930s still haven't been broken yet. No other natural disaster in the recorded U.S history has affected so much of the country for so long.

In summary, natural disasters in both China and the U.S have their own characteristics. On one hand, China has a more noticeable monsoon climate; the flood and drought disasters are the most typical types of agricultural disasters; earthquake and typhoon both cause tremendous damages. On the other hand, hurricane/tornado, flood, and earthquake are the major natural disasters of the United States. Because of the influence of the "Circum-Pacific Earthquake Belt", earthquakes have caused destructive damages to both countries. However, China is also affected by the "Mediterranean Sea—the Himalayas earthquake Belt". In addition, China has more complex terrains and landforms; therefore, the earthquakes in China are more complex and could cause more destructive damages. Floods take place frequently on the plains of both countries. However, the limited areas of plains in China are often the most densely settled places and locations for agriculture, population, city and the economy. Therefore, floods and droughts in China are more damaging than those of the U.S. Finally, the coastal areas of both countries are affected by hurricanes (typhoons). Since China's coastal areas have a denser population concentration, they are more vulnerable to hurricanes.

2.2.5 *Physical Zonations*

2.2.5.1 Physical Zonation of China

Regional differentiation is the theoretical basis for physical zonation. This results in the phenomenon that under the impact of solar radiation and the Earth's internal energy, the natural environment and its present components some degrees of consistencies apply to certain directions but are different for other directions. The most basic and common differentiation applies to phenomena of zonal and non-zonal change. The latter can be divided into humidity zonation (longitudinal zonation) and vertical zonation (altitudinal zonation).

As the field of geography developed in China since the 1950s, the fundamental materials and resources for geographic research have been gradually improved. Accordingly, various zonation schemes have been put forward by many Chinese geographers. The one proposed by Huang Bingwei in 1959 turned out to be the most representative scheme. It is comprehensive and well designed, and has been included into the "Physiographical Atlas of the People's Republic of China" after its revision.

The basic principles for this zonation are that the high-level unit follows the zonal principles but the low-level unit follows the non-zonal principles. The system

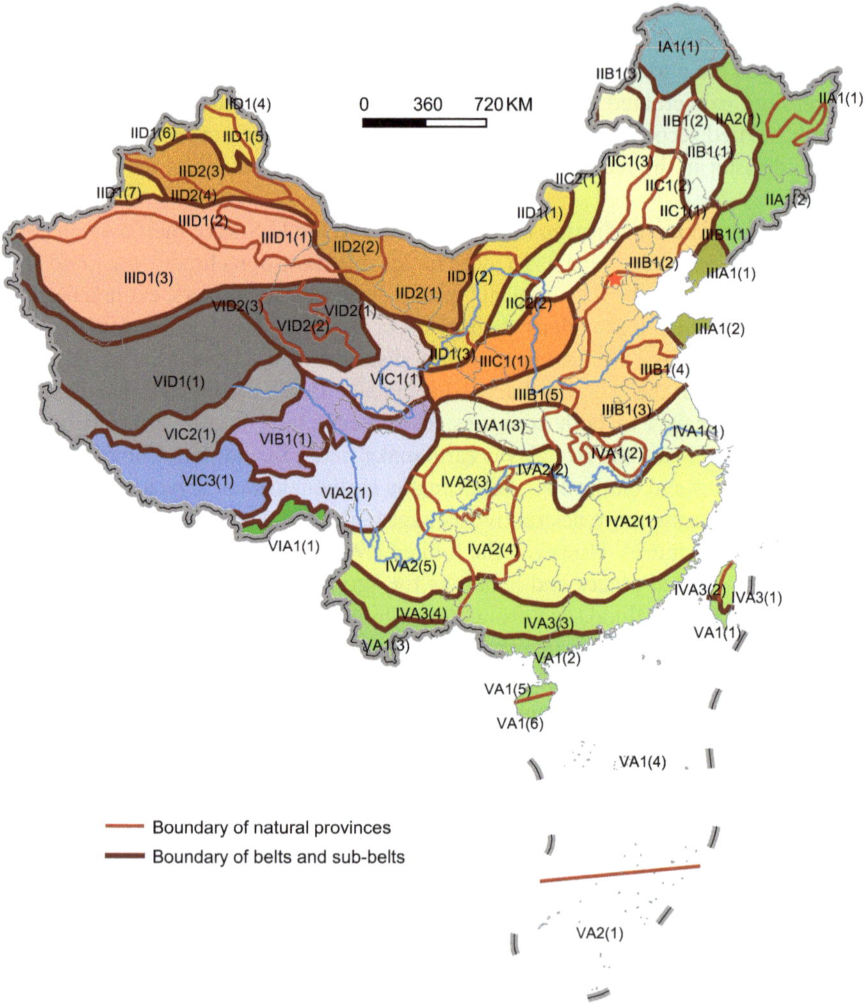

Fig. 2.15 Physical zonation of China (From Physiographical Atlas of the People's Republic of China, 1965)

divides the country into zones from the top to the bottom, based mainly on the method of the superposition of the leading symbols. The scheme divides the nation into three natural areas (the eastern monsoon area, the Northwest Arid Areas and the Qinghai-Tibet Plateau area), 6 heat zones (the equatorial zone, the tropical zone, the subtropical zone, the warm temperate zone, the temperate zone and the cold temperate zone), 18 physical regions and sub-regions, 28 physical belts and sub-belts and 90 physical provinces (Fig. 2.15, Table 2.1).

The Eastern Monsoon Area The boundary of the Eastern Monsoon Area is the east of the Eurasian continent, with low-flat topography, deep soil and notable

Table 2.1 Physical zonation of China

Heat zones	Zonal vegetation
Cold temperate zone	IA1 the bright coniferous forest in cold temperate zone (wet)
Temperate zone	IIAl the needle broadleaf mixed forest in temperate zone (wet)
	IIA2 the sylvosteppe in temperate zone (wet)
	IIB1 the meadow in temperate zone (sub-wet)
	IIC1 the grassland with dark chestnut soil in temperate zone (sub-wet)
	IIC2 the grassland with light chestnut soil in temperate zone (sub-wet)
	IID1 the semi-desert in temperate zone (sub-wet)
	IID2 the desert in temperate zone (sub-wet)
Warm temperate zone	IIIA1 the broadleaved deciduous forest in warm temperate zone (wet)
	IIIBl the sun-dry broad leaved deciduous forest in warm temperate zone (sub-wet)
	IIIC1 the steppe in warm temperate zone (sun-dry)
	IIID1 the desert in warm temperate zone (dry)
Subtropical zone	IVA1 the broadleaved deciduous forest and broad-leaved evergreen forests in the cool subtropical zone (wet)
	IVA2 the broad-leaved evergreen forests in middle subtropical zone (wet)
	IVA3 the broad-leaved evergreen forests in warm subtropical zone (wet)
Tropical zone	VAl the selva and the monsoon forest (wet)
	VA2 Tropical equatorial rainforest zone (wet)
Cold temperate zone in Qinghai-Tibet plateau	VIAl the broad-leaved evergreen forests in the Qinghai-Tibet Plateau (wet)
	VIA2 Coniferous forest and meadow in the Qinghai-Tibet Plateau (wet)
	VIB1 meadow in the Qinghai-Tibet Plateau (sub wet)
	VIC1 meadow, steppe, forest in the Qinghai-Tibet Plateau (sub-dry)
	VIC2 alpine steppe in the Qinghai-Tibet Plateau (sub-dry)
	VIC3 meadow, steppe in the Qinghai-Tibet Plateau (sub-dry)
	VIDl desert and semidesert in the Qinghai-Tibet Plateau (dry)
	VID2 desert and semidesert in the Qinghai-Tibet Plateau (dry)

monsoon influence. Its wind direction and rainfall vary with the seasons. In summer, the southeast monsoon comes from the ocean which brings plenty of rainfall so that the climate is humid and the hydrographic network is complex. From the south to north, the forest vegetation and soil are different because of the difference in temperature. In addition, influenced by the extensive and long-term human activities, most of the natural forests are no longer in existence and almost all of the arable lands have been reclaimed as farmland. This region, with the high density cities, towns and rural settlements everywhere, is an area where population density is largest and economy is the most active.

The Northwest Arid Area It is located near the center of the Eurasian continent. Since there is less influence from the southeast monsoon and fewer mountains, it has an arid or semiarid climate. There are few rivers, and they are mainly inland rivers. The vegetation types mainly include desert, desert steppe and steppe, mountain forests and alpine meadows only in the higher mountain elevations. The soil is barren and the salinization common. In the vast basins, there are many mobile and half-mobile sand dunes, which are the main source of sandstorms. In this area, there are fewer human activities compared to the eastern monsoon area. Due to excessive reclamation and overgrazing, frequent sand storms and the grassland degradation are the major problems.

The Qinghai-Tibet Plateau Area It is the largest plateau of the world with an average elevation of more than 4,000 m, thin air, low and cold temperature, intense radiation and strong winds. In the hinterland where there is less moisture coming from the outside, the temperature is so low that there are still many glaciers exist in the high mountains. Due to the harsh environment, plants and animals are rare and the vegetation mainly includes the cold desert and the alpine grassland. In the southeast, canyons are cut deep and the vertical vegetation belt spectrum mainly consists of sub-alpine forests and alpine meadows. In this area, only small populations live and there are even depopulated zones. Human activities are mainly limited to some valley areas.

There are 4 steps in the comprehensive zonation system of China (Table 2.2). The first step divides China into 6 heat zones and 1 extremely cold region according to the surface temperature; the second divides China into 18 natural are-as according to the moisture content; based on the divisions from heat and moisture, the third divides China into natural regions according to the conditions of the soils and the vegetations; the last step further divides China into natural provinces according to the biological and climatic conditions of the natural zones.

2.2.6 Physical Zonation of the U.S.

The physical zonation of the U.S. is based mainly on the ecological features, which classify and divide the continental U.S into various zones and sub-zones according to the characteristics of the ecosystems and various environmental factors that drive the differentiations of their complex components. The zones from this system represent the combinations of the ecosystems with similar functions. It enhances the management of lands from a single isolated resource to the entire catalogue of resources within the ecosystem for the goal of achieving the ecological monitory, biological conservation, and sustainable development. At present, the most commonly used one is the Robert Bailey system, which divides North America into ecoregions and sub-ecoregions based mainly on their climatic conditions and their corresponding vegetations. In reference to the Koeppen climatic classification system, Robert divided the North America into 4 large natural areas and 15 divisions. The conterminous U.S is mainly located in the moist temperate area and the arid area (Table 2.3).

The Humid Temperate Area The main controlling factors of the areas in mid latitude (30 °N to 60 °N) are the tropical and polar air masses. This area is influenced by hurricanes where the majority of precipitations comes from the rising

Table 2.2 The index system for physical zonation in China (Huang 1993)

System	Zonation basis			Significance and description
Natural area (3) (level 0)	Eastern monsoon region, northwest arid area, the Qinghai-Tibetal pine areas; integrated features			Reflect then on-zonal The hue for background (not on map)
Heat zones and the sub-bands (6) (level 0)		Accumulated temperature ≥10 °C	Average temperature of the coldest month	Zonal feature Landscape and agricultural cropping Code with I–VI (not on map)
	Boreal	<1,700 °C	<−28 °C	
	Temperate	1,700–3,200 °C	−28–−8 °C	
	Warm	3,200–4,500 °C	−8−0 °C	
	Subtropical	4,500–8,000 °C	0–16 °C	
	Tropical	>8,000 °C	>16 °C	
	Maximum elevation of plateau farming			
Natural areas, sub-areas(18) (level 1)	Heat zones as basis, similar water and heat mix Dryness: wet < 1 Sub-1 1.2 (1.5) Semi-arid 1.2 (1.5) ~ 2 Drought > 2			Hydrothermal general grid Determine the direction of land use and agriculture Coded with A–D
Sub-zones of natural areas (28) (level 2)	Combination of water-heat as basis Soil (sub-categories), vegetation (formations Programmed) and same land use			Basic unit of zonation Determine land use structure, crop varieties, cropping Coded 1–3 (color representation)

moist air within the hurricane air masses. The seasonal variation of temperature and precipitation is apparent. The seasonal variation of the energy and the temperature is larger than the diurnal variation. The middle latitude area's climate has the distinctive winter, that the tropical climate does not have. The vegetations mainly include the temperate deciduous broadleaf forest, the coniferous forest, and the evergreen trees.

The Dry Area The basic characteristic of dry climates is that the loss of the surface water through evaporation is greater than the precipitation it receives each year. Generally, the dry climates can be divided into two areas: the arid desert and the arid prairie. The semi-arid prairie is a transitional zone between the desert zone and the moist zone. The boundary line between the arid zone and the semi-arid zone is arbitrary, which is usually defined by using one-half of the rainfall between the humid area and the semi-arid prairie. Of all the climates, the dry climate is the most widespread, occupying 1/4 of the Earth's land area.

Table 2.3 The physical zonation of the conterminous U.S. (Bailey 2002)

Area	Division	Equivalent Koppen climates	Zonal vegetation	Principal zonal soil type
Humid temperate Area	Warmer continental	Dcb	Mixed evergreen and deciduous-coniferous forests	Gray brown podozolic
	Hot continental	Dca	Broadleaf forest	Gray brown podozolic
	Subtropical	Cf	Broadleaf-coniferous evergreen, coniferous-broadleaf semi-evergreen forests	Red and yellow podzolic
	Marine	Do	Mixed forests	Brown forest and gray-brown podozolic
	Prairie	Cf, Dca, Dcb	Forest-steppes and prairies, savannas	Prairie soil, chernozems, and chestnut-brown soil
	Mediterranean	Cs	Dry steppe; hard-leaved evergreen forests, open wetland and shrub	Grassland soil
Dry Area	Tropical-subtropical steppe	BSh	Open woodland and semi-deserts, steppes	Sierozem, and brown soil
	Tropical- subtropical desert	BWh	Semi-deserts, deserts	Sierozem, desert soil
	Temperate steppe	BSk	Steppes, dry steppes	Sierozem, desert
	Temperate desert	BWk	Semi-desert, desert	Sierozem, desert

2.3 Environmental Issues and Problems

2.3.1 China's Status Quo and Action Plans

2.3.1.1 Air Pollution

Sulfur dioxide (SO_2), nitrogen oxides (NOx) and particulate matter (PM) are accessed in 325 prefecture-level (and above) cities and 113 major cities according to *the Ambient Air Quality Standard* (GB 3095–1996). It can be concluded from the reports that the atmospheric quality and acid rain pollution status of cities nationwide have stayed steady but are still in heavy contamination of the air in the larger urbanized areas, in particular in the three key areas and regional clusters Beijing-Tianjin-Hebei regional cluster, the Yangtze River Delta and the Pearl River Delta, in 2012.

Beijing can serve as an example. The nation's capital has a semi-dry monsoon influenced humid continental climate with a high natural dust precipitation. In recent decades, with the acceleration of industrialization and urbanization processes, there are more than twenty million tons of coal burned, emission from more than five million vehicles and more than 150 million square meter of urban construction area every year in Beijing. All this contributes significantly to the atmospheric pollution. Beijing has suffered frequently from haze and smog since 2013. The air contains a larger density of PM 2.5 particles—those smaller than 2.5 microns in diameter, which can penetrate deep into the lungs—which mainly comes from coal firing and motor vehicle emissions. In addition, the unusual sustainable steady atmosphere and high aerosol concentration make for the continuous accumulation and agglomeration of aerosols, thus increase the persistence of haze and smog. In January 2013, there were only 4 days of clear weather with five times of severe haze pollution processes in Beijing-Tianjin-Hebei regional cluster. Beijing is now struggling to keep a clean air situation as only the wind in winter and the rain in summer can kick out the haze and smog.

The Beijing Government published the *Clear Air Action Plan from 2013 to 2017* in order to significantly improve the air quality which focuses on prevention and control of PM2.5 and promotes emission reduction of multi-contaminants. On Sep 12th, 2013, the Airborne Pollution Prevention and Control Action Plan (2013–2017) unveiled by the State Council sets the goals for the nation's 338 cities. For the Beijing- Tianjin- Hebei regional cluster, concentration levels of PM2.5 particles must be cut by 25 % by 2017 from 2012 level, under the plan. In this plan, ten measures are published: (1) Reduce contaminant emission (coal consumption from 68 to 65 %; removal of 15 million heavy polluting motor vehicles); (2) Strictly control the new capacity of energy-intensive and heavy-pollution industries (such as a reduction of steel production capacity); (3) Promote cleaner production and develop public traffic; (4) Accelerate the restructuring of energy industry (by increasing the supply of natural gas and non-fossil fuels); (5) Enhance energy efficiency and environmental protection indexes constraints; (6) Promote new system of simulation and restriction of emission reduction and energy conservation; (7) Push industrial transformation and upgrading by laws and standards; (8) Format a mechanism of joint prevention and control of Beijing-Tianjin-Hebei mega-region, Yangtze River Delta mega-region and Pearl River Delta mega-region in order to work on PM2.5 pollution in populated areas and major cities; (9) Severe pollution weather should be taken consideration by local government's public emergency management; (10) Build the social standard of behavior of "share the same breathe and work together", everyone participates and plays their respective duties. "Western countries have spent decades trying to improve their air quality, and are still trying. What China is doing here is trying its best to make improvements happen as soon as possible, but we also need to be realistic about the hardships ahead and prepare for a protracted war against pollution. It's a war that will involve every single member of the public," said Chai Fahe, Vice-President of the Chinese Research Academy of Environmental Sciences (China Daily 2013).

2.3.1.2 Water Pollution

The monitoring of fresh water includes surface water (rivers, lakes and reservoirs) and underground water in China. In *Environmental Quality Standards for Surface Water* (GB 3838–2002), surface water is divided into five classes according to its purpose for use and protection target.

Class I: mainly for source of water and national nature protection areas
Class II: mainly for class one protection areas for centralized potable water source, protection areas for rare fishes, spawn ground for fishes and shrimps etc.
Class III: mainly for class two protection areas for centralized potable water source, protection areas for general fishes and swimming areas
Class IV: mainly for general industrial water areas and entertainment water areas not directly touched by body
Class V: mainly for farmland water areas and water areas for general landscape requirement.

If there are several classes of function in the same water areas, the highest class of function will be taken for classification. Different classes of function have different standards accordingly.

In 2012, the country's state-monitored surface water sections were in a light pollution situation. Among the state-monitored surface sections of the Yangtze River, Yellow River, Pearl River, Songhua River, Huaihe River, Haihe River, Liaohe River, the rivers in Zhejiang Province and Fujian Province, the rivers in Northwestern China and the rivers in Southwestern China, the percentage of Class I–III is 68.9 %, Class IV–V is 20.9 %, and the rest is below Class V (see List of Classes I to V). Major indexes are chemical oxygen demand (COD), 5 days' biochemical oxygen demand (BOD5) and permanganate index.

Among 62 state-monitored lakes and reservoirs, the percentage of Class I–III is 61.3 %, Class IV–V is 27.4 %, and the rest is below Class V. Major indexes are total phosphorus (TP), chemical oxygen demand (COD), 5 days' biochemical oxygen demand (BOD5) and permanganate index.

Lake Taihu can serve as an example. Taihu Lake, located south of the Yangtze River Delta, is one of the five largest freshwater lakes in China. Since the mid-twentieth century, the lake was polluted gradually with nitrogen and phosphorus materials. Large areas of the lake were in eutrophication, and the quality of freshwater kept deteriorating. In recent decades, the quality level of freshwater was downgraded from mainly Class II to mainly Class IV, and the level of eutrophication upgraded from mainly mesotrophic and mid-eutrophication to severe eutrophication. The Cyanobacteria Event in Taihu Lake happened in May, 2007. A toxic-algae bloom in Wuxi, Jiangsu Province, caused a lack of freshwater for the whole city. The bottled water was sold out in every store and shop. By now, the water quality at Taihu Lake has improved significantly through technical measures and policies such as cutting pollution sources, salvaging algae, water diversion, dredging and ecological recovery.

2.3.2 *Status Quo and Action Plans in the U.S.*

The United States of America has been experiencing various kinds of environmental problems. Among them, air pollution and water pollution are the most common and widespread ones.

2.3.2.1 Air Pollution

In general, air pollution in the U.S. consists of particulate pollutants and gaseous pollutants such as sulfur dioxide (SO_2), nitrogen oxides (NOx), ozone (O_3), carbon monoxide (CO), volatile organic compounds (VOC), certain toxic air pollutants, and gaseous metals from fossil fuel fired power plants, industrial facilities, petroleum refineries, and automobiles. Particle pollution (PM2.5 and PM10) includes a mixture of compounds that are from mechanical processes like construction, mining, agriculture, and chemical processes like burning fuels and emitting gases. Generally, they are grouped into five categories: sulfate, nitrate, elemental carbon, organic carbon, and crustal material. Currently in the nation, electric utilities emit over 60 % of national SO_2 emissions, agricultural operations emit over 80 % of national NH3 emissions, solvent uses contribute about 50 % of the national VOC emissions, and automobiles pump out approximately 60 % of national CO emissions.

With less than 5 % of the world population but the leading consumer nation in the world, the United States is the second largest air polluter, behind only China, and produces nearly 7,000 million metric tons of CO_2 in 2011, accounting for 17–18 % of the total greenhouse pollutants worldwide. With the number of automobiles approaching 200 million, automobiles are the single largest primary source of air pollution, emitting over half of the carbon monoxide, a third of the nitrogen oxides, and almost a quarter of the hydrocarbons into the atmosphere. In the United States, on average, four out of five Americans own a car (cars per capita was 0.799 in 2011). Americans' passion for cars has steadily increased over the years. When the Clean Air Act was issued in 1970, only 5 % of U.S. household owned 3 or more cars. This number, however, has increased to 19.1 % with an average of 2.1 vehicles for every family in 2011.

Secondary pollutants such as ground-level ozone formed from automobile pollutants have caused heavy smoggy skies and harmful dirty air in many US cities, especially in California, where many of the most-polluted cities in the US are heavily concentrated. With warm climate, curved topography, and high traffic volume, it has created a perfect condition for air pollution to generate, linger, and spread. For example, Los Angeles provides the classic example of photochemical smog and thermal inversions being created. According to the California Department of

Motor Vehicles (DMV), as of January 1, 2007, there were 6,675,888 automobiles, commercial vehicles and motorcycles registered in the County of Los Angeles. In the City of Los Angeles alone, there were a total number of 2,499,764 registered vehicles. In addition, people traveling on the roadways in Los Angeles/Orange counties experience the highest hours of delay (93 hours) among the metropolitan areas in the nation. When the enormous amount of harmful gases comes out from the massive number of automobiles, the city's coastal location and topographic setting have trapped and confined those harmful gases in a broad, mountain valley. When cooler air from the ocean slides under the warmer urban air, an inversion layer typically forms around 500 m altitude. Since the surrounding mountain ranges are higher than the inversion altitude, the valley tends to hold in the inversion layer and the pollution under it.

In the U.S., air quality has improved significantly since the Clean Air Act was created in 1970. As of 2010, ozone levels across the country had dropped 13 % since 2000, while particle pollution was 24 % lower, according to the American Lung Association, which began to monitor air quality in 2000. A study led by researchers at Harvard School of Public Health (HSPH) has found beneficial effects between reduced air pollution levels and improved life expectancy in 545 counties in the U.S. from 2000 to 2007. However, despite the progress made for cleaner air, roughly half the people in the United States (50.3 %) live in counties that have at times unhealthy levels of either ozone or particle pollution. Poor air quality has caused significant health risks to people with lung diseases like chronic obstructive pulmonary disease(COPD) and asthma, people with heart disease, diabetics and those who work or exercise outside. On average, air pollution is responsible in the U.S. for an estimated 50,000 premature deaths each year and costs from air pollution-related illness are estimated at $150 billion per year.

2.3.2.2 Water Pollution

Water pollution in the U.S. comes from both "point source" (PS) and "nonpoint source" (NPS). While the sources of water pollutants are almost infinite, point source pollution mainly includes those easily identified as outlet pipe, ditch, channel, tunnel, and conduit from individual industrial and sewage treatment plants. They are discrete, visible, and confined, therefore, are relatively easy to control. Nonpoint source pollution, on the other hand, comes from many dispersed sources; consequently, it is extremely difficult to control. Nonpoint source pollution comes from agricultural practices, residual areas, urban runoff, mining operations, and atmospheric deposition. When rainfall or snowmelt runoff moves over and through the ground, it takes away natural and human-made pollutants into lakes, rivers, wetlands, coastal waters and ground waters. The surface runoff from storm water in urban area is particularly detrimental since all sorts of debris such as sediment, oil, fertilizer, and pesticide residues, as well as organic residues from vegetation, animal droppings, and garbage can be flushed into streams, lakes, and oceans when it drains into the ditches and pipes quickly. Every year, enormous amounts of oil, paint, lead, grease, and toxic chemicals from automobiles have been washed off from the countless miles of dirty surfaces of paved streets and highways, parking lots, garages, and buildings roofs and lawns.

Agricultural runoff is another leading source of water pollution for rivers and streams, lakes, and bays. According to the latest National Water Quality Inventory, agriculture has contributed to 60 % of the impaired river miles and half of the impaired lake areas surveyed by states, territories, and tribes. The substantially increased use of nitrogen, phosphorus and other nutrient fertilizers for boosting crop yields has led to an sharp increase in the amount of contaminants being washed down to the water system around the country. Today, in the United States, most of the phosphorous and organic nitrogen pollution in the water comes from agricultural sources. Of the hundreds of millions pounds of almost 25,000 kinds of pesticide products such as herbicides, fungicides, and insecticides used in the agricultural fields, less than 1 % have actually reached the pests. The majority ends up as chemical residues in the nation's rivers, streams, and water bodies. Livestock farms that house hundreds of thousands of pigs, chickens, or cows, produce vast amounts of manure and urine; they also contribute excessive amounts of microbes, nitrate pollution and drug-resistant bacteria to water supplies. Those nutrients have threatened the water quality across the country and often cause algae blooms that cause oxygen insufficiency in the water and destroy aquatic life. For example, a broad "dead zone" that stretches over several thousand miles in the Gulf of Mexico has been forming due to the inadequate level of oxygen to support aquatic life because of the algae blooms from nutrients in animal waste.

Water pollution has become a serious problem around the country. According to a recent US wide assessment study of the water system by EPA, nearly about 44 % of the nation's streams, 64 % of the lake areas, and 30 % of the bay and estuarine areas are not clean enough for fishing and swimming. Another recent study indicated that only 28 % of the nation's streams have top healthy biological living conditions in their region. Each year, nearly 20 million Americans fall ill from waterborne parasites, viruses or bacteria that are from the animal waste. Drinking water contaminated with nitrates near animal plants such as hog factories has increased the occurrences of blue baby syndrome, caused deaths in infants, spontaneous abortions, and disease outbreaks. For instance, in the coastal waters in North Carolina, bacteria and viruses from animal waste have been linked to the death of more than one billion fish.

In recent years, the United States has made tremendous advances in controlling point source pollution from industries and sewage treatment plants. Meanwhile, tremendous amounts of efforts have been put into addressing the nonpoint source water pollution issue. In 1987, the country established the Nonpoint Source Management Program by the Clean Water Act Amendments; in 1990, the Coastal Nonpoint Pollution Program was established by the Coastal Zone Act Reauthorization Amendments. Various programs have been initiated from the local governments to deal with water pollution particularly NPS problems. In addition, many public and private groups have developed initiatives and education activities for pollution prevention, pollution reduction, and monitoring. However, those efforts and achievements could be in jeopardy as we push further into those previously inaccessible places where human could not physically go but rely strictly on technology for resources such as oil. The potential disaster that can be brought forth into our water systems sometimes can be catastrophic and beyond human being's ability to control and manage. For example, when the BP Deepwater Horizon oil well exploration which took place over a mile below the ocean's surface, occurred on April 20, 2010, nobody could go to site to stop the spill before millions of gallons of crude oil had been released into the Gulf of Mexico. The damage that this disaster brought upon the ocean, beaches, shores and wildlife will probably never be fully restorable.

2.4 The Comparison of the Physical Geography

2.4.1 Physical Geography

2.4.1.1 Comparison of Landforms

In China, the topography decreases from west to east in a ladder-like fashion, so most of the rivers originate from the west and flow eastward. This feature has resulted in a gradient development of the water resource. The huge differences in surface reliefs and the very rugged terrains have imposed big obstacles for agricultural productions and transportation infrastructure constructions. The topography of the United States is high in the east and west but low in the middle, which leads to the north-south alignment of most mountains and river networks. The central low and flat Great Plains makes it easy for the two sides to connect with each other, therefore, the difficulty for the development of transportation and resource is relatively minimal.

The two countries share some other common features as well. First, the topography of the two countries has an enormous impact on its respective climate and other natural environments, but it is more evident in China. Both countries have majestic mountains (the Himalayas and the Rocky Mountains respectively) in the west. In China, the uplift of the Qinghai-Tibet Plateau blocks the warm and moist air from the Indian Ocean, which forces the subtropical high pressure system to move further north, causing the northward shifting of the arid and semi-arid areas. The soaring and broad Cordillera Mountains which is composed of a serious of mountain ranges, intermontane basins and plateaus in the western United States has created many barriers to the eastward flows of the Pacific air masses. In the east, the north-south trending Appalachian Mountains, although not very high and wide, is enough to block the warm and humid air masses from the Atlantic Ocean to move westward. In addition, The mountains on the east and west sides have confined polar air mass from Canada, the tropical air mass from the Gulf of Mexico, and the Atlantic air masses to the continent's central portion. This has led to the violent collisions of those air masses with each other, usually causing extreme weathers such as hurricanes and tornados. Furthermore, the vast flat Great Plains does not cause any problems for the extremely rigid air masses from the Arctic to move southward to bring cold weathers to the immense central and the southeastern plains in the winter.

Second, the landscape types in both countries have affected the agricultural production and the distribution of population and cities significantly. Overall, the populations in the two countries are mostly concentrated in the central and eastern plains. In addition, both countries' economies are generally more developed in the East than in the West. In China, this situation is more apparent. The dominating landform in the United States is plain. The central Great Plains, accounting for about 1/2 of the total land, is very suitable for large-scale farming and the development of transportation systems. While in China, plateaus and mountains are the leading landforms, and plains only occupy about 12 % of its total land area. The high quality farmlands are scarce and scattered, not appropriate for large scale farming by the machinery. This has created greater challenges for farming, and the development of urban and transportation systems.

2.4.1.2 Comparison of Climates

China borders the largest continent (the Eurasia Continent) in the west, and the largest ocean (the Pacific Ocean) in the east. It is a country with the most typical monsoon climate. The United States touches the Pacific Ocean in the west and the Atlantic Ocean in the east. In the U.S, the monsoon climate is not that noticeable, but it has the Mediterranean and the temperate marine climates that China lacks. Additionally, the climates in the two countries are significantly affected by mountain ranges (such as the Himalayas and the Rocky Mountains), but it is more obvious in China. The magnificent Qinghai-Tiber Plateau has created a big thermal contract between the ocean and the land, leading to hotter summer and colder winter.

From the view point of the climatic characteristics, the inlands in both countries are dry and cold in the winter and hot and humid in the summer. The coastal areas are warm and humid, with abundant rainfall. However, China has a more noticeable seasonal variation and ocean-land contrast. As for temperature and precipitation, both countries are warm and moist in the summer, cold and dry in the winter, and generally declining precipitation from southeast to northwest. Nevertheless, China's continentality of climate is more significant and arid and semi-arid area is broader. Although the two countries' west are arid areas, the extent and the scope of the dryness in western China are much more extreme than those of the United States, with a large area of them belonging to the extremely dry lands. For the regional differences in precipitation, the U.S West Coast is affected more by the oceanic currents. Although there is a difference in the amount of precipitation between the East and the West, the disparity is quite small. On the contrary, China' west is located deep inland and is isolated by towering mountains; consequently, the west and east have a significant difference in the amounts of precipitation they receive.

For climate, both countries have various and complex types of climate, and have broad latitudinal zonal distributions of temperate, subtropical, tropical, and highland climate. However, their differences lie in that China has the world's most typical monsoon climate, and its eastern region is dominated by the monsoon climate, including the tropical monsoon climate, the sub-tropical monsoon climate, the temperate monsoon climate, the temperate continental climate and the highland climate. China's northwestern area is located far away from the ocean and is controlled by the continental air mass; consequently, it has temperate continental climate, except for the highland climate in the area of the Tibetan Plateau. Meanwhile, most parts of the United States belong to the temperate climate and the subtropical climate, with an only exception in the south tip of Florida with a tropical climate. In the United States, the temperate continental climate covers the most extensive area. While the southeast coast has a humid subtropical monsoon climate, the west coast has Mediterranean climate and the temperate continental climate, and limited areas in the western Cordillera Ranges have the highland climate. Therefore, the climate in each of the two countries has its own features.

2.4.1.3 Comparison of Water System and Surface Landscape Zones

Affected by the terrains, most of the rivers in China flow from west to east and finally drain into the Pacific Ocean. On the other hand, due to the existence of two drainage divides in the United States, the Rocky Mountains and the Appalachian Mountains, there exist three river systems, including the one flowing from west to east into the Atlantic Ocean, the one flowing from east to west into the Pacific Ocean, and the one running from north to south into the Gulf of Mexico. In addition, the Great Lakes are a unique feature of the United States water system.

The landscapes in both countries have significant latitudinal and longitudinal zonal features. However, China's topography is more towering and rugged, which has created typical three-dimensional zonal landscapes, i.e., more altitudinal zonal features for the vegetations and the soils. Besides, China's zonation is more obvious from the ocean to the inland, reflected by changes from the arid lands in the northwest to the wet lands in the southeast and the complex changes in the agricultural zones. Affected particularly by the so-called "The world's third pole", the Qinghai-Tibetan Plateau, and the large size population, China has become one of the countries in the world with the largest differences in regional landscapes and human-land systems.

2.4.2 Natural Disasters

China and the U.S are two very different countries. China is a large agricultural country with high population density, accounting for only 6.5 % of the world land area but 20 % of the world total population. The fast paces of industrialization and urbanization have generated a big pressure to its land and its fragile ecosystem. On the other hand, the U.S. occupies 6.4 % of the world land area but only 5 % of the world population. It has highly modernized industry and agriculture, and extremely

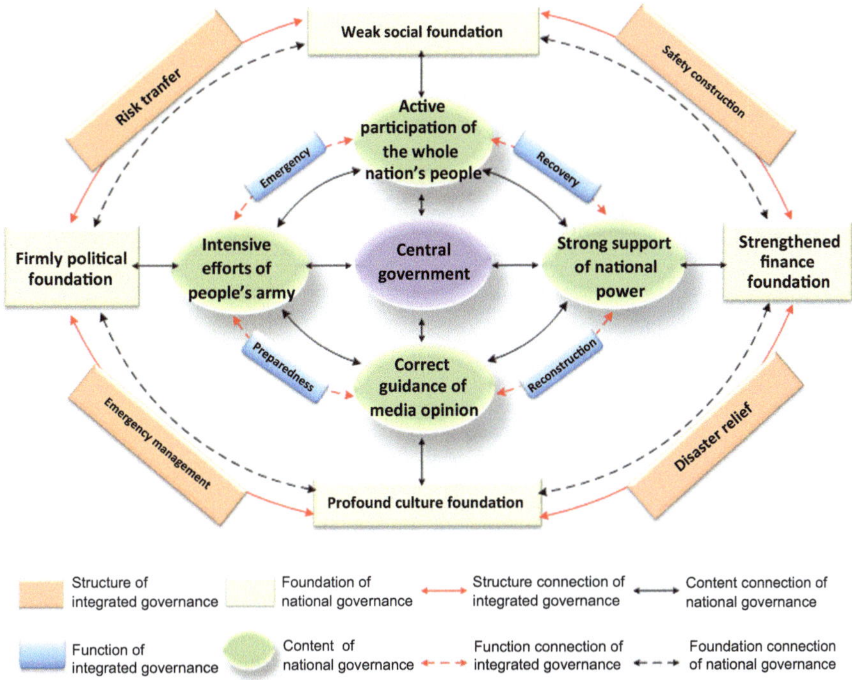

Fig. 2.16 The Chinese government disaster prevention model

developed science and technology. Therefore, when facing the natural disasters of the same intensity, because of the differences in the vulnerabilities of their social ecosystems and the political systems, the two countries have very different ways of dealing with the disaster prevention and mitigation.

Under the leadership of the State Council's Emergency Management Office and the National Disaster Reduction Committee, China implements a disaster prevention system (model) of "Central Leadership, Divisional Responsibility, and the Combination of the Hierarchical Management and the Local Management", as well as the watchful idea of "Prevention First, and One Entity for Prevention, Fighting and Rescue". The structure of the national comprehensive disaster prevention system is composed of the "Safe Fortification, Disaster Relief and Aid, Emergency Management and Risk Transfer". The function of the national comprehensive hazard prevention system is composed of the "Disaster Preparedness, Emergency Response, Recovery and Reconstruction" (Fig. 2.16).

In the United States, under the current federal system, the highest authority of the disaster management is the Department of Homeland Security. The Federal Emergency Management Agency (FEMA) within this Department is the executive agency responsible for the coordination of the cross-region cross-branch operations after an incident. The responsibilities of FEMA include the emergency prevention,

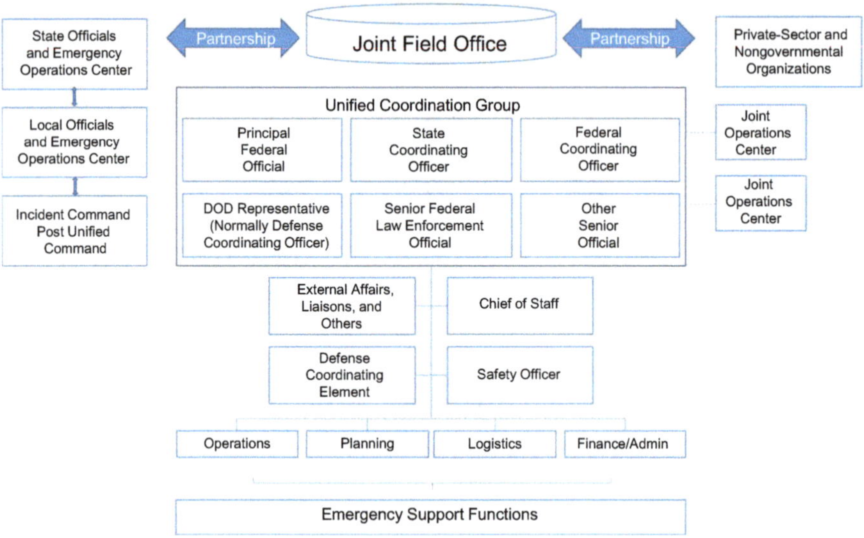

Fig. 2.17 The American government disaster prevention model (National Response Framework of the Homeland Security of America)

the emergency preparedness, the emergency response, the disaster recovery and reconstruction, the protection of public facilities, and the reduction of casualties and property losses. Emergencies are managed at the most-local level possible. However, if the emergency is terrorist related or an "Incident of National Significance", it will become a national event. Under the emergency cases, on behalf of the president, FEMA can coordinate various rescue efforts from the state and local governments, federal authorities, Red Cross, private sectors, nongovernmental organizations, and volunteers to respond to and recover and reconstruct from the disasters. The United States has made a number of response laws and regulations for natural disasters and emergencies, which become the legal basis for the federal government to develop, implement, and support the disaster mitigation plan (Fig. 2.17).

In general, the Chinese government implements a powerful "top—down" disaster management model with a high effect but low efficiency in the utilization of the resources. The US also has a nationwide federal coordination management structure, but it mainly takes the "bottom up" approach to engage in the states as the main components and pays great attentions to the resources from the nongovernment organizations, communities, and individuals. Although its effect of dealing with the disaster is low, the efficiency of using the various resources is high. Therefore, in the prevention of the natural disasters, especially the catastrophic ones, the most efficient effective approach is to combine the "top-level (nationwide) " model from China and the "bottom (local)" model from the United States or other Western countries together. Efforts should be put on strengthening the construction of the legal system, the mechanism, and the various capabilities. It also should exercise the disaster insurance actively, especially the catastrophe insurance, and transfer the risk through effective means to reduce the disaster losses. Built upon this, the two countries should learn from each other to improve their systems and abilities for disaster prevention and mitigation.

2.4.3 *Physical Geography Regionalization*

2.4.3.1 The Uplift of the Qinghai-Tibet Plateau and Its Impact on the Regional Climatic Differences

The substantial uplift of the Qinghai-Tibet Plateau over an immense extent was the most significant regional event during the Earth's evolution history since the Cenozoic in China. The rapid uplift of this large plateau not only changed the circulation patterns of the airs on the plateau itself, it also affected the configuration of the modern East Monsoon, intensified the downdraft of the airs in the Northwest. It has completely changed China's regional climatic structures and the physical geographical processes. Generally, when the altitude of the Qinghai-Tibet Plateau reached 3,000 m, its dynamic and thermal functions to the air movements were powerful enough to disrupt the atmospheric circulation patterns in East Asia.

Qinghai-Tibet Plateau, with a land area of nearly 2,500,000 km^2 and accounting for a quarter of China's total land size, is located at 25 °N–40 °N, a transitional region between the Westerlies and the Subtropical High-Pressure Belt. It's more than 4,500 m high altitude embraces up to one third of the Troposphere in the middle latitude area during the winter. It stretches across over 31 longitudinal degrees, approximately 2,700 km long in the east-west orientation, and 15 latitudinal degrees, nearly 1,400 km wide (1/3 of the Westerlies) in the north-south direction. Such a massive obstacle has blocked the ramification process of the Westerlies and the movement of the warm and humid air masses into Asia's interior. Additionally, it has a significant impact on the southward expansion of the winter winds and on the desertification of Asia's interior.

The Dynamic Functions of the Qinghai-Tibetan Plateau The Qinghai-Tibet Plateau plays an important role in blocking the movements of the airs near the surface. In winter, the cold air accumulates on the northern Qinghai-Tibet Plateau and then divides into two parts. While one moves into the Tarim Basin along the Al Jinshan Mountains, the other flows westward or northwestward along the Qilian Mountains into the Hexi Corridor and then continue further south to form the winter wind channel, enhancing the strength of winter wind in the southeastern area. In the summer, on the way to flow northward from the Bay of Bombay, when the Southwest Monsoon meets with the Qinghai-Tibetan Plateau, it is divided into two branches: one along the Himalayas to become an easterly and the other moves along the direction of the mountains into the China's southwest to strengthen the water vapor circulation there. It increases the precipitation in the periphery of the plateau and exacerbates the dry condition in the inner plateau due to the rain shadow effect. In the west part of the Qinghai-Tibet Plateau, when the Westerly is blocked by the plateau in the winter, it is divided into the northern and southern branches, with the former being much stronger than the latter. The southern branch moves around the south side of the plateau to become a southwesterly, enhancing the dry and hot southwestward flowing air. The northern branch becomes a southwesterly at the north of the plateau. It then bypasses northern Xinjiang to become a northwesterly, further strengthening the strength of the winter wind. In the north of the plateau, a ridge is formed where the subsiding air prevails, further exacerbating the desertification processes in the Northwest. Overall, because of the obstruction from the Qinghai-Tibet Plateau, after the processes of ramification, bypassing, converging and eastward moving, the Westerly has become the strongest wind in the Northern Hemisphere.

The Thermal Functions of the Tibetan Plateau The great thermal differences between the surface of Qinghai-Tibetan Plateau surface and the freely flowing air masses with the same altitude in the atmosphere have produced significant thermal effects on the atmospheric circulations. In summer, like a fireplace thrusting into the atmosphere, the plateau heats up the air near the surface to force it to ascend, while pulling up more airs from the Indian Ocean as supplies to fuel this process. Consequently, it brings abundant monsoon rainfalls. On the contrary, in the winter, the huge plateau is just like a colossal piece of ice, cooling the air above it and forcing the air to flock to the Indian Ocean. It enhances the strength of the southward moving cold air from the North, making it a strong winter wind.

In short, the uplift of the Qinghai-Tibetan Plateau has important implications on forming and developing of the Eastern Asia Monsoon. The uplift of the plateau can be divided into three main phases. In the first phase, about 10–9 Ma ago, the Asian Monsoon began to form; in the second phase about 3.6–2.6 Ma ago, the uplift of the plateau accelerated and the Asian Monsoon in the winter and the summer increased their strengths at the same time; in the last phase since 2.6 Ma ago, with the continuing uplift of the plateau, the variability of the summer and winter monsoon increased and the winter monsoon strengthened. With the uplift of the Qinghai-Tibet Plateau, China's climate can be divided into three unique regions (Table 2.4), which constitute the foundation in shaping China's macro natural landscapes.

2.4.3.2 Reasons for the Climatic Regionalization in the United States

The regionalization of the climate and the physical environment in the U.S is closely associated with the land-sea distribution and the latitudinal zonation. The southeast coast is mainly the subtropical monsoon climate with largely the subtropical evergreen broadleaf forest. As a result of the difference in the thermal properties between the land and sea, in the winter, the Northwesterly is prevailing, bringing cold and dry weather condition; in the summer, the dominating wind is the southeasterly, which brings high temperature and abundant rainfall. The southwest coast (mainly the California coast) has a Mediterranean climate. Due to the control of the subtropical high pressure system in the summer, the weather is hot and dry; in the winter the mid-latitude Westerlies brings warm and humid air to this area that covered by subtropical evergreen conifers. The Northeast Coast is mostly the

Table 2.4 The impact on China's regional climate from the uplift of the Qinghai-Tibet Plateau

Evolution trend of regions	Original background	Major impact
Eastern region -the formation and strengthening of East Asian monsoon system	Ocean-land monsoon, wetter	Southward expansion of the winter wind–strengthening the power The plateau increased the ocean-land monsoon strength– big annual air temperature range, precipitation concentrated in summer The ramification of the Westerly and its disappearance to some extent control the monsoon's path(October/May) Compared with other parts of the same latitude, the temperate zone in the monsoon zone shifted to the south
Northwest -drought	Secluded inland, more arid, with desert, desert steppe, steppe-based vegetation	Accelerated the descending of the accumulation of cold air Water vapor from the Indian Ocean was blocked Drought intensified, extent expanded
Qinghai-Tibet Plateau Area -Cold and arid	Plateau surface <2,000 m still sub-tropical savanna, steppe climate	The plateau blocked the water vapor from the Indian Ocean and the Pacific Ocean, interior becoming dry Plateau cooling, becoming colder Plateau monsoon

temperate marine climate with temperate deciduous broadleaved forest, influenced by the Westerlies all year long. The warm and humid air taken by the wind from ocean to land brings mild, humid, and rainy weather to the area. In addition, situated in the high latitude area and affected by the frigid tundra climate, the state of Alaska is extremely cold and dry.

The giant ranges (for example the Rocky Mountains and the Appalachian Mountains) also affect the regionalization of the climate and the physical environment in the U.S greatly. The Rocky Mountains is an important climate boundary in the North America continent. It blocks the eastward intrusion of the Polar Air Mass from the Pacific Ocean, the westward move of the Polar Air Mass from Canada, and the tropical Air Mass from the Gulf of Mexico, leading to the great differences between the East and the West in many aspects, such as the distribution of precipitation and temperature. Furthermore, the Rocky Mountains are also an important drainage divide for the river systems. The Appalachian Mountains in the

East prevents the Atlantic water vapor from traveling further into the interior. It has certain influences on the development of the climate zones in the East. However, Because of its relatively low altitude and continuity, its influences to the climate of the U.S are inferior to the Rocky Mountains to the climate of the U.S or the Himalaya Mountains to the climate of China.

The Evolution of the Modern American Topography Outline The outline of the modern American landscapes is formed after a series of events over a long geological history. The core of the North America continent is the ancient land before the Cambrian. With the tectonic movements from the interactions between the continent and its adjacent plates, the outline was developed gradually. During the Archean period, after a series of tectonic movements and collisions, most part of the Great Lakes area tended to rise slowly over a long period so that the ancient crystalline base rocks were widely exposed and a huge hardened and stable continental land mass—the Canadian Shield was gradually formed, as part of the Laurasia. In the south and west of the Canadian Shield, the subsidence was the main tectonic movement with many transgression and regression activities. In the Precambrian, the crystalline rock basement was covered by a deep post-deposition, forming the central platform which covered most of the American central plain. In the eastern central platform, the internal low plains were mainly made of sediments from the Paleozoic, while in the west the Great Plains was made mainly of the Cenozoic sedimentary layers that covered the Paleozoic sedimentary layers.

During the Early Cambrian, the old Atlantic Ocean was formed between the ancient North America and Old Europe, and the old Appalachian Ocean was formed between the ancient North America and the ancient Africa. During the Devonian, with their collisions, the ancient North America and the Old Europe plated developed into the ancient Europe—America continent. After the Devonian, the ancient Gondwana rotated clockwise, causing the ancient Africa and the ancient Europe—America continents to gradually move closer to each other, and finally the two collided in the late Carboniferous sutures, forming the northern Appalachians. Through a lengthy process of the plate activities, the Appalachian geosyncline set off three large-scale mountain orogenic events to form a series of NE—SW trending parallel folds and thrusts. After the creation of the mountains, it went through several long processes of peneplain stages. During the Triassic, the north-south faulting occurred. In the Tertiary, the mountains rose again, and a new erosion process began, resulting in the foot platforms, the Blue Mountains, the Ridge Valley area and the Appalachian Plateau from the southeast to the northwest.

The large-scale upheaval of the Cordillera started from the Mesozoic. The reopening of the Jurassic Atlantic drove the North American Plate moved westward to cause serious of tectonic movements to create the Alaska Range, Cascade Mountains, Sierra Nevada Mountains, and mountains in California Peninsula and so on. From the late Cretaceous to the early Tertiary, the expediting retrograde movement of the plate triggered a serious of large-scale mountain orogenesis in the eastern Cordillera geosynclines, leading to the uplift of the Rocky Mountains, the significant rise of the Colorado Plateau, along with the strong magmatic intrusions and fault activities. Since the Oligocene, the North American Plate continued to move westward to battle directly with the Pacific Plate. This tectonic event started a new wave of Alpine movements which still last today and led to the formation of the coastal mountains along the Pacific Rim, the folded Sierra Nevada Mountain belts, the significant uplift of the Colorado Plateau, and the block faults within the Great Basin.

The Evolution of the Modern Climate Since the breakup the Pangea, the three major geological structures that shaped the climate in the United States include the erosion of the Appalachian Mountains, the uplift of the West Cordillera and it accompanying volcanic activities, and the uplift of the northeastern margins of the North America continent. When the North America was separated from the Gondwana, the landform was high and active in the east but relatively low and stable in the west. However, as the Appalachian Mountains continued to be eroded away, its impact on the climate reduced greatly. On the other hand, with the continuous uplift of the Cordillera's west margins, its influence on the climate had been increasing. During the past 5 million years alone, generally the western region increased by more than 500–2,500 m, which had a profound impact on this region's climate and biological production. During the Pleistocene, most mountains had reached to 3,000 m high and were covered by snow.

This role exchange between the east and the west had a significant impact on the transformation of the heat and moisture in the continental United States, which led to an extreme climate of the West being getting drier and the Central being more vulnerable to the tropical and polar air masses. The uplift of the northeastern edge in North America, to a large extent, led to the development of the Cenozoic glaciers in this region. During the Cenozoic, this region was uplifted by another 1,500–2,000 m. When the mountains and plateaus have been raised to a certain altitude, it formed the climate conditions for the growth of glaciers.

Due to the obstruction from the West Cordillera, the warm and wet air masses from the Pacific Ocean were sharply uplifted to change their characteristics. In addition, the dry and cold air masses originated from the frigid northern cold land and the Arctic ice and snow sheets, because of no natural barriers, can sweep freely across the central continent, reaching to the Gulf of Mexico in the winter. With the north-south alignment of mountains on both sides, the central lowlands becomes an ideal place for the transformation of the tropical wet warm air mass and the polar dry cold air mass. The collision of these two air masses often produced extreme weathers such as severe thunderstorm and cyclones. Other mountains from the West Cordillera Mountains also have a great impact on the climate. For example, the mountains on the California Coast are high enough to prevent the winter storms from the Pacific from moving eastward, resulting in a 100 km long semi-arid region in the west of the Central Valley. The west of the lofty Sierra Nevada Mountain is humid temperate climate, but its east side is a semi-arid climate. There are many similar changes everywhere in the Cordillera, such as the Great Basin with a temperate desert climate even if it is located between 35 °N to 45 °N.

2.5 Conclusions

China and the U.S. are two large countries in terms of land area in the world. They are both located in large continents, with thousands of kilometers from north to south and from east to west. Their vast land areas determine the variety of landforms, climates, vegetation patterns and natural disasters in each country. Their geographical locations in the respective continent and the relative location to oceans entail the similarities and differences in terms of their physical geography. Physical geographical conditions cast in most cases strong influences on human activities and therefore influence regional human-environment interactions. These activities and interactions, e.g. population distribution, agricultural and industrial production, will be elaborated in the following chapters.

References

Air Quality Trends. http://www.epa.gov/airtrends/.

Air Trends Design Values. http://www.epa.gov/air/airtrends/values.html.

Antony, R. O. (Eds.). (2002). *The physical geography of North America*. New York: Oxford University Press.

Bailey, R. G. (1983). Delineation of ecosystem regions. *Environmental Management, 7*, 365–373.

Bailey, R. G. (1988). *Ecogeographic analysis: A guide to the ecological division of land for resource management* (18 p). Washington: USDA Forest Service. Misc Publ 1465.

Bailey, R. G. (2002). *Ecoregion-Based Design for Sustainability*. New York: Springer-Verlag.

Birdsall, S. S., Palka, E. J., Malinowski, J., et al. (2009). *Regional landscapes of the United States and Canada* (7th edn.). Hoboken: Wiley.

Bryson, R. A., & Hare, F. K. (Eds.). (1974). *Climates of North America*. Amsterdam: Elsevier Scientific Pub Co.

China Daily (2013). *Nation gets tough on air pollution*, China Daily, Sep 13, 2013, p. 1, 4.

Chinese Academy of Sciences. (1985). *Physical geography in China Editorial Board, China Natural Geography (Zonglun)*. Beijing: Science Press.

Clean Air Status and Trends Network. http://www.epa.gov/castnet/.

Dahl, T. E. (1990). *Wetlands losses in the United States 1780's to 1980's (PDF)* (20 pp, 13 pp). 2.66 MB, About PDF. Washington, D.C.: U.S. Department of the Interior, Fish and Wildlife Service.

Diaz, Robert J., & Rosenberg, R. (15 Aug 2009). Spreading dead zones and consequences for marine ecosystems. *Science*, 926–929. American Association for the Advancement of Science. 29 Jan 2009.

Editor Committee of Encyclopedia of China—China Geography Editorial Board. (1993). *Encyclopedia of China (China Geography)*. Beijing: China Encyclopedia Publishing House.

Edward, A. K., & Robert, H. B. (Eds.). (2008). *Natural hazards: Earth's Processes as hazards, disasters, and catastrophes* (2nd edn.). Upper Saddle River: Pearson Prentice Hall. (07458).

Environmental, P. A. (2006). National Emissions Inventory (NEI) Trends report: Average annual emissions, all criteria pollutants. http://www.epa.gov/ttn/chief/trends/.

EPA Monitoring Network. http://www.epa.gov/ttn/amtic/.

Huang, B. (1993). Integrated Work of Physical Geography for Sixty Years. Beijing: Science Press.

Hypoxia 101. What is hypoxia and what causes it? (29 Jan 2009). Mississippi River and Gulf of Mexico Watershed Nutrient Task Force. 2009. U.S. Environmental Protection Agency, Office of Wetlands, Oceans and Watersheds.

Impervious Surfaces and Hydrologic Balances of Watersheds. Watershed Academy Web. (19 Feb 2009). Growth and water resources. 12 Sept 2008. United States Environmental Protection Agency.

Local air quality trends. http://www.epa.gov/airtrends/where.html.

Marine Debris; Habitat Protection. (19 Feb 2009). US EPA Office of Wetlands, Oceans and Watersheds. 4 Sept 2008. United States Environmental Protection Agency.

National Core Monitoring Network. http://www.epa.gov/ttn/amtic/ncore/index.html.

Overview. Gulf of Mexico Hypoxia. (21 Aug 2008). Louisiana Universities Marine Consortium. 29 Jan 2009.

Riebsame, W. E., Changnon, S. A., & Karl, T. R. (1991). *Drought and natural resources management in the United States: Impacts and implications of the 1987–89 drought* (pp. 11–92). Colorado: Westview Press.

Shi, P. J. (2003). *Atlas of natural disaster system in China*. Beijing: Science Press.

Stedman, S., & Dahl, T. E. (2008). Status and trends of Wetlands in the coastal watersheds of the Eastern United States 1998–2004 (PDF). 36 pp, 8.7 MB, About PDF. National Oceanic and Atmospheric Administration, National Marine Fisheries Service, and U.S. Department of the Interior, Fish and Wildlife Service.

The City of Los Angeles Transportation Profile. (2009). City of Los Angeles Department of Transportation z 100 S. Main St. z 9th and 10th floors z Los Angeles, CA 90012.

Trends in ozone adjusted for weather conditions. http://www.epa.gov/airtrends/weather.html.

U.S. Environmental Protection Agency, Office of Water. (29 Jan 2009). National water quality inventory: Report to congress, 2004 Reporting Cycle (PDF) 11 pp, 74K, About PDF). Washington, D.C.: Environmental Protection Agency, 2009.

U.S. Environmental Protection Agency, Office of Wetlands, Oceans and Watersheds. (2009). Economic Benefits of Wetlands (PDF). 4 pp, 281 K, About PDF. Washington, D.C.: Environmental Protection Agency, 2006. 29 Jan 2009.

U.S. Environmental Protection Agency, Office of Wetlands, Oceans and Watersheds. (2009). Protecting Water Quality from Urban Runoff (PDF). 2 pp, 232K, About PDF. Washington, D.C.: Environmental Protection Agency, 2003. 29 Jan 2009.

U.S. Environmental Protection Agency, Office of Wetlands, Oceans and Watersheds. (2009). Wetlands: Protecting Life and Property from Flooding (PDF). 4 pp, 286 K, About PDF Washington, D.C.: Environmental Protection Agency, 2006. 29 Jan 2009.

U.S. Geological Survey (USGS) Publications on natural disasters. U.S. Department of the Interior, USA.

Vankat, O. L. (1979). *The natural vegetation of North America: An introduction*. New York: Wiley.

Wang, J. A. (2010). Geographic Atlas of China. Beijing: SinoMaps Press.

Wang, J. A. (2007). *Geographical tutorial*. Beijing: Higher Education Press.

Wang, J. A., Shi, P. J., Wang, P., et al. (2006). *The temporal pattern of natural disasters in China*. Beijing: Science Press.

Wang, J. A., Sun, H., Xu, W., et al. (2002). Spatio-temporal change of drought disaster in China in recent fifty years. *Journal of Natural Disasters, 11*(2), 1–6.

Wetland Functions and Values. (12 Sept 2008). Watershed Academy Web. Environmental protection agency, Office of Wetlands, Oceans and Watersheds. 29 Jan 2009.

Zhou, L. (1993). Theories and Practice of Chinese Agricultural Zonation. Hefei: USTC Press.

Chapter 3
Population/Ethnic Geographies of China and the U.S.

Lucius Hallet IV, Jing'ai Wang and Rudi Hartmann

3.1 Population Geography of China

China is the world's most populous country. The population increased dramatically from 1949 to 2010 (from 542 million in 1949 to1.37 billion in 2010) (Table 3.1). Due to the imbalance of regional economic development, China has an unequal population distribution with high population densities in the urbanized eastern areas and very sparse population densities in the less urbanized western regions. The division can be drawn along the Hu-Huangyong line (named after Chinese geographer Hu Huanyong; Hu 1990; Hu 1935; Hu 1986) from northeast to southwest. The highly populated metropolitan areas, such as the Pearl River Delta, the Yangtze River Delta, the Beijing—Tianjin megaregion, are in sharp contrast to the sparsely populated rural and pastoral areas, such as the Qinghai-Tibetan Plateau and the northwest interior basin.

In China, censuses have been taken six times at 1953, 1964, 1982, 1900, 2000 and 2010. The other years' data is obtained by spot-check and recorded in the Chinese Statistical Yearbook.

3.1.1 General Features of the Population in China

China has a large population base that has seen slight changes to its proportion within the world's total population. In recent decades, however, natural growth rates

L. Hallet IV (✉)
Department of Geography, Western Michigan University, Kalamazoo, USA
e-mail: Lucius.hallett@uwmich.edu

J. Wang
School of Geography, Beijing Normal University, Haidian District, Beijing, China
e-mail: jwang@bnu.edu.cn

R. Hartmann
Department of Geography and Environmental Sciences,
University of Colorado Denver, Denver, Colorado, USA
e-mail: Rudi.Hartmann@ucdenver.edu

R. Hartmann, J. Wang (eds.), *A Comparative Geography of China and the U.S.*,
GeoJournal Library 109, DOI 10.1007/978-94-017-8792-5_3,

Table 3.1 Chinese population volume and proportion within the world population. (National Bureau of Statistics of China 2011; UNFPA 2010)

	1949	1960	1970	1980	1990	2000	2005	2010
World population (billion)	2.436	3.027	3.634	4.415	5.284	6.000	6.477	6.909
China population (billion)	0.542	0.662	0.830	0.987	1.134	1.295	1.308	1.370
Proportion of world (%)	22.24	21.87	22.84	22.36	21.46	21.58	20.19	19.83

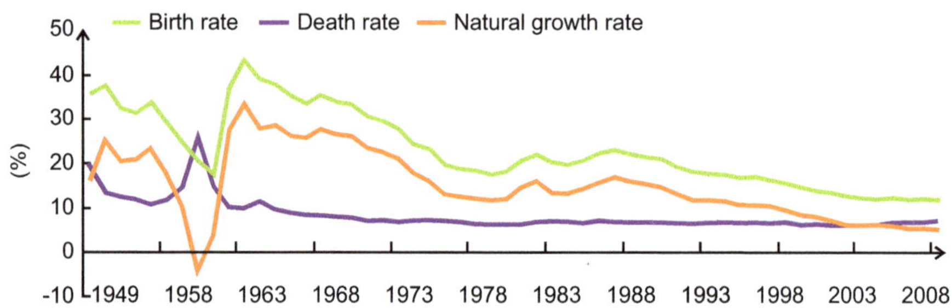

Fig. 3.1 China's population birth rate, death rate and natural growth rate from 1949 to 2008. Note: birth rate, death rate and natural growth rate in 2009 are 11.95 %, 7.08 % and 4.87 %, in 2010 are 11.90 %, 7.11 % and 4.79 %. (National Statistics Bureau of China 2010)

decreased significantly because of the implemented family planning policies, and China's percentage in the total world population has fallen from 22.24 % in 1949 to 19.83 % in 2010.

The birth rate, death rate and natural growth rate are three key indicators of natural changes to a population. The death rate of China's population decreased during past years except for 1960; the trend in China's birth rate is consistent with the natural growth rate (Fig. 3.1), which means that the birth rate continues to play an important role for natural population growth. Nevertheless, the birth rate has been significantly reduced through implementation of family planning policies to achieve the goal of controlling excessive population growth in China.

An analysis of the fluctuations of Chinese population growth for the time period 1948–2010 (see Fig. 3.1) shows that the Chinese population has been gradually transformed from high birth rate, low death rate and high natural increase to low birth rate, low death rate and low natural increase. Four distinct stages can be distinguished: (1) The time period 1949 to 1958 saw a first peak in population growth, with the annual birth rate exceeding 3 % and an annual average natural growth rate up to 2.24 % resulting in an annual net population increase of 13.11 million. (2) From 1958 to 1961 marked by a drop in population growth as the 1961 birth rate was only 1.8 %, the average annual natural growth rate was 0.46 %. The only year with a decrease of the population was 1960 (at the end of the period of the "Great Leap Forward Movement"). Due to severe natural disasters and a declining food production the Chinese population experienced slow growth. (3) From 1962 to 1973 the number of the population soared from 670 to 890 million, with an annual net increase of 19.46 million; the average

Table 3.2 Gender ratio based on census in China

Year	1953	1964	1982	1990	2000	2005	2010
Gender ratio	107.56	105.46	106.30	106.60	106.74	102.19	105.2

annual natural population increase rate reached 2.56%. The year 1963 set a new record with 3.35%. (4) 1974 to the present, with the birth rate and the natural growth rate decreasing synchronously. In 1998 the natural population growth rate dropped to below 1.0%. In the following years it continued to further drop and reached 0.48% in 2010. This has been a remarkable trend primarily caused by the continued implementation of the family planning policies; it resulted not only in successfully curbing excessive population growth in China but also represented an important contribution to the control of the world population growth.

3.1.2 Population Structure

China's population structure can be examined from a variety of aspects including gender ratio, age structure, educational structure and urban-rural population structure. Gender and age are two major demographic attributes of the population; the former can be used to indicate the gender ratio (the corresponding number of males per 100 females), the latter to express the proportion of the different age groups in the population, e.g. in form of a population pyramid.

3.1.2.1 Gender Structure

Usually, the gender ratio is less than 100 in developed countries and less than 105 among most developing countries. In China it has remained at high levels (above 105) from the second census in 1964 to the sixth census in 2010 (Table 3.2). During the years of 2000–2010, the overall gender ratio of China's population has seen fluctuations in a downward trend, from 106.74 in 2000 down to 105.2 in 2010.

In addition, there are significant rural-urban differences in the gender ratio. The rural gender ratio is generally higher than in the urban regions due to Chinese traditional values as expressed in a male preference for a child. Family planning policies have had also an influence on the preferred gender of a child, since the allowed number of children in a family is limited (usually to one or to two) and many parents prefer to have a boy to succeed their "family lines" by all means, thus compounding the imbalance in the gender ratio.

3.1.2.2 Age Structure

During the past 50 years age structure in China has gradually and consistently changed as to the proportion of the following three age groups or cohorts: (1) 0 to

Table 3.3 Changes in the age structure of China's population (National Bureau of Statistics of China 2012)

Year	Persons ages 14 and below (%)	Persons ages 15 to 64 (%)	Persons ages 65 and above (%)
1953	36.28	59.31	4.41
1964	40.69	55.75	3.56
1982	33.59	61.50	4.91
1990	27.69	66.74	5.57
2000	22.89	70.15	6.96
2005	20.27	72.04	7.69
2010	16.60	74.50	8.90

14 years (2) 15 to 64 years and (3) 65 years and over (Table 3.3). During the period of 1953 to 2010, the proportion of children has overall declined; the proportion of the age group 15 to 64 years has increased significantly, and the proportion of the elderly ages 65 and above has been growing. During 1953 to 1964, the age group 0 to 14 increased sharply and China's age structure clearly reflected the type of a younger population. However, by 1982 the proportion of the adults had risen to 61.50 %, and that of the elderly to 4.91 % which was an indication that China had begun the transition from a young to an adult type population. The population pyramids (in Fig. 3.2) clearly show this trend. In 2000, the children proportion in the total population dropped to 22.89 %, the adult proportion reached 70.15 %, while the elderly proportion increased to 6.69 % indicating that the age structure of China's population was transitioning to an older population type. By 2010, children of China's population dropped to below 20 %, while the proportion of elderly population increased to 8.9 %, reflecting the completed transition to the type of an older population in China.

With a completed transition of the population age structure to an older type of population, China is facing, on the one hand, the heavy burden of an aging population comparable to the situation in most developed countries—though China's average income levels are not as high as those in most developed countries. China's population is aging faster and the proportion of 65 and older population group is getting larger. On the other hand, the labor force (15 ~ 64 years) in China has reached 72.7 %, with the world's most abundant labor resources in place. Thus, China is experiencing a historically unprecedented "demographic dividend" period which also means that China will have to face increasing employment pressure as the current demographic dividend will gradually erode in the future.

3.1.2.3 Educational Structure

The biggest change in the educational structure of China's population is that the illiterate proportion declined significantly over the past several decades. In 1949, 80 % of population above the age of 15 was illiterate, 62.4 % in 1964. The rate of illiterate persons in China fell to 22.2 % in 1990, to 9.1 % in 2000, and even to 7.1 %

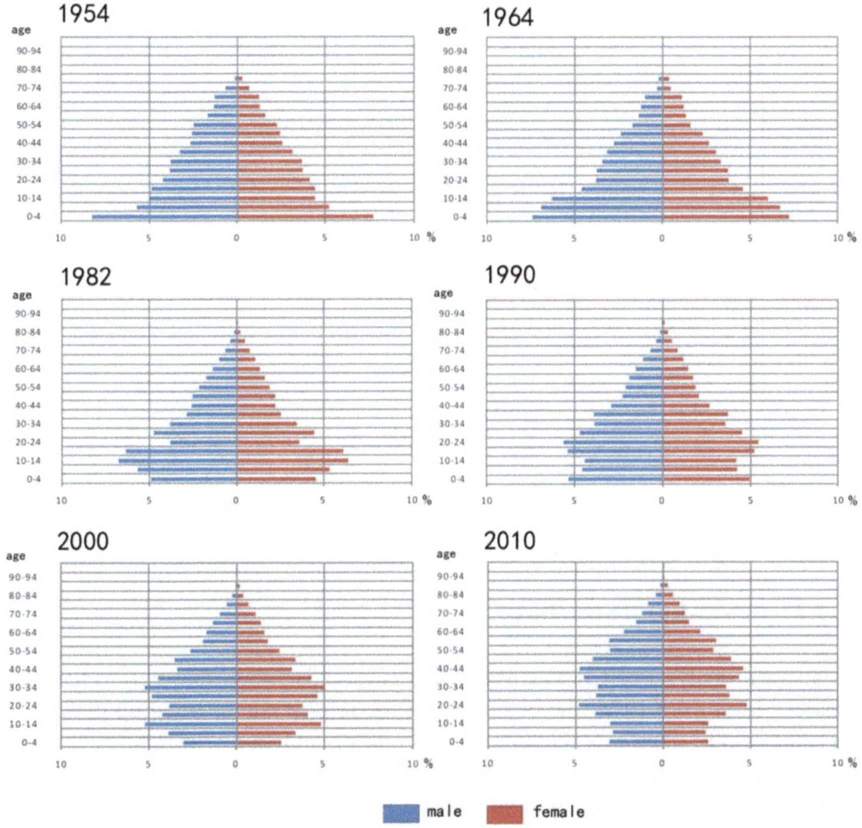

Fig. 3.2 China's population age-gender pyramids. (National census data 1955, 1965, 1983, 1991, 2000 and 2010)

in 2009. Three quarter of the illiterate population lives in western rural areas such as Tibet, Qinghai, Guizhou, Gansu, Ningxia and Yunnan.

Because of the country's large population base the increasing educational attainment in China marks a significant contribution to the improvement of the overall educational status of the world's population. Compared with developed countries, the proportion of the population in China receiving higher levels of educational attainment was low. Though, the proportion of the Chinese population participating in primary education was very large. However, the proportion of the Chinese population receiving degrees from a university and from high school increased considerably. From 1964 to 2010, for instance, the number of people (per one hundred thousand) with a university degree rose from 416 to 8,930, a degree from senior high school from 1,319 to 14,032 and a degree from junior high school from 4,680 to 38,788. The proportion of the population with some high school education increased ten times over this time period. However, the proportion of persons with a higher education degree was still very small, only 8.9 %.

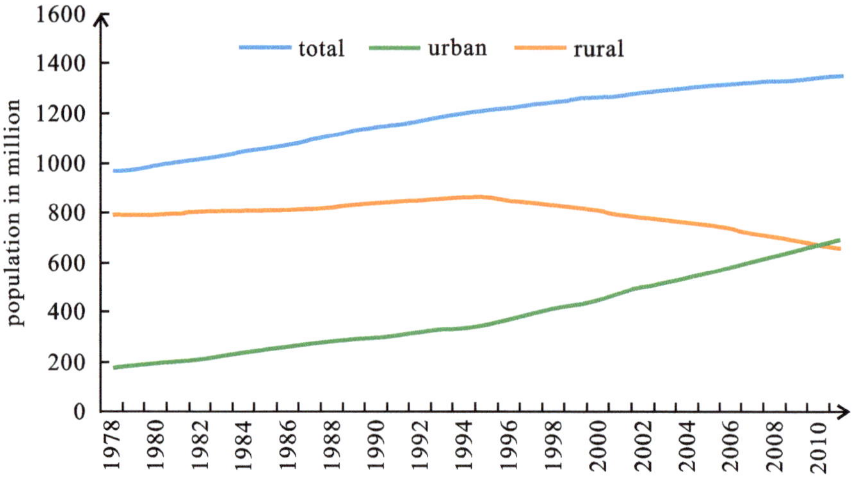

Fig. 3.3 Rural and urban population in China since the reform and opening up period (1978). (National Bureau of Statistics over the years)

3.1.2.4 Urban-Rural Structure

A statistical analysis of China's population (Fig. 3.3) shows that during the past decades years, from 1978 (the beginning of the reform and opening up period) to 2011, the overall trend was an increase of China's population. The rapid pace in increase, however, slowed after 1990. Along with the total population growth, China's urban population was growing rapidly (from 172.45 million in 1978 to 690.79 million in 2011). The rural population rose significantly before 1990, but since 1990 the proportion of rural population in the total population first stagnated then declined steadily. Thus, the year 1990 marks a turning point for the growth of China's population. Figure 3.4 represents the urban population.

3.1.3 Population Distribution

China's vast total population shows enormous differences among regions. Population density in the southeast is very high, while it is very low in the northwest. The "Hu Huanyong population boundary" divides China into two parts, the southeast and the northwest. The southeast accounting only for 42.9% of the total area captures 95.4% of the country's population, only accounting for 42.9% of the total area, while the northwest with 57.1% of total area has only 5.6% of the population. Although China's population has developed from a few thousands of people in ancient prehistoric times to more than 1.3 billion people today, the "Hu Huanyong division line" has always been the population boundary in China (Fig. 3.5). Moreover, China's population distribution is closely related to elevation and topography, with the following characteristics:

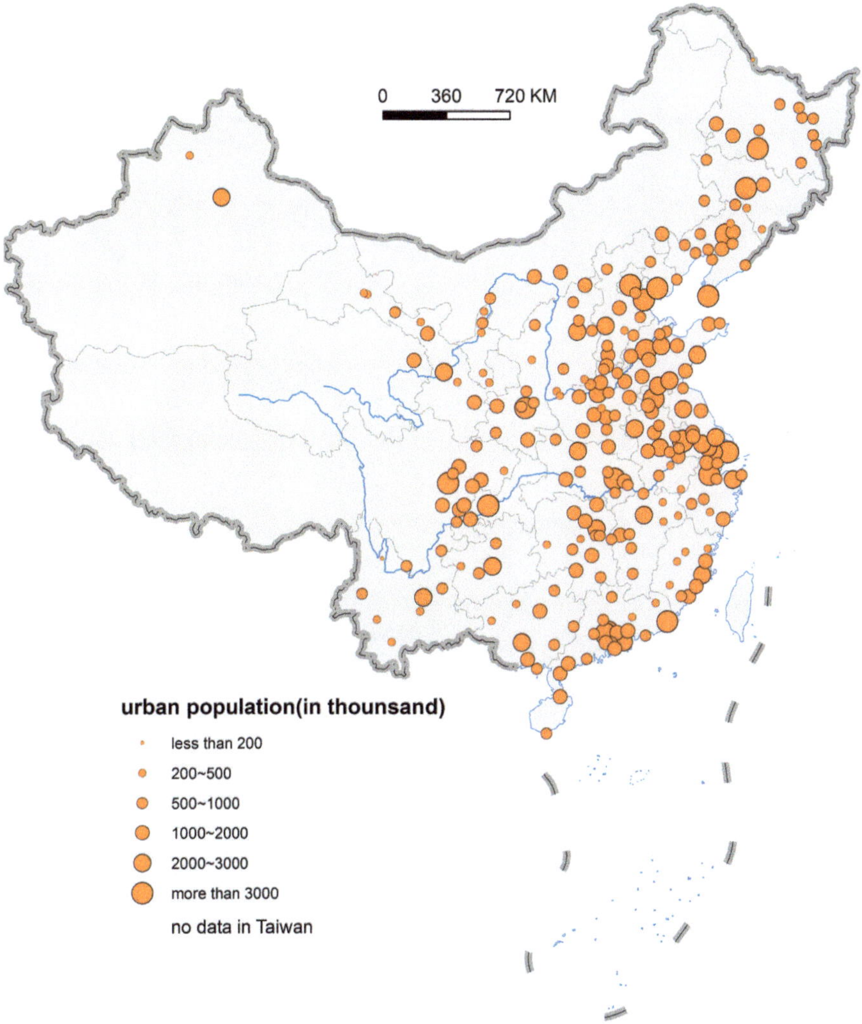

Fig. 3.4 Urban population in China (2010)

Dense populations in low lying areas (plains) and sparse populations in the mountains and plateaus

China's population density decreases rapidly with rising elevation (Table 3.4), from nearly 70 % of the population living in areas below 200 m above sea level to 2 % or so of the population living at higher elevations of more than 2,000 m. The population density is 400 person/km^2 for the first elevation range while it decreases to less than 7 persons/km^2 for the latter which is closely related to a restricted potential of the higher terrain for agricultural production, transportation infrastructure and urban development. Plains are more suitable for farming and provide better

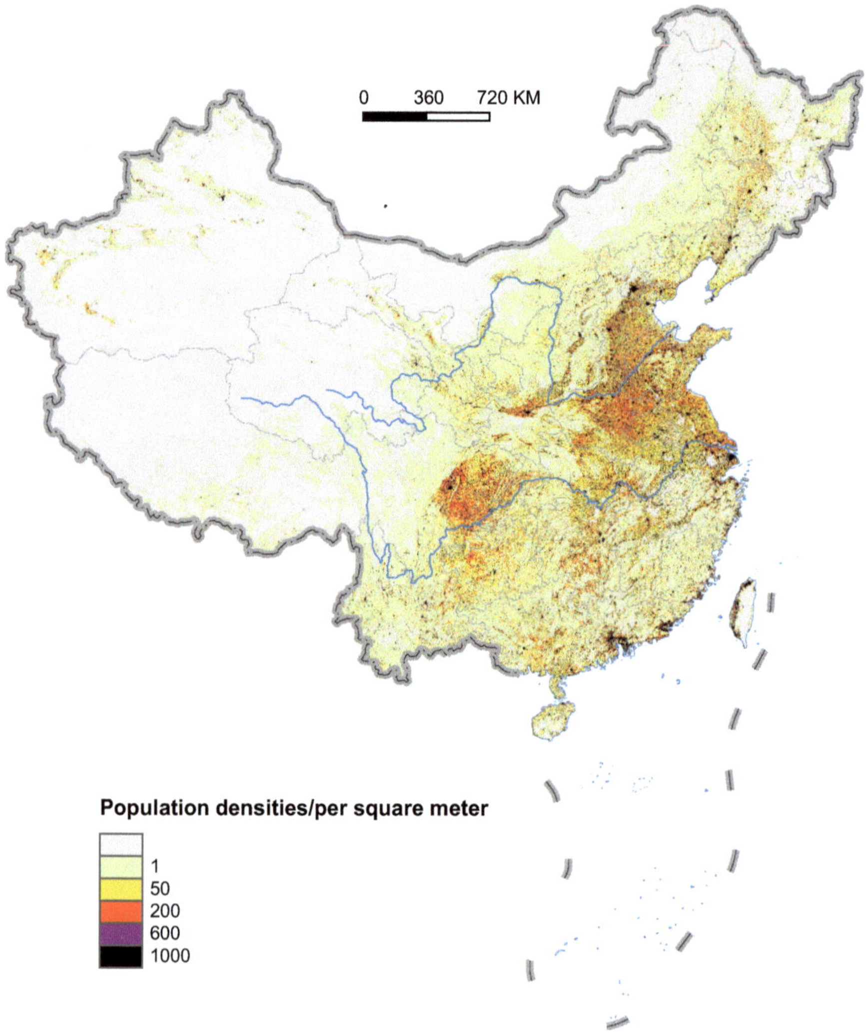

Fig. 3.5 Population densities (2010)

conditions for improvements in the transportation infrastructure and for urban construction. On the plateaus, due to the restrictions of the terrain, water and weather conditions, only animal husbandry can be successfully developed which confines the size of the population.

Higher population densities in warm and humid regions and lower in cold and dry areas

China's population is relatively sparse in areas with cold or dry climates such as the Qinghai-Tibet Plateau, Inner Mongolia Plateau, the northwest and the interior dry zone. The cold and dry climate conditions limit agricultural development and

Table 3.4 Population density changes with altitude. (The fifth national census 2001)

Altitude (m)	<200	200~500	500~1000	1000~2,000	2000~4000
Potulation proportion (%)	68.1	18.9	9.1	9.1	2.1
Area proportion (%)	15.0	10.2	16.9	25.0	32.9
Population density (person/km^2)	419.2	194.1	55.0	37.7	6.7

manufacturing as well as the use of some industrial precision instruments which have all the effect of severely impeding the processes of industrialization and urbanization, and consequently allowing less population growth.

Distribution of the population is closely related to cultivated land and urban areas

The spatial variation of population density in China is directly related to the distribution of farmland and to urban development. Population is more numerous in cultivated land areas, and less numerous in forest and grassland areas. These types and qualities of lands have greatly restricted agricultural production. In areas with arable land more surplus food can be produced to satisfy the needs of larger urban populations so that the regional population density increases. By contrast, in forest and grassland areas the population density is relatively low. The development of agriculture in these areas is restricted by barren soil making it difficult to sustain a larger population.

3.1.4 Spatial Variations of China's Population

3.1.4.1 The Spatial Variation of Population Density

Since 1949, China's five national censuses showed that the macro pattern with high population densities in the southeast and sparse population densities in the northwest divided by the "Hu-Huanyong population line" continues to prevail. There is little spatial expansion of the sparsely populated areas in the northwest, while the eastern areas with already relatively high population densities have seen significant expansions.

3.1.4.2 Driving Forces for Urbanization and Population Migration

Since the beginning of the reform and opening up period (1978) population migration and mobility are mostly related to the processes of industrialization and urbanization. The developed areas and the areas with high levels of urbanization are the main destination zones for the migrant population, while the less developed provinces (though with large populations) are the main origin areas of migration in China. 58.3 % of the national mobile population is concentrated in the coastal areas, 25 % in the central areas, and 16 % in the west. On a provincial level, 17.5 % of the floating population (without valid residential permit) lives now in Guangdong and

39.5% in Jiangsu, Zhejiang, Shandong and Liaoning. If we examine the number of mobile populations in the total residents of host areas the following picture and general proportion evolves: in the regions along the coast one migrant for every six residents, in Beijing and Shanghai one migrant for every three residents and in the central and western regions one migrant for every 12 persons. The "workers flood" (migrant workers seeking jobs without having the proper Hukou household registration) comes mainly from 6 provinces which have abundant labor resources and fewer opportunities for local non-agricultural employment. These provinces include Sichuan, Hunan, Anhui, Hubei, Henan and Jiangxi accounting for about half of the total rural migrant labor force on the move.

3.2 Ethnic Geography of China

3.2.1 Ethnic Composition of China's Population

According to the fifth national census (2000), China has 56 nationalities. On the mainland there are 1,159.40 million Han people, which make up 91.59% of the total population. Minority populations make up 106.43 million people or 8.41% of the total population. Compared with the fourth national census (1990), the Han population increased by 116.92 million or by 11.22%; the minority population increased by 15.23 million or by 16.70%. Among the 55 ethnic minorities the Zhuang are the largest population, with 16.17 million; the smallest minority group is the Lhoba, with only 2,965 people. The greater relative growth of the minority populations is mainly due to relaxed standards of the population restriction policies. There is also the group of non-Chinese foreigners who became citizens of the People's Republic of China and other nations not identified, of totally about 0.75 million accounting for 0.06% of the total population (Table 3.5).

The Han people as China's main nation not only makes up a very large proportion of the population of the country but is also by far the leading ethnic group worldwide (among 2,000 ethnic groups in the world). The Han are roughly 5 times the number of the Hindustan, the world's second-largest ethnic group, or equivalent to the total population of the following seven largest nations including the Hindustan, the Americans in the United States, Russian, Bengali, Japanese, Brazilians and Germans. Among the other 55 minorities, the Hui and the Manchu use the same language as the Han, namely Mandarin Chinese; the other 53 nations use their own native languages. 29 ethnic groups use languages of the Tibetan language family, mainly in the south, and 17 ethnic groups use a language of the Altaic language family, mainly in the north. Three ethnic groups use a language of the South Asia language family, two groups use a language of the Indo-European language family and the Gaoshan speak an Austronesia language. There is one ethnic group whose affiliation to a language family is inconclusive. The phenomenon of the linguistic exchanges between nations is quite common while Chinese Mandarin is the common nationally spoken language.

Table 3.5 Ethnic groups and their Number within the Chinese Population. (The fifth national census 2001)

Groups	Population	Groups	Population	Groups	Population
Achang	33,936	Hui	9,816,802	Qiang	306,072
Bai	1,858,063	Gaoshan	400,000	Salar	104,503
Bonan	16,505	Jino	20,899	She	709,592
Blang	91,882	Gin	22,517	Sui	406,902
Buyei	2,971,460	Jingpo	132,143	Tajik	41,028
Korean	1,923,842	Kirgiz	160,823	Tatar	4,890
Daur	132,394	Lahu	453,705	Tujia	8,028,133
Dai	1,158,989	Li	1,247,814	Tu	241,198
De'ang	17,935	Lisu	634,912	Va	396,610
Dongxiang	513,805	Lhoba	2,965	Uyghur	8,399,393
Dong	2,960,293	Manchu	10,682,263	Uzbek	12,370
Derung	7,426	Maonan	107,166	Xibe	188,824
Russians	15,609	Monba	8,923	Yao	2,637,421
Oroqen	8,196	Mongol	5,813,947	Yi	7,762,286
Ewenki	30,505	Miao	8,940,116	Yugur	13,719
Gelao	579,357	Mulao	207,352	Tibetan	5,416,021
Hani	1,439,673	Naxi	308,839	Zhuang	16,178,811
Kazak	1,250,458	Nu	28,759	Han	1,159,400,000
Hezhen	4,640	Pumi	33,600		

3.2.2 Distribution of Chinese Ethnic Groups

A unique geographical area is one important indicator for the formation of every nation. In the long process of the national exchanges between groups and national integration China's various ethnic minorities have attained a unique geographic distribution.

Han Chinese settlements are more concentrated in the eastern half, whereas minority settlements areas are more concentrated in the western half

The provinces with the highest percentage of Han Chinese are located in the east (with 99.73 % Han in Jiangxi, 99.71 % Han in Shanxi, 99.67 % Han in Jiangsu, etc.), while ethnic minorities live mainly in the following 5 autonomous regions and their surrounding areas: Inner Mongolia, Xinjiang, Tibet, Guangxi and Ningxia. Mongols live largely in Inner Mongolia. Uygur and Kazakh settlements are concentrated in Xinjiang while Tibetans live largely in Tibet (92.2% of the Tibetan population, with the remaining population living in Qinghai, Sichuan, Yunnan and other provinces). The main Zhuang settlement area is in Guangxi while Ningxia is inhabited by the Hui. The distribution of ethnic minorities in the provinces of Yunnan, Guizhou, Qinghai, Gansu, Jilin, Sichuan is relatively large (Fig. 3.6). Yunnan Province has the most ethnic minorities, a total of 22. Thus, most of the settlement areas of the minorities are distributed in the northwest and the southwest, which is an important factor for the population distribution pattern in China.

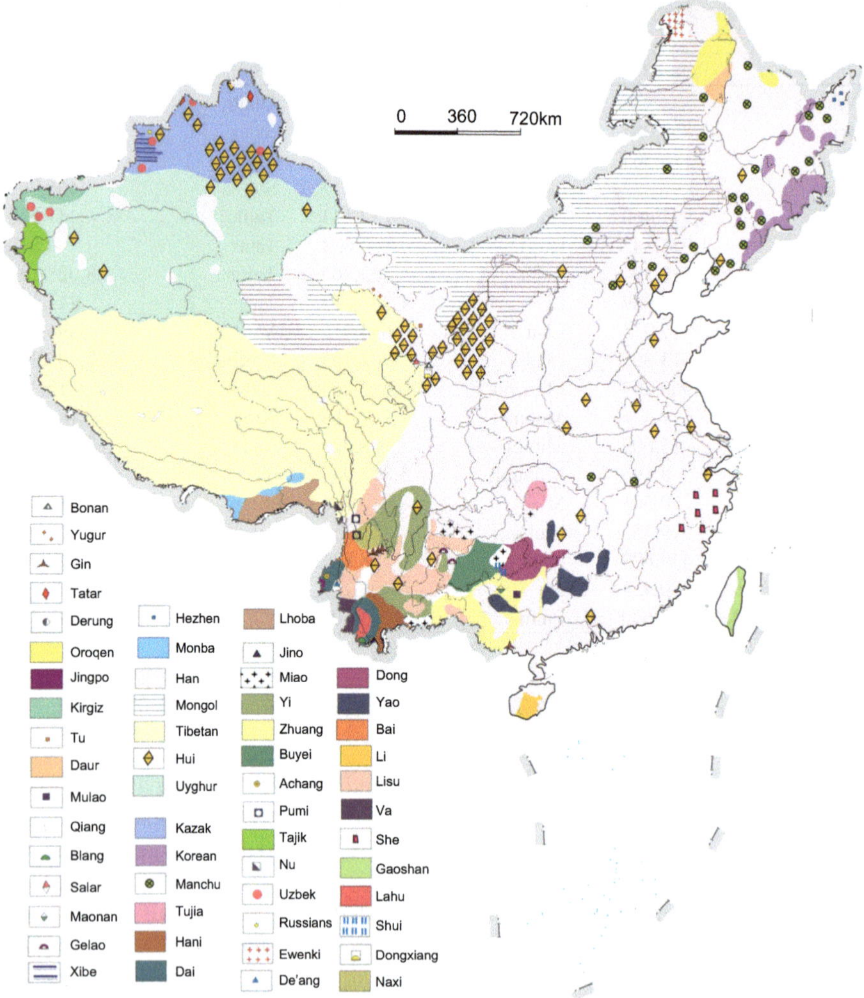

Fig. 3.6 Regional Distribution of the 56 Ethnic Group of China. (Rong and Tang 2004)

The Han live mainly is in the central coastal areas, while the ethnic minorities are mainly distributed in border areas. More than 30 ethnic minorities in China have their same ethnic group living in adjacent locations at neighboring countries, such as the Kazakhs of Xinjiang, China and the Kazakh population of the neighboring country of Kazakhstan. They have remained in close contact with frequent gatherings of relatives and friends at, for instance, weddings and holiday celebrations. As a result, trade relations with neighboring countries have developed rapidly.

The distribution of racial/ethnic groups is widely dispersed and mixed with small settlements and living conditions intertwined

In China, the more than 100 million of minority populations are widely distributed in many parts of the country. After decades of national contact, migration, integration and development a basic spatial pattern has evolved which has been characterized by the contrast of "big mixed, small settlements". All of the 23 provinces, 5 autonomous regions and 4 municipalities in China have settlement areas where ethnic minority groups live. Even on a county level, often settlement units are found where two or more ethnic groups live. While China's total ethnic minority population is relatively small (compared to the Han) its distribution is widely dispersed, as the example of the Hui and the Manchu shows. The Hui population is distributed over more than 2,000 cities and counties characterized by a large dispersion and a small concentration in the Ningxia Hui Autonomous Region. During the Qing dynasty (1644 ~ 1911) the then leading Manchu minority were dispersed all over China. In our times, the Manchu population is distributed over 2095 cities and counties, though still relatively concentrated in the three northeastern provinces of Liaoning, Jilin and Heilongjiang. Those areas saw the formation of fixed settlement areas after the long-term political developments and associated migration processes. Other examples are the Mongols and the Tibetans. 73 % of the Mongolian population lives now in Inner Mongolia while 70 % of the Tibetan population resides in Tibet and Sichuan. Some of the smaller nations are basically concentrated in a limited number of counties or even a few townships, such as the Keno, Qiang, Brown, Mao Nan and others. Due to their migration history, certain nations are concentrated in several areas which are far apart from each other. For example, one part of the Xibe group lives in the Ili Kazak Autonomous Prefecture in Xinjiang while another part of this ethnic minority group has made its home in a few counties of Liaoning Province, several thousand kilometers away.

Han Chinese settlements in the plains and hills, ethnic minorities distributed in mountains and plateaus

Due to historical factors, the Han are basically distributed in the plains and hilly areas and ethnic minorities are mostly distributed in the mountains and plateaus. The Tibetan settlement area is on the Tibetan Plateau. Topographically, the main Han Chinese neighborhoods are found in the third stepped unit to the east of the line Daxing'an Ranges—Taihang Mountains- Wushan Mountains—Xuefeng Mountains, with the exception of the Changbai Mountains. In the second stepped zone the Han Chinese and the majority of the minorities have their livelihood. The Han Chinese people are distributed at a lower altitude, while ethnic minorities live usually in higher elevations. In the transition zone between the second step and the first step are found the neighborhoods of most minority groups. This area is the most complex in ethnic and linguistic composition. The nations with an agriculture-oriented livelihood and a more developed economy are distributed in the lower zone while the livestock-based regions in the higher zone. The shifting cultivation-oriented nations with the least developed economy are found in the highest zone.

3.3 Population Geography in the U.S.

Who are the Americans and where do they live? Perez and Hirschman summarize this best (2009, p. 1): "The racial and ethnic makeup of the American people is in flux." Throughout the course of this chapter we will examine the patterns and distribution of the people who make up the population of the United States, where they are found, how they differentiate themselves from others and the potential growth and changes coming in the next millennium. We will try to identify certain trends, both recent as well as long-term ones, impacting the way we view the United States currently. Included in this analysis will be the distribution of social preferences, different ethnicities, the rural/urban divide as well as many other population variables.

One of the main problems in any attempt to place a definitive value on the population is the difficulty in gathering actual numbers. The United States Government takes a census of its citizens every 10 years. The most recent census was taken in the early months of 2010 with the data only partially available at the time of this publication. The maps and tables that constitute the data from which most of this chapter is written are from the 2000 Census (U.S. Census Bureau 2000), the 2010 Census (U.S. Census Bureau 2010), the American Community Survey, Census.gov and other US government publications (see bibliography). For a country in such a state of 'flux' as the US this multiple availability of data, both factual from 2000 and 2010 and extrapolated otherwise, poses a dilemma as to how to best summarize the information while remaining true to the most recent actual numbers.

3.3.1 General Features of Population in the U.S.

According to the Central Intelligence Agency[1] as of June 2012 the estimated total population of the US was 313,847,465 people. During that time period, the US Census Bureau's Data Finders Population Clock link said that there were 313,719,945 Americans. This makes the US the third largest population in the world behind China and India. There is one birth every 8 seconds, one death every 14 seconds, one international migrant (net) every 44 seconds, thus a net gain of one person every 13 seconds.[2] Results of the 2010 census revealed that the total population increased by roughly 9 % from the 2000 census. Only the age group 25 to 44 years decreased (about 3.4 % against the 2000 census).

3.3.2 Population Structure

The median age for all those 313 million Americans was 37.2 years with females averaging 38.5 and males 35.8. The female portion of the total population is 50.8 %

[1] www.cia.gov/library/publications/the-world-factbook/geos/us.html.

[2] www.census.gov/population/www/popclockus.html.

and since females also have a longer life expectancy there is variability in the numbers. The male portion of the population grew by 9.9 % from 2000 to 2010 while females increased by 9.5 %. The age group from 18–65, what is considered the working portion of the population, contained 194.3 million people, comprising 62.9 % of the total population. Those 65 year and older totaled almost 13 % of the population with males numbering 16,910,232 and females 22,571,696. It must be noted that there are more males than females under the age of 14 and many more women than men over the age of 65.

3.3.2.1 Age Structure of Population

The figure Population by Sex and Age along the right hand side of Fig. 3.7 is a population pyramid showing the gender/population distribution of the American population from the Census Bureau. Separated into 5-year cohorts from birth, there are two distinct population categories of special interest: ages 45–49 and 50–54. These represent in 2010 people born in the mid 1950's to the mid 1960's. The slight increases in those aged 15–19, 20–24 and 25–29 represent the children of that phase. The aftermath of World War II and the Korean conflict, as well as the liberalizing of the American mind set in the 1960 fostered a population boom reflected in these findings. At the top of the pyramid note that women over the age of 65 represent a larger portion of society than men do as they have a longer life expectancy. This is very much reflected in the ages 80+ and may be of special interest to those who study the American population in the future.

3.3.2.2 Population in Labor Force

Estimates from the U.S. Census Bureau suggest that in 2010 there are roughly 237 million Americans 16 years or older in the Civilian non-institutional population (ftp.bls.gov/pub/special.requests/lf/aat3.txt). Of those, 139 million are employed, representing 58 % of the total population. Almost 15 million are unemployed, or 9.6 % and 83 million are not in the labor force as it is defined. Table 3.6 breaks that population labor force down further.

Those in the population working have a per capita income of $47,400. While this represents a very large income, especially contrasted with China's, it still ranks 10 in the world behind such countries as Luxembourg, Qatar, Norway and Kuwait (CIA World Fact book). The population works in multiple areas: 0.7 % are in farming, forestry and fishing; 20.3 % are in manufacturing, extraction transportation and crafts; 37.3 % are in managerial, professional and technical jobs; 24.2 % work in sales and office occupation and 17 % work in other types of services. Thus the majority of the workforce, roughly 78 % are in professional or service jobs. All of these jobs contribute differently to the economy: agriculture accounts for 1.2 %, industry for 22.2 % and services 76.7 % (Census Bureau, CIA World Fact Book).

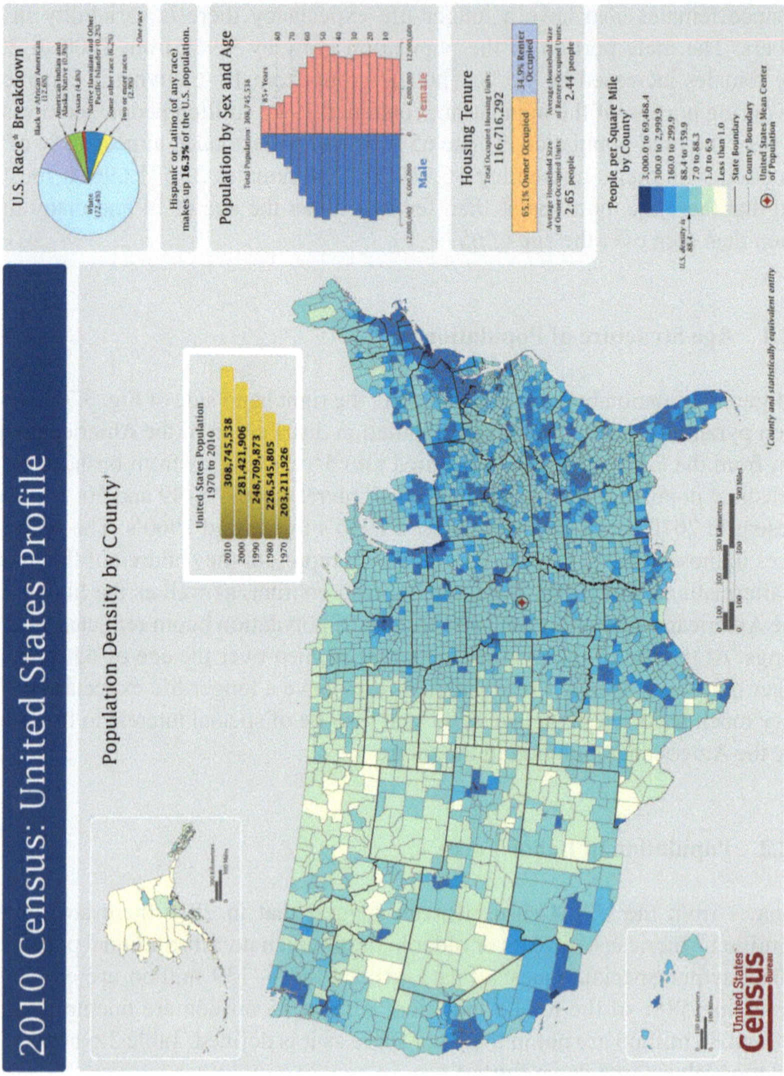

Fig. 3.7 2010 Census: United States Profile, Population Density by County

3.3.2.3 Social, Educational and Economic Populations

There are 116.7 million total households across the country with family populations numbering 262 million (U.S. Census Bureau, 2010 Census Summary File 1, factfinder.census.gov). Of these households, 77.5 million have children under 18 years of age. Non-family households number 39.2 million with households 65 years and older numbering 10.4 million reflecting the high number of household with single members of either gender.

Table 3.6 Civilian Non-institutional Population. (numbers in millions, each category represents % of available population, not total US population, ftp.bls.gov/pub/special.requests/lf/aat3.txt)

Ageprofile	Total population in labor	Employed total	Percent of total population in labor	Unemployed total	Percent of labor force
16–19	16.9	4.3	26	1.5	26
20–24	21	12.6	60	2.3	15.5
25–34	40.9	30.2	73	3.3	10
35–44	40	30.6	76	2.7	8.1
45–54	44.2	33.1	75	2.7	7.7
55–64	35.8	21.6	60	1.6	7.1
65	38.7	6.2	16.2	0.5	6.7

The growth rate of the population as a whole (2011 estimate) is around .96% which ranks the United States right in the middle of world countries, with Zimbabwe at 4% highest and the Northern Mariana Islands the lowest at −4%. China by comparison is at.49% growth rate. This is mediated of course by the large population of China to begin with. The United States has a birth rate of 13.83/1000 live births and a death rate of 8.38/1000 population. The birth rate is also right in the middle for countries with Niger first with 50/1000 and Japan second lowest at 7.3/1000. China's birth rate is near to that with 12.29/1000. The Chinese death rate is also lower at 7.03/1000 population (all data from www.CIA.gov/library/publication/the-world-factbook/geos/us.html).The United States as a whole is an educated society. The percent of high school graduates or higher of the total population is 84.6%. Twenty-seven percent of the population has a bachelor's degree or higher. Table 3.7 below highlights some of the educational charicteristics of the population. Although the population over the age of 25 is estimated to be under 200 million, 27% of those people have a bachelors degree or higher. This suggests that the population as a whole is very literate as only 6.4% of those have less than a 9th grade education.

3.3.3 Population Distribution

The population of the U.S. is growing in different places, but it is also displaying different growth rates. Regional growth in the South and the West is increasing far faster than that of the Midwest or the Northeast parts of the country. The South and the West represent slightly less than two/thirds of the total population. In the U.S., California, Texas and New York are the most populated states followed closely by Florida, Illinois, Pennsylvania, and Ohio. It is when one looks at individual counties within states where the different distribution patterns begin to emerge. It is the urban areas where most of the population is centered. Thus, a state such as New York or Georgia may have a large population but that population is centered, in the case of Georgia, in the counties around Atlanta and in the case of New York, in several cities along Lake Ontario, along the Interstate 90 corridor and down the

Table 3.7 Educational Attainment, Population 25 years and older (in millions) (U.S. Census Bureau, 2005–2009 American Community Survey)

Total population	197,440,193					
Less than 9th grade	9–12th, no diploma	High school graduate	Some college	Associates degree	Bachelor's degree	Graduate or professional
12.5	17.8	57.8	40.1	14.6	34.3	19.9
Percentage of total population						
6.4	9.1	29.3	20.3	7.4	17.4	10.1

Hudson River from Albany to New York City (Population Distribution and Change 2010 Census Briefs).

There are seven states with populations between 10,000,000 and 37,253,000: California, Texas, New York, Florida, Illinois, Ohio, and Pennsylvania. There are seven states whose populations are very low, between 563,000 and 999,999: Alaska, Delaware, Vermont, North and South Dakota, Montana and Wyoming. Of the states with populations between 5 and 10 million only Arizona, Colorado and Washington are located in the western part of the country, the remainder: Missouri, Wisconsin, Michigan, Indiana, Massachusetts, New Jersey, Maryland, Virginia, North Carolina, Tennessee, Kentucky, Minnesota and Georgia are considered to be eastern. The remaining states have populations that fall between 1 million and 5 million. The largest numeric growth is seen in Texas but as a percentage of total state populations, the western states of Nevada, Arizona, Utah, and Idaho are seeing their numbers increase the most. The only state losing population is that of Michigan where populations have decreased from 2000 to 2010 by .6 % (Population Distribution and Change 2010 Census Briefs).

3.3.3.1 Population Density

In Fig. 3.7, the map itself represents population density for counties in the U.S. It clearly shows how much of the population is concentrated in certain counties within particular states. This pattern will emerge in other forms as well as indicated below. These heavily populated counties, numbering between 500,000 and 9,520,000 are the sites of the United States' larger metropolitan areas. The population density of the United States is quite heavily weighted to the eastern, western and northern portions of the country. These are the original areas of inhabitation, populated by Western Europeans first, especially in the east, and then successive waves from Europe and Africa after and they trend towards the migratory patterns of later disbursement as well, with heavy populations across the middle tier of New York, the southern Pennsylvania areas and then up through the upper portions of Midwestern states such as Ohio, Indiana, and Illinois and the southern portion of Michigan and then around Lake Michigan in the region of Chicago. There are high densities along the southern portions of the Appalachian Mountain chain towards Atlanta, Georgia where people moved out from North Carolina and South Carolina as well. Moving farther west however the urban areas of larger cities begin to show the highest densities. Cities such as Denver, Colorado; Salt Lake City, Utah; Phoenix, Arizona; and the corridor from Seattle, Washington down through Portland, Salem and Eugene, Oregon. Figure 3.7 also shows how the eastern portion of the US tends to have large tracts of high densities but the western portion shows many rural areas with very low densities. The people per square mile by state indicate that the highest state density is in the New York-New England area with Alaska, Montana and Wyoming bringing up the rear with the lowest. Much of the Central United States has densities ranging from only 7 to 79.5 people. Along with Washington and California having high densities per state, this suggests that the United States is a bi-coastal populated country. The entire eastern portion of the country has population densities above the national average of 79/sq. mile showing just how little population there is in the central portions of the country. The very small population of Alaska, contrasted with its very large land area also contributes to the discrepancy.

There are several distinctive metropolitan areas around the country. The New York-Newark area is home to 19 million people; the Los Angeles-Long Beach area has 12 million; Chicago has 9.4 million; Dallas-Fort Worth 6.3 million; and the Philadelphia region at 5.9 million (Population Distribution and Change 2010 Census Briefs). Also indicated from the figure are two areas of lesser populated areas: the Rocky Mountains of Nevada, western Colorado, New Mexico, Utah and central Idaho; and the western parts of Texas, Oklahoma, Kansas, Nebraska, all of North and South Dakota and eastern Montana. These are the great agricultural areas of the midwest, home to corn, soy, wheat cereal crops as well as feedlot production of pork and cattle.

3.3.3.2 Urban Population and the Rural Dilemma

The United States has largely an urban population. The percentage of state population in metropolitan area counties shows that the US as a whole is 80 % urban. Seventeen states are above the average having more citizens who live in cities. Only Alaska, Idaho, Montana, Wyoming, North and South Dakota, Iowa, Arkansas, Mississippi, Kentucky, West Virginia, Vermont, and Maine are less than 50 % urban in their total population. While it would seem logical to suggest that the states with the highest total population, states such as Texas and California, would be those with the highest urban populations, it has to be noted that these are also among the largest states in total area so the urbanization is even more apparent. Thus both Texas and California are very urbanized and, as noted in other chapters in this book, these states are agriculturally highly productive. This suggests that the populations are for the most part living in the great cities of those states and that the land is being used for large scale agricultural practices that do not require large human labor inputs. The pressures upon the land for expanding urban areas is slightly alleviated due to the aggrandizing of farmland and the expansion of farm size with its corresponding flight to urban areas of those populations (Table 3.8).

Table 3.8 Population Change for the Ten most Populous Statistical Areas: 2000 to 2010. (U.S. Census Bureau 2010; Census and Census 2000)

Metropolitan statistical area	Population		Change	
	2000	2010	Number	Percent
New York-Northern New Jersey-Long Island, NY-NJ-PA	18,323,002	18,897,109	574,107	3.1
Los Angeles-Long Beach-Santa Ana, CA	12,365,627	12,828,837	463,210	3.7
Chicago-Joliet-Naperville, IL-IN-WI	9,098,316	9,461,105	362,789	4.0
Dallas-Fort Worth-Arlington, TX	5,161,544	6,371,773	1,210,229	23.4
Philadelphia-Camden-Wilmington, PA-NJ-DE-MD	5,687,147	5,965,343	278,196	4.9
Houston-Sugar Land-Baytown, TX	4,715,407	5,946,800	1,231,393	26.1
Washington-Arlington-Alexandria, DC-VA-MD-WV	4,796,183	5,582,170	785,987	16.4
Miami-Fort Lauderdale-Pompano Beach, FL	5,007,564	5,564,635	557,071	11.1
Atlanta-Sandy Springs-Marietta, GA	4,247,981	5,268,860	1,020,879	24.0
Boston-Cambridge-Quincy, MA-NH	4,391,344	4,552,402	161,058	3.7

For information on confidentiality protection, nonsamplingerror, and definitions, see www.census.gov/prod/cen2010/doc/pl94-171.pdf. The full names of the metropolitan statistical areas are shown in this table: abbreviated versions of the names are shown in the text.

Also note that the number of counties that were metropolitan in both 1990 and 2000 (those in light blue) and those that were metropolitan in 2000 and nonmetropolitan in 1990 (those in dark blue) far outnumber those that were nonmetropolitan in 2000 and metropolitan in 1990. Not only is the population of America growing, it is also rapidly urbanizing.

3.4 Racial/Ethnic Geography in the U.S.

Race is a complex and frustrating topic as it relates to the population of the United States. Generally speaking Americans are divided into whites, blacks, Hispanics, Asians but at many levels 'most Americans, except for recent immigrants, are probably descended from multiple geographic, ethnic, and racial origins' (Perez and Hirschman 2009, p. 2). The projections do not take into consideration the mixing of racial and ethnic identities through intermarriage, as it is difficult to build consensus on the categories of those involved. There is also the distinction between identity

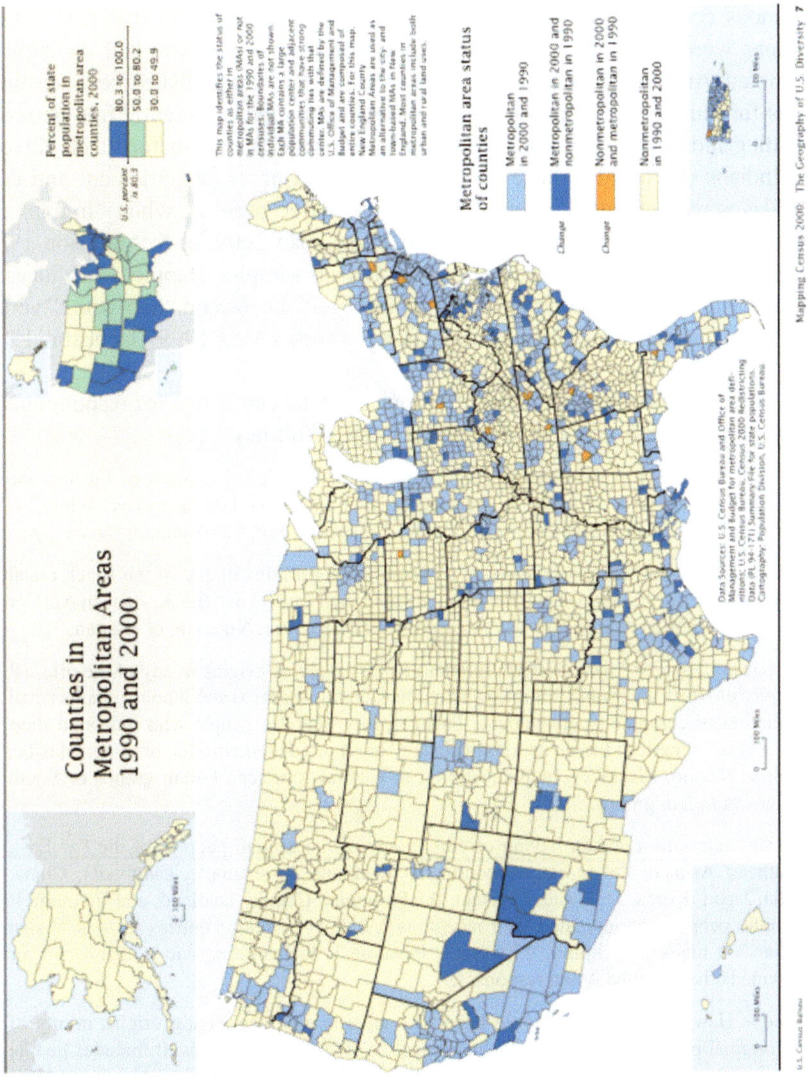

Fig. 3.8 Urban population in the U.S.

and ancestry as it pertains to how one self-identifies (Perez and Hirschman 2009). Race and ethnicity are seen in different contexts by both social scientists and in everyday life (Lee and Bean 2004) (Fig. 3.8).

Census data can only be interpreted from the numbers submitted and those of race depend upon the choices of those who have multiple racial and ethnic origins (Edmonston et al. 2002, p. 249). Many of the newly arriving immigrants do not see themselves as any particular color, nor are they viewed as such by others (Lee and Bean 2004). The first census was conducted in 1790 making the US the longest continuous census taking country (Farley and Haaga 2005). In that first census questions were raised about the composition of society and who had the right to be counted. Article 1, Section of the US Constitution distinguished between three groups for purposes of taxation and Congressional apportionment: "free persons" (including indentured servants), "other persons" (a euphemism for black slaves), and "Indians not taxed" (those living beyond areas of white settlement and control) Slaves were counted as only three-fifths of free persons, while "Indians not taxed" were not counted at all (Perez and Hirschman 2009, p. 5; Anderson 1988, p. 9; Klinker and Smith 1999, p. 25). Later Census samples grappled with ideas of mixed race make-up, percentage of blood purity, and the idea on "non-white" versus "white" in immigrant cultures coming from Europe (Alba 1999; Jacobson 1998; Lee and Bean 2004).

The 2010 Census had specific definitions of racial categories for respondents to self-identify through. The form suggested them as follows:

> "White" refers to a person having origins in any of the original peoples of Europe, the Middle East, or North Africa. It includes people who indicated their race(s) as "White" or reported entries such as Irish, German, Italian, Lebanese, Arab, Moroccan, or Caucasian.
>
> "Black or African American" refers to a person having origins in any of the Black racial groups of Africa. It includes people who indicated their race(s) as "Black, African Am., or Negro" or reported entries such as African American, Kenyan, Nigerian, or Haitian.
>
> American Indian or Alaska Native: refers to a person having origins in any of the original peoples of North and South America (including Central America) and who maintains tribal affiliation or community attachment. This category includes people who indicated their races(s) as "American Indian or Alaska Native" or reported their enrolled or principal tribe, such as Navajo, Blackfeet, Inupiat, Yup'ik, or Central American Indian groups or South American Indian groups.
>
> "Asian" refers to a person having origins in any of the origian peoples of the Far East, Southeast Asia, or the Indian subcontinent, including, for example, Cambodia, China, India, Japan, Korea, Malaysia, Pakistan, the Philippine Islands, Thailand, and Vietnam. It includes people who indicated their race(s) as "Asian" or reported entries such as "Asian Indian," "Chinese," "Filipino," "Korean" "Japanese," "Vietnamese," and "Other Asian" or provided other detailed Asian responses.
>
> "Native Hawaiian or Other Pacific Islander" refers to a person having origins in any of the original peoples of Hawaii, Guam, Samoa, or other Pacific Islands. It includes people who indicated their race(s) as "Pacific Islander" or reported entries such as "Native Hawaiian," "Guamanian or Chamorro," "Samoan," and "Other Pacific Islander" or provided other detailed Pacific Islander responses.
>
> "Some Other Race" includes all other responses not included in the White, Black or African American, American Indian or Alaska Native, Asian, and Native Hawaiian or Other Pacific Islander reace categories described above. Respondents reporting entries such as multiracial, mixed, interracial, or a Hispanic or Latino group (for example, Mexican, Puerto Rican, Cuban, or Spanish) in response to the race question are included in this category. (Overview of Race and Hispanic Origin: 2010 issued March 2011, page 3).

As you can see by the categories, it is very difficult for the Census Bureau to get specific numbers of national origins for groups of people when, for example, Chinese and Indian natives are grouped together as Asian. It makes for distinctions between

Hispanic origins difficult since they are lumped together as "other." The different populations coming to America can be seen however in the differences between the original census and 2010. The U.S. has become very complex racially and how to count various groups impacts almost impossible. As a part of the total the various groups fall out as follows: One Race has 299 million, White has 223 million, Black or African American 38 million, American Indian and Alaska Native 2.9 million, Asian 14.6 million, Native Hawaiian and Other Pacific Islander 540,000, Some Other Race 19.1 million and Two or More Races at 9 million.

The difficulties of total population growth are best seen in the category of Hispanic: total population of the country as a whole grew by 27 million from 2000. Much of this came from those reporting themselves as not white alone and either Hispanic or Latino as their ethnic origin. The Hispanic group grew from 35.3 million in 2000 to 50.5 million in 2010, thus over half of the total growth of the U.S. occurred in this group. But what does that say about individual groups within that category? It is a rhetorical question of course but it has vast implication within American racial and ethnic political distinctions.

3.4.1 Distribution

Figure 3.9 below shows the counties with the highest percentage of total population for a single group from the six census groups. In the upper right corner is a representation of the states with the highest percentage. Only Hawaii with Asian, Washington DC with Black, and Puerto Rico with Hispanic are not White, not Hispanic.

As mentioned earlier in the chapter, definitions of race make large sweeping generalization problematic when defining where Americans are. Figure 3.9 shows the six census 2000 classifications on race to map 'groups with highest percent of county population.' It suggests that the southern border with Mexico states of California, Arizona, New Mexico and Texas have counties where the highest percentage of the population is Hispanic or Latino. This also suggest that there is not as much dispersal of Hispanic or Latino people in the rest of the country, thus prognostication of Hispanic peoples increasing all over the country may be slightly flawed. There are portions of Arkansas, Louisiana and Mississippi along the Mississippi delta; Central portions of Alabama, Georgia and South Carolina, and areas along the North Carolina/Virginia border where Black or African American are the highest populations within counties. This would suggest that the large migrations of blacks from the rural south to the urban north might not have impacted population as much as previously thought. This data only emerges at the county level as the state level is not detailed enough for analysis.

Figure 3.9 above shows which minority group is the highest percentage of total county population. This map excludes White, not Hispanic as a group. From Texas, Kansas, and Nebraska west Hispanic or Latino dominates with the exceptions of American Indian and Alaska Native areas indicating tribal land holdings. The southeastern portion is predominantly black, especially Louisiana, Mississippi, and

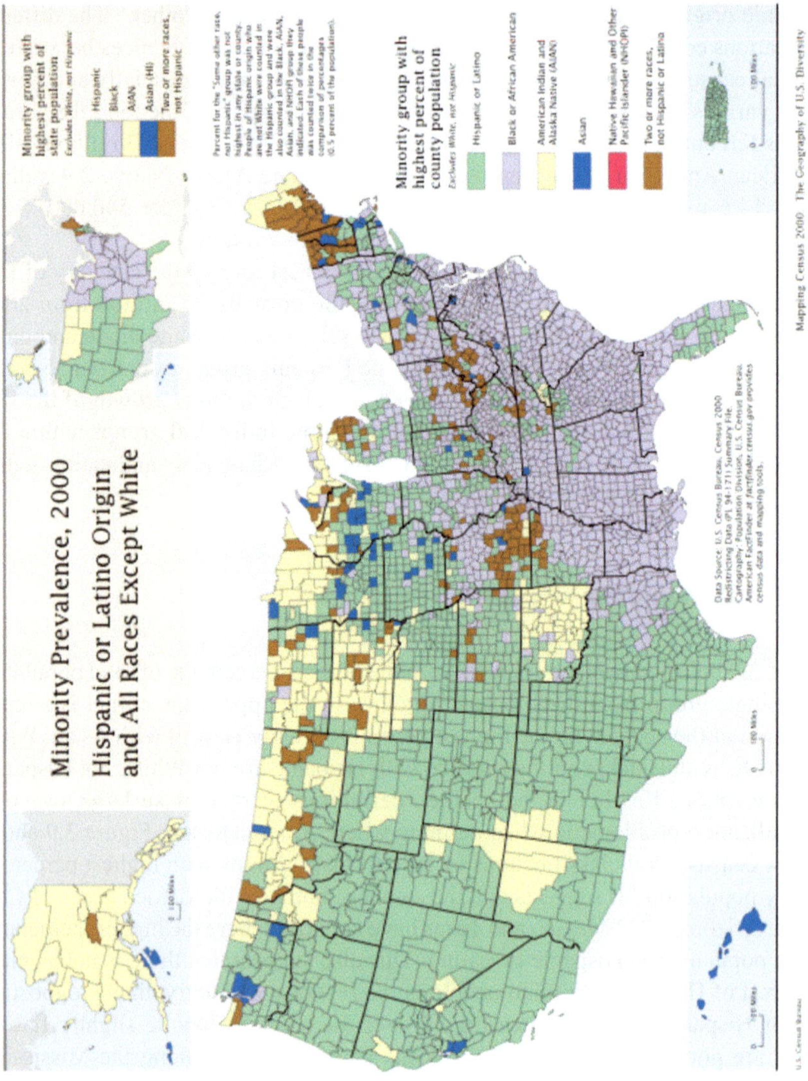

Fig. 3.9 Minority Prevalence. (http://2010.census.gov/2010census/data/index/php)

South Carolina, while southern Missouri and Vermont, New Hampshire and Maine are mostly Two or more races, not Hispanic or Latino. The counties colored as dark blue represent Asian as the group with the highest percent and shows some historical facts such as the area around San Francisco and the island of Hawaii but some interesting later immigrations from countries such as Vietnam and Laos as well as they impact such places as Minnesota, Wisconsin and Iowa.

3.4.2 Migration and Immigration

The United States is a country founded upon immigration into and in-migration within its borders. While America is a highly mobile society with nearly '1 in 7 people changing residence each year' (Census Atlas 2007, p. 108), it is also one that has increased in population through numerous immigrations from the rest of the world. The country is in its fourth immigration wave. This is called the 'Globalization Wave' and had a population impact from 'foreign-born' residents of 31 million by 2000 (Kritz and Gurak 2005, p. 259, see Martin and Midgley 2003). Most immigrants coming to the US between 1995 and 2000 lived in one of six states: California, Florida, Illinois, New Jersey, New York, and Texas as social networks link newcomers to those who are already established. All of these states had over 1 million foreign-born residents (Census Atlas 2007). By far the largest population of foreign-born people came from Mexico, over 9 million. The next two countries of origin were China and the Philippines at roughly 1.5 million followed by India, Vietnam Cuba, Korea, Canada, El Salvador and Germany at just under 1 million.[3] The largest percent of foreign born as a percentage of total population live in California, Florida, central Washington State, western and southern Texas, the metropolitan New York City, Chicago, Philadelphia, and Boston areas.

Historically, net domestic migration filled in much of the country as people migrated from their original entry ports. The largest number of people originally coming was from Germany, over 50 million. Next, the Irish at 37 million, African Americans and the English at 27 million, followed by those who labeled themselves as American at just fewer than 20 million. Mexicans numbered 18 million followed by Italian's, Polish, French, American Indian at 8 million, and then Scottish, Dutch Norwegian, Scotch-Irish and Swedish, all around or under 5 million. As the total population grow, however, the percentages of particular ancestry's decrease. For instance, German's decreased from 23 % in the 1990 Census to 16 % in 2000 and stayed that way through 2008. Similarly, the Irish, English and African American percentages of the total population decreased reflecting the sizable immigrations coming from other places such as Mexico, China, and smaller populations such as Brazilian, Albanian, and Honduran (Census Atlas 2007).

3.4.3 Trends and Future Projections

The net growth of the Unites States population can be attributed to three primary causes—the total fertility rate for women of 2.06 children per lifetime, a net immigrant growth rate of 4.18/1000 population and a life expectancy rate of 78.37 for the total population. These life expectancy rates are 75.92 for men and 80.93 for women, thus women live longer and contribute to the total longer. These three

[3] China includes those who responded China, Hong Kong, Taiwan, or the Paracel Islands. Korea includes those who responded Korea, North Korea, or South Korea.

factors are critical in an understanding of where the population will go in the future. The major trend is that of an increasing population for the for-seeable future since the fertility rate and net immigration rate will naturally increase the number of people (Shrestha and Heisler 2011). By the year 2050 it is anticipated that the population will number almost 440 million.[4] The fertility rate of American women ages 15–44 shows a small decline over time although the age group 40–44 slightly increased in 2010.

A second trend is that the American population is getting older. The crude death rate has remained relatively consistent since 1950, hovering between 8.1 and 9.7 per 1,000 persons. This is an inverse relationship when looking at an aging population. One would expect the death rate to go up as a population ages, but the fact that the rate is consistent over time means that people are living longer as a general rule and contributing to the growth of the population longer. The effects of this aging are not well known when looking towards that future. Will health care costs escalate? Is there a general decline in an aging populations contribution to the overall economic health of the country? Are there enough people to fill the jobs of those leaving the work force?

The third and perhaps most significant of the trends in the United States population is the effect of the net immigration rates currently experienced by the various different states and how this will change the population structure as it currently stands. This is because 'major racial and ethnic groups are aging at different rats, depending upon fertility, mortality, and immigration within these groups' (Shrestha and Heisler 2011). The most mobile of all age groups are young adults who move for various reasons such as economic opportunity, lack of future quality of life possibilities or potential violence due to political beliefs. The United States favors the entire immigration of these families, not just the individual so, due to different ideas on birth control, fertility rates and family size, where they settle will directly impact population figures.

As mentioned earlier in this chapter, the changing definition of race on the 2000 and 2010 Census has created many new catagories of self-identification for racial and ethnic persons. This is making the United States much more diversified racially and ethnically. Hispanics are now the number one minority, with almost 13% of total population in 2000 growing to approximately 30% by the year 2050 (Shrestha and Heisler 2011). But what to make of this since there are so many different ways for someone of Latino origin to identify with? (See Chap. 3.3). What is does suggest is that in the next 40 years, the United States will become much more racially diverse as it allows its citizens to identify themselves in catagories other than simply American.

[4] U.S. Census Bureau, Table 2. Projections of the Population by Selected Age Groups and Sex for the United States: 2010 to 2050 (NP2008-T2), issued August.

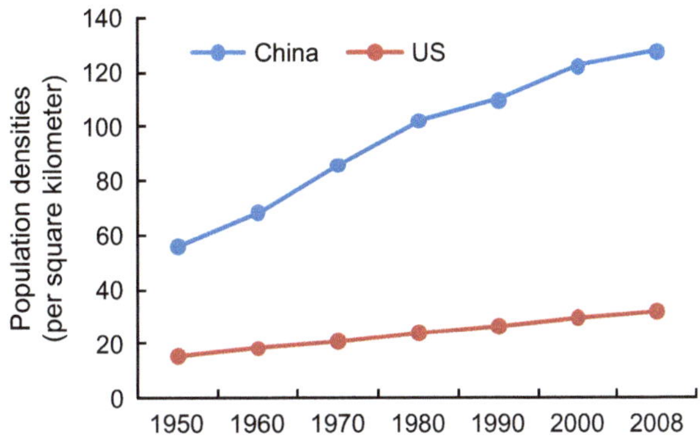

Fig. 3.10 Comparison of population densities changes between China and the U.S. (1950–2008) (Calculated with the relevant data from "China Statistical Yearbook" and the USA statistical yearbook-1949–2008)

3.5 Comparisons of Population Geography of China and the U.S.

3.5.1 Comparisons of the Changes in the Population Patterns of China and the U.S. (1950–2008)

Population densities and population patterns of China and the U.S. have been changing over time (Fig. 3.10). In Fig. 3.11, the map represents how the densities changed in these two countries for recent sixty years. Obviously, the densities kept increasing, but the growth rate in China is much higher than in the U.S.. China's population density rose to 127.61 per km^2 in 2008, more than doubling from 55.78 per km^2 in 1958. It is expected that the rate would slightly decrease in the future since the rate slowed down after 1980. The population density of the U.S. rose to 31.58 per km^2 in 2008, nearly two times the 15.19 per km^2 it was in 1958; it showed a slow but steady growth. The graphs below indicate the gap in population densities between the two countries which is more of a chasm.

From Fig. 3.11, the conclusion can be safely drawn that there were few basic changes in the population distributions of these two countries over time periods. In general, there was a much higher population density in the east than that in the west. However, compared to 1950, population density grew more rapidly in the east of China, and the maximum population density of China had exceeded that of the U.S. by six times by 1960. In the next decade, there were not many variations in the population distributions except a further increase in numbers of people in both countries. In the following decade 1970 to 1980, tremendous population growth took place in the east of both China and America. Apart from that, the population in California, which lives on the west coast of the U.S., increased substantially and disproportionately resulting in a bi-coastal population distribution in the U.S.. In the following decade 1980 to 1990, the average population density in China expanded to over 100 people per square kilometers while there was an increase to only 25.84 people per square kilometers in the U.S. in the same period. By 2000,

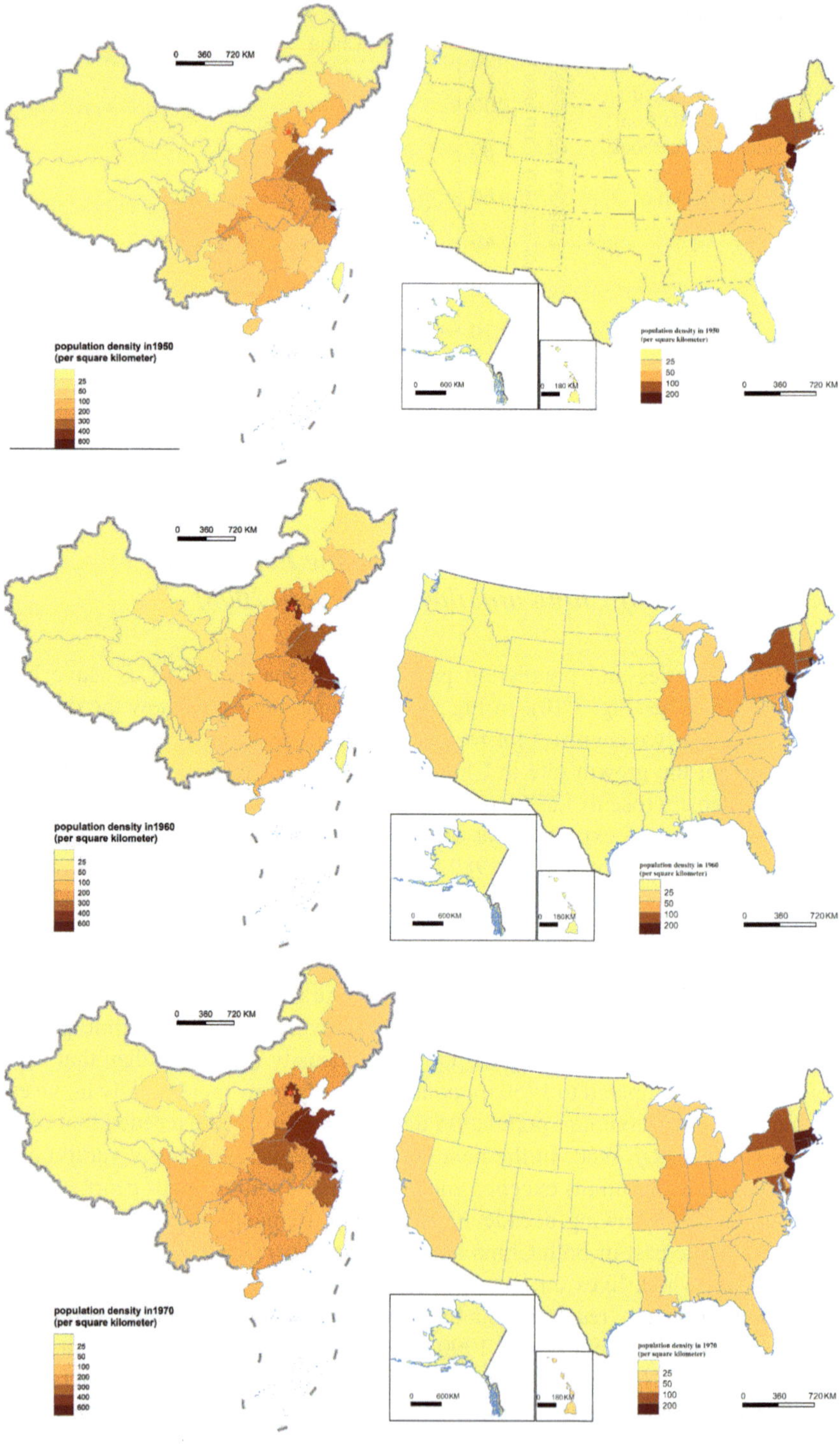
0 360 720 KM
population density in1950
(per square kilometer)
25
50
100
200
300
400
600
population density in 1950
(per square kilometer)
25
50
100
200
0 600 KM
0 180 KM
0 360 720 KM
0 360 720 KM
population density in1960
(per square kilometer)
25
50
100
200
300
400
600
population density in 1960
(per square kilometer)
25
50
100
200
0 600KM
0 180KM
0 360 720KM
0 360 720 KM
population density in1970
(per square kilometer)
25
50
100
200
300
400
600
population density in 1970
(per square kilometer)
25
50
100
200
0 600KM
0 180KM
0 360 720KM

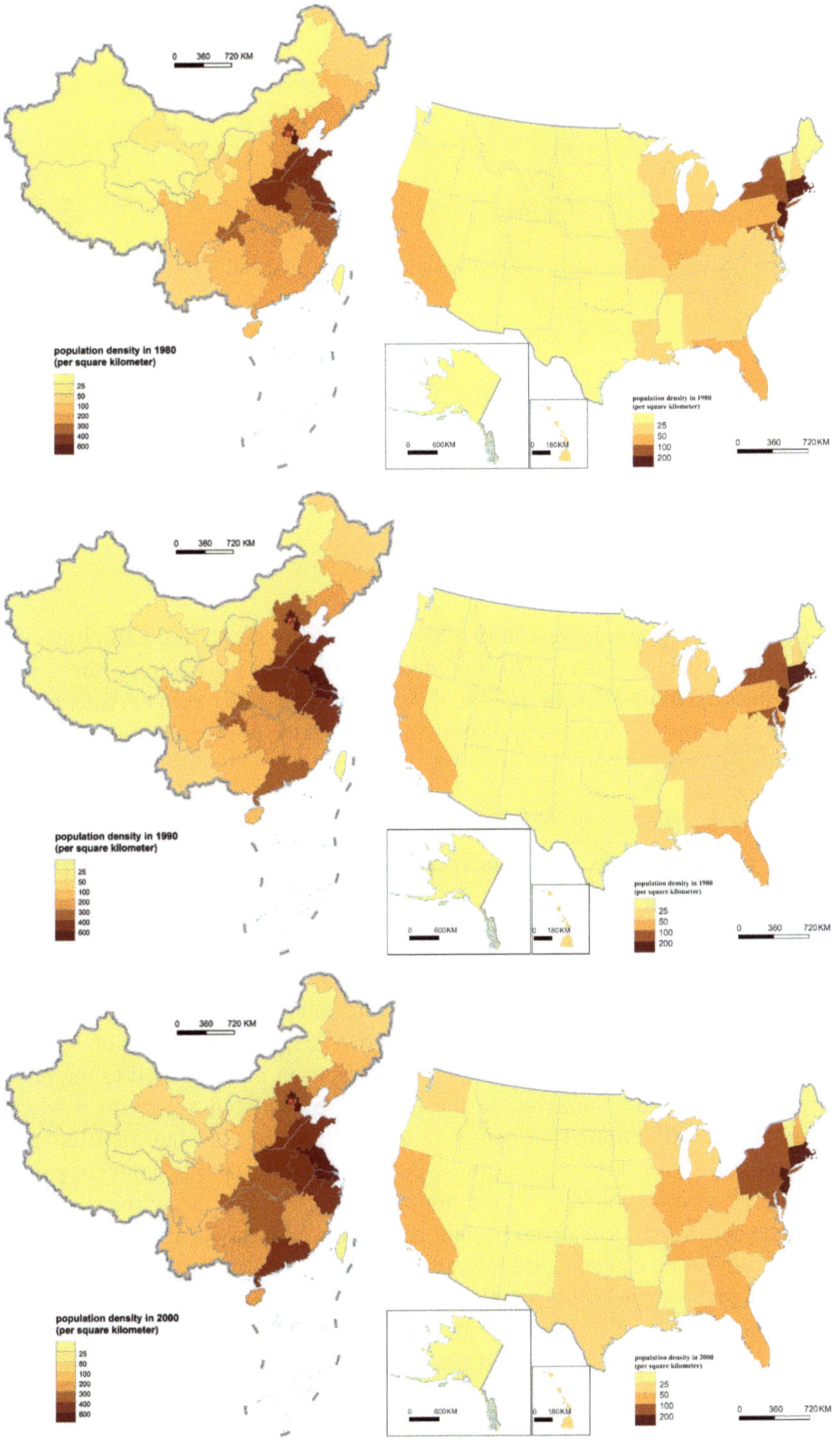
0 360 720 KM
population density in 1980
(per square kilometer)
25
50
100
200
300
400
600
0 600KM
0 180KM
population density in 1980
(per square kilometer)
25
50
100
200
0 360 720KM
0 360 720 KM
population density in 1990
(per square kilometer)
25
50
100
200
300
400
600
0 600KM
0 180KM
population density in 1980
(per square kilometer)
25
50
100
200
0 360 720KM
0 360 720 KM
population density in 2000
(per square kilometer)
25
50
100
200
300
400
600
0 600KM
0 180KM
population density in 2000
(per square kilometer)
25
50
100
200
0 360 720KM

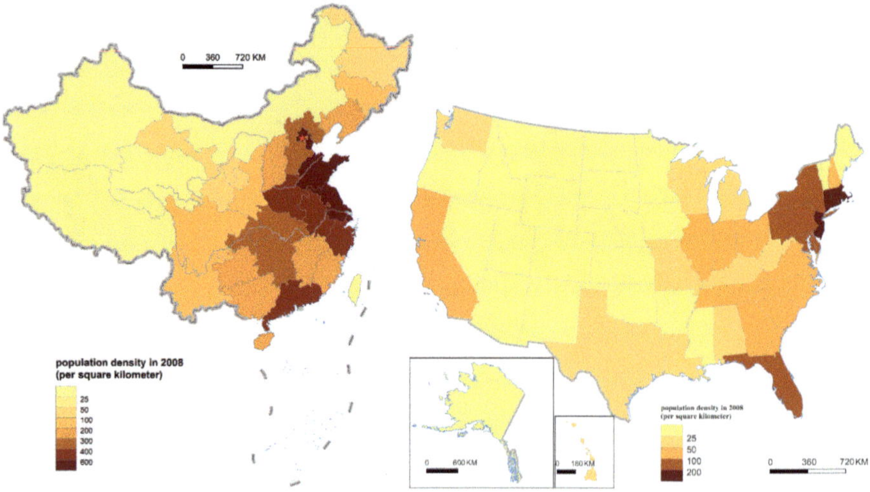

Fig. 3.11 The series maps of population density in China and the U.S. (From top to bottom: 1950, 1960, 1970, 1980, 1990, 2000, 2008)

the Chinese population increased in a higher rate compared to the U.S., especially in provinces such as Shanxi, Guangxi and Guizhou. On the contrary, the year of 2008 witnessed more obvious growth of the population in the west of America than in China, thus further displaying disparities between the population distributions of these two countries.

3.5.2 Comparisons of Driving Forces of Population Patterns in China and the U.S.

Population densities of China and the U.S. are both relatively high in the east and low in the interior west (although they are relatively high along the Pacific Coast to the far west in the U.S.). China shows a more significant uneven pattern between east and west regions. Much of this can be attributed to the physical geography of the countries. In the U.S. there has been a general westward progression of the population center (a generalized ideal of the distribution of the population over time—sometimes called the demographic center) beginning with the first white settlers. As the population of the U.S. grew during its two hundred year history, this demographic center moved westward as well. It moved from near the state of Delaware towards its present location in Missouri. It is the expansion and movement of people across the country from east to west, as well as the growth of the Pacific coast states of California, Oregon and Washington as well as bordering Arizona, that has filled in the country's population center (Knox and Marston 2010).

The topography of China is high in west and low in east, mainly formed by mountains and plateaus. There are very limited areas of plains topography (about 12 % of the total land area) concentrated in the eastern region. Therefore, most of China's population is concentrated in the limited areas of the eastern plains. The division of the "Hu Huangyong line" forming a northeast to southwest line (see Fig. 3.5), shows the high density population and higher levels of urbanization in the east, and the sparse population and lower level of urbanization in the west. Because of the high-density and highly uneven distribution of population, China is confronted with a harsh problem of human-environmental contradiction and significant environmental problems such as ecological fragility and pollution concentrated in urban areas.

Spatial patterns of the U.S. population distributions mainly depend on its historical immigrations, domestic migrations, etc. A large number of Europeans immigrated into the United States in the 1820s up to the 1920s; they were mainly concentrated in the East, although there were significant populations coming up from colonial Mexico to populate the west coast. During the expansion of these populations into the rest of the country they followed pathways westward consistent with source, for example, New Englanders typically immigrated westward along the northern border with British Canada and settled in the northern areas of the country. After the Second World War, Asia and Latin America gradually replaced Europeans as the main sources of immigrants to the United States, and much of these people flowed to the southwest and urban regions, (more on spatial mobility in the U.S. with maps) so the immigrant flows had somewhat re-shaped the original spatial pattern of the U.S. population. Internal migration is another driving force for the U.S. population pattern, the so called "Sun Belt" zones of south and west gradually replacing the frozen northeast as a major destination for the migrants, especially since the mid-19th century. These "westward movements" and the process of national land exploitation, the concentration in the central plains area of a large number of agricultural populations and the migrations of foreign immigrants flowing into other regions of the United States have changed the shape of the country to one where the bulk of the population is in large metropolitan areas (see Chap. 7 in this volume).

3.5.3 Historical and Cultural Differences Between the Population of China and the U.S.

China and the U.S. are typical representatives of the ancient Oriental civilization and the modern Western civilization, respectively. They both possess large populations with high cultural/ethnic/racial diversities. China is an ancient agricultural country with a history of more than five thousand years of settlement reflecting significant internal and regional features. The United States is representative of a modern Western civilization featured by an immigrant culture of "multi-ethnic blending" the so-called "melting pot," restricted in a historical perspective to less than 300 years. This is a difference that must not be understated. The different historical relationships between the citizens of each country and their physical landscape make for a large part of the differences between the two countries.

When it comes to the sources of the population, there are 56 ethnic groups in China and about 99 % of the populations are natives. On the contrary for the U.S., only about 1 % of the population is Native American or Indigenous, 99 % of the populations are immigrants of some form who arrived after the discovery of the western hemisphere. So compared to the United States, the Chinese nation is relatively conservative, advocating the concepts of collectivism and entireness, while American culture emphasizes individual values, democracy and freedom, personal promotion and competition. Thus, the social systems of China and the U.S. can be summarized as "Top down" and "Bottom up" respectively.

3.5.4 Comparisons of Population Policies and Schemes in China and the U.S.

China's population policies/schemes include the household registration system (Hukou) and the birth control policy. The U.S. population policies/schemes are mainly based on the immigration policies during different periods and relatively sound social security system, etc.

Firstly, China's household registration system to some extent has weakened the free movements of population and economic factors, and has aggravated the "dual structure" and wealth gaps between the urban and rural areas. This has resulted in an inequality of urban and rural areas in terms of housing, education, health care, social insurance, among many. There is no similar system in the United States and this allows people to move freely. The Chinese government has also introduced policies to encourage and support people to move to sparsely populated areas in central and western regions. In the U.S. there were significant movements of people to rural areas but they were primarily driven by either economic gain, as in the various gold rushes, or land grabs, as in the opening of previously held Native American lands.

Secondly, the population policy of "birth control" is another striking feature of China. This is a national basic policy implemented to control the natural population growth; the policy has not only effectively controlled the excessive growth of China's population, but also has made an important contribution to the total control of the world's population. However, it resulted in a family structure of "4-2-1" which means heavy burdens of supporting the aged, the problem of loneliness for single child and so on. In contrast, the United States is the world's largest country to receive immigrants, so the population policy of the U.S. is mainly based on its immigration policy. Immigrants were entitled to free entry initially, but later, especially since 1982, there has been a certain limit for immigration. The "quota immigration policy," based on selective principles was implemented in 1982. Especially after World War II, the U.S. has adopted a "talent import policy"to attract the elite worldwide, which successfully help the U.S. becoming a leader in scientific and technical fields.

Finally, when it comes to the population pyramid structure, the proportion of China's labor force (age of 15 to 64 years) is very large in its total population, with the most abundant labor resources in the world, witnessing an unprecedented "demographic dividend" period. But the large numbers of workers may also cause great pressures on employments, what's worse, in contrast with the United States and other developed countries, China's aging trend is faster, larger, and the level of the national economy of China is not as high. Thus, China might experience a danger situation, that of "getting old before rich". By contrast, the U.S.'s immigration policies allow absorption of elite talents from all over the world, which means the U.S. is enjoying an international demographic dividend. Although it also suffers a potential population aging problem, the various U.S. social systems, such as medicare and retirement programs provides a more sound social security system compared to China, which is a guarantee for the public well-being.

3.6 Conclusions

China and the U.S. are both large countries in the world in terms of population, ranking first and third in the world. The distribution of their large populations in their territory is extremely uneven. In China it shows a southeastern-to-northwestern gradient, while in the U.S. it is mostly an east-middle-west disparity. The uneven distribution of population is a result of the distribution of physical geographical conditions and socioeconomic activities, as well as years of domestic migration. Population dynamics in both countries show the dynamics of human-environment interaction from the historical perspective. Both countries have quite a number of ethnic groups with one leading group each. They have seen conflicts and tensions over racial, ethnic and cultural issues in the distant and recent past. China's large and aging population is creating serious challenges to its economic development, ecological service and environmental qualities. The Chinese government is trying to overcome this difficulty by continuing the baby planning policy, which on the other hand increase the potential risk of population issues due to the single-child-family problems. By contrast, the U.S has other types of problems in form of an aging population and by attracting immigrating elites and other groups from all around the world which can take advantage of the existing social security system. Population, people, as one of the key player in the human-environment system, exhibits so many common but critical differences, and this is the basis for understanding the social, economic, and urbanization issues in the following chapters.

References

Alba, R. (1999). Immigration and the American realities of assimilation and multiculturalism. *Sociological Forum, 14*, 3–25.

Anderson, M. J. (1988). *The American census: A social history*. New Haven: Yale University Press.

Du, X. R., & Tang J. J. (2004). *Atlas of China*. Beijing: SinoMaps Press.

Edmonston, B., Lee, S. M., Passel, J. S. (2002). Recent trends in intermarriage and immigration and their effects on the future racial composition of the US population. In P. Joel & C. W. Mary (eds.), *The new race question: How the census counts multiracial individuals* (pp. 227–255). New York: Russell Sage Foundation.

Hu, C. Q. (1990). *Huanyong Hu's anthology of population Geography*. Beijing: China Financial and Economic Publishing House.

Hu, H. Y. (1935). Distribution of population in China. *Acta Geographica Sinica, 2*(2), 33–74.

Hu, H. Y. (1986). *Abridged edition of population Geography of China*. Chongqing: Chongqing Publish Group.

Jacobson, M. F. (1998). *Whiteness of a different color: European immigrants and the Alchemy of race Cambridge*. London: Harvard University Press.

Klinker, P. A., & Smith, R. M. (1999). *The unsteady march: The rise and decline of racial inequality in America*. Chicago: University of Chicago Press.

Kritz, M. M., & Gurak, D. T. (2005). In R. Farley & J. Haaga (eds.), *The American people: Census 2000* (pp. 259–301). New York: Russell Sage Foundation.

Lee, J., & Bean, F. D. (2004). America's changing color lines: Immigration, race/ethnicity and multiracial identification. *Annual Review of Sociology, 30*, 221–241.

Martin, P., & Midgely, E. (2003). Immigration: Shaping and reshaping America. *Population Bulletin, 58*(2), 11–14.

Perez, A. D., & Hirschman, C. (2009). The changing racial and ethnic composition of the US Population: Emerging American Identities. *Population and Development Review, 35*, 1–51.

Shi, P. J. (2003). *Atlas of natural disaster system of China*. Beijing: Science Press.

Shrestha, L. B., & Heisler, E. J. (2011). The changing demographic profile of the United States. In CRS Report for Congress 7–5700. http://www.crs.gov RL32701. Accessed 31 March 2011.

U.S. Census Bureau. (2000). 2000 Census Summary File 1.

U.S. Census Bureau. (2010). 2010 Census Summary File 1.

U.S. Census Bureau. (2010). Overview of race and Hispanic origin: 2010, 2010 Census Briefs.

Wang, J. A. (1998). Mode selection of distribution of Chinese population and sustainable development. *Journal of Beijing Normal University (Social Science Edition), 1*, 72–80.

Wang, J. A., Shi, P. J., Yi, X. S., et al. (2008). The regionalization of urban natural disasters in China. *Nat Hazards, 44*, 169–179.

Wu, C. P., & Xie, N. (2011). 1980–2010: Retrospect and prospect of China's population policy for 30 years. *Gansu Social Sciences, 1*, 1–5.

Xu, W. X. (2007). An important research on contemporary China population policy: A comment of the research on contemporary China population policy. *Fujian Tribune (The Humanities & Social Sciences Monthly), 3*, 138–139.

Zhang, S. Y. (2004). An analysis on migration situation in China based on birth place. *Market & Demographic Analysis, 10*(3):1–19.

Zhang, L. P. (2008). New changes of the American migration in the Postwar Era and its implication for China. *Journal of Shanxi University (Philosophy and Social Science Edition), 4*, 96–101.

Zhang, W. X. (2001). The recent processes of population Geography in America. *Human Geography, 1*, 6–10.

Zhao, J., & Chen, C. K. (1999). *China Geography*. Beijing: Higher Education Press.

Chapter 4
Agriculture and Food Production in China and the U.S.

Mark Leipnik, Yun Su, Robert Lane and Xinyue Ye

4.1 Introduction

This chapter discusses the main agricultural characteristics of the People's Republic of China (China) and the United States of America (the U.S. or often just "America"). Since these two nations are the two most important agricultural producers on the earth they merit a detailed comparative analysis. The comparison of these two nation's agriculture is not done on a point by point basis because for example the U.S. does not produce tea, silk or cashmere nor rely on rice as a dietary staple and China does not raise bison or avocados. Nor does each nation have a similar geographic distribution of arable lands, although the countries are of similar size. However, the following discussion does cover the spatial distribution and production of important crops raised in each nation and the major issues facing each country in the agricultural sphere. Where there are directly comparable issues and characteristics such as the similarity of the Loess Plateau in northern China and Palouse Plateau in eastern Washington State in the U.S. or the existence of a corn belt in both China and in the U.S. it is noted. There are in fact many similarities between agriculture in the U.S. and China since both nations are the number one, two or three leading producers for a wide range of

M. Leipnik (✉)
Department of Geography and Geology,
Sam Houston State University, Huntsville, USA
e-mail: geo_mrl@shsu.edu

Y. Su
School of Geography, Beijing Normal University, Beijing, China
e-mail: suyun@bnu.edu.cn

R. Lane
Department of Agricultural and Industrial Science,
Sam Houston State University, Huntsville, USA
e-mail: blane@shsu.edu

X. Ye
Department of Geography, Kent State University, Kent, USA
e-mail: xye5@kent.edu

R. Hartmann, J. Wang (eds.), *A Comparative Geography of China and the U.S.*,
GeoJournal Library 109, DOI 10.1007/978-94-017-8792-5_4,

commodities including corn, soy beans, wheat, eggs, pork, cattle and cotton. Both nations face similar issues with loss of high quality farm lands to urbanization and water supply and quality issues related to groundwater overdraft and non-point source pollution. Perhaps the most salient difference is that in China about 300 million people are engaged in agriculture on only about 7 % of the world's arable land and produce about 19 % of the world's food, supplying most of the needs of about 22 % of the world's people. In America, about 2 million people are farmers, ranchers or fishermen (but only 960,000 are solely employed in it). These generally independent farmers have access to about 9 % of the world's arable land (the most of any country, China is third after India) and use it to produce about 11 % of the world's food. This bounty greatly exceed the needs of about 4.5 % of the world's people, the surplus provides a significant proportion of the world's exports of many key crops like wheat, corn and soy beans. From these differences flow many other important characteristics such as the extensive nature and export orientation of U.S. agriculture and the intensive and self-reliant character of Chinese farms.

4.2 Agriculture in China

4.2.1 The Long History of China's Agriculture

4.2.1.1 Primitive Agriculture and Traditional Agriculture

The origins of China's agriculture can be traced back 10,000 years. Starting about seven or eight thousand years ago through about three thousand years ago, primitive agriculture was quite widespread. At first, agriculture in China mainly consisted of food gathering and hunting. Farm production formed a pattern that was wheat-dominated in Northern China and rice paddy-dominated in Southern China. Domestic rearing of livestock acted as a sideline to farming. The earliest livestock raised were dogs, pigs, chickens, and water buffalo. Gradually, the society formed a structure of configurative farming and animal husbandry, and produced a stable settled village life based on agriculture. Because of the fertile soil and warm dry climate and availability of rivers for irrigation, the Loess Plateau of the Yellow River (similar in soil and high potential for wheat production to the Palouse Plateau of eastern Washington State) became the earliest region where primitive agriculture emerged. (Lu 1984)

During the "Spring and Autumn" Period (770 BC–221 BC), China entered into the era of traditional agricultural. The characteristics and advantages of traditional agriculture is that by fertilization and intensive cultivation, crop yield per unit of land area is continuously improved with recycling of nutrients and maintenance of soil fertility. With the universal adoption of iron tools and the plow, there was a revolutionary change in China in that human labor was replaced by animal power. With traditional cultivation methods, the land was used with some fields kept fallow in rotation with those used to grow crops. Green manure crops, especially legumes, were used in this crop rotation.

During this period, China built many famous water conservation and irrigation projects, such as the Dujiangyan Dam in Sichuan Province, the Karezin (called Qanats in Iran and Afghanistan) system in Xinjiang Province, the Great Canal, etc. Karezin are sub-surface channels used to provide a reliable supply of water for human settlements and irrigation in hot, arid and semi-arid climates. The Great Canal is the longest and earliest canal in the world. These projects not only prevented floods by controlling water flow, but also ensured irrigation by adjusting water distribution, which made a great contribution to the development of Chinese ancient agriculture (Zhou 2000).

At the same time, the theory of agriculture was developing rapidly, and people had more knowledge of soil and crops. With horticulture and animal husbandry developing to a considerable extent, man began to alter the natural environment and develop a variety of modes of agriculture.

At the earlier stage of the Traditional Agricultural era, Chinese agricultural production was centered on the Yellow River Basin. However, in the middle Tang Dynasty (618 AD–907 AD), because of war in the north and construction of irrigation projects in the south, the agricultural production center started to shift to the south. After the beginning of the Song Dynasty (960 AD–1279 AD), the middle and lower reaches of the Yangtze River became the new agricultural production center (Zhao and Chen 1999).

4.2.1.2 Modern Agriculture

In order to meet the need of improved living standards and national development, traditional agriculture gradually was transformed into modern agriculture. In comparison with traditional agriculture, modern agriculture was no longer confined to farming and aquaculture. It has extended to include genetics, food processing, and information services, etc. Its continuous development is dependent on investment in new technology. With modern management theory and methods, production efficiency has improved by leaps and bounds.

Since the founding of the People's Republic of China in 1949, with the reclamation of land, redistribution of land to landless peasants, the building of hydraulic engineering works, the use of quality seeds, use of synthetic fertilizers and employment of advanced technologies, the efficiency of land utilization in China has greatly improved. It is a great achievement that China, with only limited and specialized imports of food, feeds 22 % of the world's population with only 7 % of the world's cultivated land. Thus a land once repeatedly scourged by famines (most seriously in 875, 1333, 1810, 1846, 1850, 1873, 1876–1879, 1896, 1907, 1911, 1928–1930 and 1959–1961) that killed hundreds of millions of people and helped seal the fate of several dynasties including the Qing Dynasty in 1912, is now self-sufficient in most foods and a net exporter of some agricultural products. (Sun 2003)

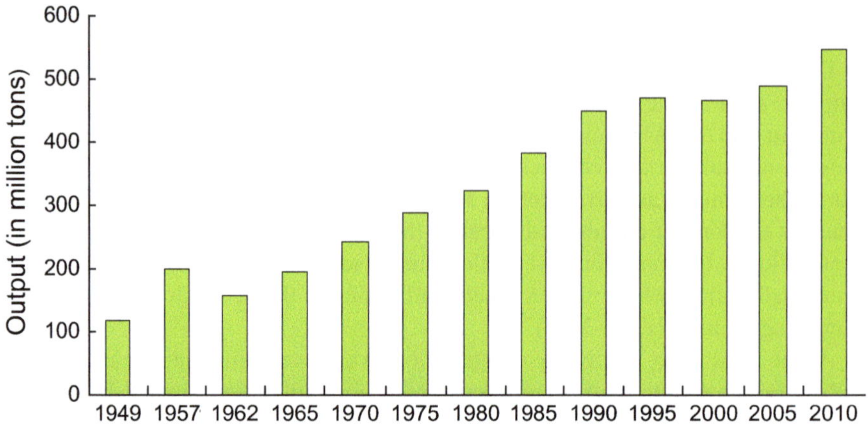

Fig. 4.1 China's grain output change (1949–2010)

4.2.2 The Development and Characteristics of China's Agriculture

4.2.2.1 The Status of Agriculture

Agriculture is an important foundation of the national economy in China. Since ancient times there has been the argument "Food is the first necessity of man; grain is the first necessity of food". With the largest population in the world, China has a huge demand for food and food issues relate to social stability and the improvement of people's living standards, so that China attaches great importance to agriculture, especially grain production.

The Chinese Government adheres to a food security policy that is based on domestic self-sufficiency; the deficiency in grain production is made up by the proper use of import and export controls. The emphasis on the improvement of agricultural efficiency relies on scientific and technological progress and a national policy that firmly maintains the primary position of agriculture in the national economy. In the past 30 years, more than 20 % of the world's growth in production of major agricultural commodities has taken place in China. As shown in Fig. 4.1, China's annual grain production has increased from 300 million metric t (in 1978) to 546 million metric t (in 2010) (National Bureau of Statistics of China 2011). The contribution of food aid to other countries by China ranked only after those of the United States and the European Union. This fact makes China an important force in safeguarding the world's food security. Due to frequent floods and droughts, there have been many periods of famine in Chinese history. But now the security of food supplies makes these famines a historical memory. Since the 1980s, China's grain self-sufficiency rate has always been maintained at more than 95 %, which has made a great contribution to safeguarding the world's food supply and stabilizing international grain prices.

Table 4.1 China's agricultural output and proportion of world's total output (2010)

Major Agricultural Product	Output (thousand metric tons)	Proportion of World's Total Output(estimate %)	Rank of World's Output
Grain[a]	546,410	22.00	1
Pork	50,712	47.34	1
Beef	6,531	9.98	3
Chicken Eggs	27,627	40.60	1
Vegetables	14,887	49.00	1
Cotton	5,961	24.10	1

[a] "Grain" includes "cereals", beans and tubers and "cereal" includes rice, wheat, corn, barley, millet and oats.

With the rapid development of China's agriculture, the output of major agricultural products has grown substantially. Many of China's agricultural products occupy an important position in the world. The production of pork, rice, vegetables, eggs, wheat, chicken, beef and cotton ranks among the highest in the world, as shown in Table 4.1. Along with the advance of globalization, the scale of China's import and export of agricultural products is constantly expanding. It has become one of the major exporting countries for fruit (ranked 1st), garlic (ranked 1st), tomato sauce (ranked 2nd), vegetables (ranked 1st), and tea (ranked 3rd) and the main importing country for soybeans, palm oil, cotton, and corn (National Bureau of Statistics of China 2011).

However, with economic development, China's industrial structure has changed significantly, which indicates that what once was a primarily agricultural country has been transformed into not only an industrial country, but arguably the world's leading manufacturer. The share of gross domestic product (GDP) and the proportion of the labor force employed in agriculture has decreased significantly (Fig. 4.2) since 1952. During the early days of the People's Republic of China (China), the proportion of primary (agriculture and related) industry in overall economy was as high as 51 %. However, with the development of modern export oriented manufacturing, the proportion of primary industry has declined and has fluctuated. In the early 1980s, the proportion of primary industry was about 30 %, with the proportion declining about 10% every 10 years after that. (China Compendium of Statistics 1949–2008, 2010). At present, the proportion of primary industry (including agriculture, forestry, animal husbandry, fisheries and related services, but excluding the mining and salt industries) accounts for only 10.1 % of the total economy in 2010.

Along with the decline of the agricultural component relative to the total size of the entire national economy, the proportion of agricultural employees in the total population is declining. Among the remaining agricultural workers, a considerable number have begun to work away from their homes and have migrated to urban centers to seek employment in manufacturing, construction and service sectors. They spend the slack farming season doing other work. They move to relatively-developed areas and work there in construction, textiles or services industries, which

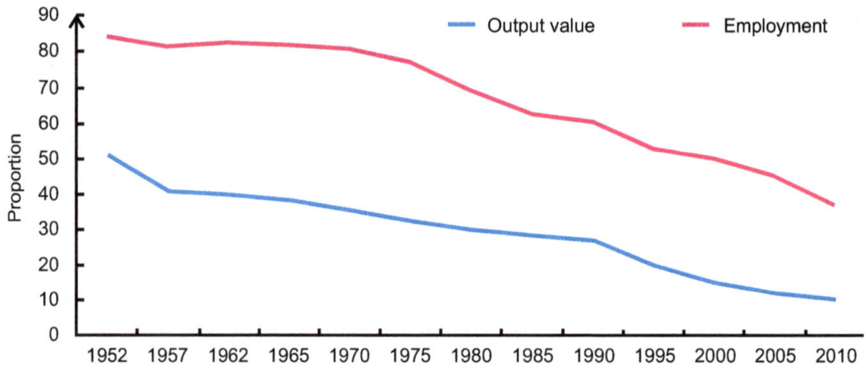

Fig. 4.2 Proportion of output value and employment of China's agricultural industry

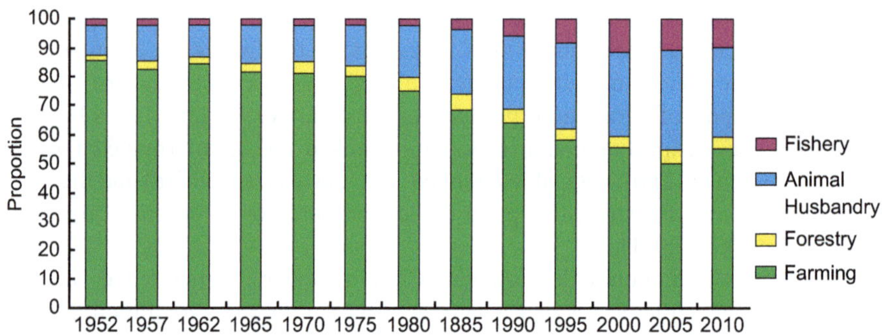

Fig. 4.3 China's proportions of agriculture, forestry, animal husbandry and fisheries

belong to secondary or tertiary industries. They have formed a migrant worker stream that has flowed out of agricultural regions into economically developed cities and/or eastern and southern areas of the country.

4.2.2.2 Changes in the Agricultural Structure

Accompanied by changes in the national economic structure, the internal structure of the agriculture sector itself has undergone changes (Fig. 4.3). Traditional monoculture agriculture has been transformed into a more complex mix in which agriculture, forestry, animal husbandry and fisheries coexist and develop. Farming accounts for a smaller proportion of the gross output value, while non-crop growing activities share is greater out of a growing value of total production.

Within the farming sector, the proportion of cereal crop acreage is declining, and the proportion of cash crops (such as cotton, vegetables, fruits or oil-seeds) and other crop acreage has increased accordingly. The internal production mix of cereal crops has changed at the same time: corn production has increased, wheat and rice production have fallen in relative terms, and soybean production has fluctuated. (Liu 1996)

From a national perspective, the proportion of the economy devoted to farming has continued to decline and efficient and economical crop production is becoming the bright spot of this structural adjustment. The traditional timber-production-dominated forestry has started to focus on ecological conservation with sustained yield, multiple use and eco-tourism related forest management practices. With the profit margin for animal products like pork and chicken higher, the proportion of animal husbandry has increased. Fisheries (particularly aquaculture) has developed rapidly and changed from "wild-caught" fishing to "fish farming". In 2010, aquaculture accounted for 71.3 % of the total fisheries production. Some types of aquaculture are export oriented such as shrimp and tilapia production.

The output value of animal husbandry has grown substantially and the production of livestock and poultry has expanded. In 1985 and 1990, egg and meat (pork, beef and mutton) production in China both ranked first in the world. At present, China's per capita meat and egg production have both exceeded the world's average. The outputs of pork and eggs account for about 40–50 % of the total world production respectively. The rapid development of animal husbandry has both prompted, and is in response to, changes in the dietary preferences of the Chinese people. In the past, the proportion of grain and other plant food based calories in the diet was up to 90 %. Over the last 60 years, meat consumption by typical Chinese has increased tenfold. Because of the development of animal husbandry, the supply of animal-based food has increased, which makes a great contribution to adjusting the Chinese people's diet and improving their health and living standards. For example, the stature of young Chinese has increased due to a higher protein diet so that in many cases children are taller than their parents and much taller than their largely rice-fed grandparents. However, there is also an increase in obesity among young people in China, although not yet on an epidemic scale as is the current situation (discussed below) in the USA where about half the population are overweight and one third are obese.

4.2.2.3 Changes in Agricultural Production Conditions

Since the reform and opening up of China starting in 1978, China's agriculture and rural economy have undergone tremendous changes and the production conditions have improved greatly. First, the level of agricultural mechanization has increased annually. According to statistics, the overall use of agricultural machinery has increased by more than seven times. The machine-sowing-and-harvesting acreage has increased every year, especially with respect to wheat, rice and soybean crops. Second, farmland irrigation, water conservation and other infrastructure has been gradually improved. The area of irrigated lands and irrigated areas that use mechanization and have access to electricity has increased significantly. Also, farms have adopted modern farm buildings and processes such as large scale greenhouses and concentrated animal feeding operations (CAFO's) that are typical of modern large scale

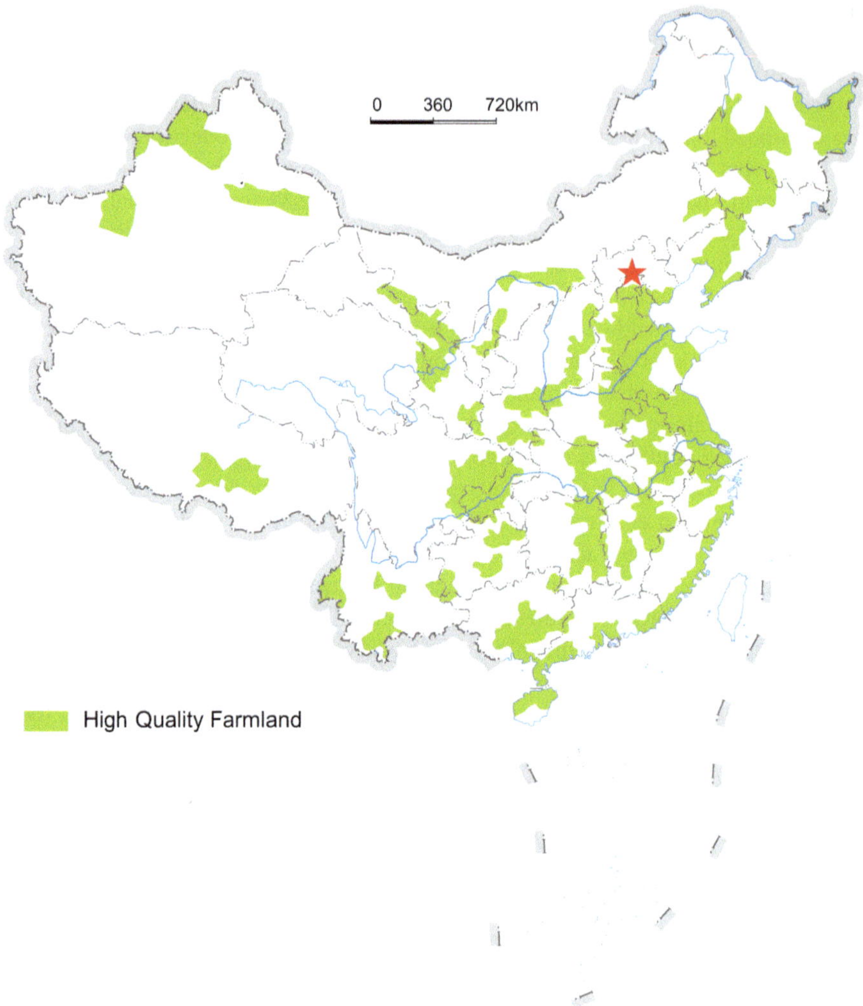

Fig. 4.4 The distribution of high-quality farmland in China

scientific agriculture and animal husbandry practices. The area occupied by greenhouses has increased substantially; mainly these greenhouses are used for growing vegetables, fruits and seedlings. (National Bureau of Statistics of China 2010)

There are about 130 million ha of cultivated land in China, ranking it fourth in the world, with high-quality land only accounting for about 30% of the total (see Fig. 4.4). In recent years, the acreage of China's arable land has been on the decline. On the one hand, with the progress of industrialization and urbanization, the spatial extent of urban areas has grown and most of this growth has been due to the conversion of cultivated land into other uses. The areas such as coastal southeast China with extensive high-quality arable lands also are areas with high population density and concentrated and rapidly-developing cities. Due to the overlap of these uses, urban sprawl has taken over many farming areas. This has led to permanent loss of arable land. Natural factors are the other main cause of loss of arable land. Land degradation due to desertification, salinization and soil erosion jointly pose a challenge to the preservation of arable land.

Nevertheless many measures are being taken to preserve the fertility of China's scarce arable lands. In different regions various targeted measures are being taken to improve soil fertility, such as use of biological or chemical modification. Crop rotation and conservation tillage is widespread in all regions. In arid and semi-arid areas of Northwest China, sand dunes are being stabilized by the use of shelterbelts. In the Northeast region, eastern plains region and coastal areas, saline-alkali lands are being reclaimed by a combination of biological and chemical methods, which includes the use of modified materials such a green manure, river mud, etc. In the Loess Plateau region, an area beset by severe soil erosion problems, planting trees and grasses, preventing excessive land conversion and reclamation of eroded lands are the main management practices being implemented. In areas such as those with steep slopes, previously plowed or grazed lands are terraced and transformed into orchards or forests and enclosed to limit erosion due to overgrazing and to limit the formation of rills and gullies in bare soils.

In the process of agricultural modernization, China has adopted a unique mechanized model. To solve the contradiction between the small scale of individual farms which are widely dispersed and the economies of scale of centralized mass production using modern farm machinery, the method of Trans-regional Operations of Farm Machines has been adopted. From the south to the north, wheat matures in a temporal sequence as the season advances. So does the appropriate time for harvest. By the trans-regional operations of combine harvesters, a small-scale farm can get temporary access to traveling mechanized harvesting technologies and skilled operators at the optimal time, thus making production both more efficient, intensive and technologically sophisticated. This trans-regional agricultural model was initiated spontaneously by farmers. Some early machine owners helped other farmers, in their village or the surrounding villages, during the busy period for harvest, which could also increase their income. Gradually, agricultural machinery cooperation was established in some places, with organized groups of farmers conducting cross-regional operations from south to north. In this process today, the Government assists in transportation and information sharing. The establishment of agricultural cooperation makes costly farm equipment more fully utilized, the rural labor force more flexible, and the farmers' incomes have also increased. Nowadays, this method has expanded from wheat harvesting to include sowing and harvesting of rice and corn, and other crops.

4.2.2.4 The Life of the Peasants

At present (in 2012), rural households are the key component of agricultural production in China. With a large population and little farmland, Chinese agricultural

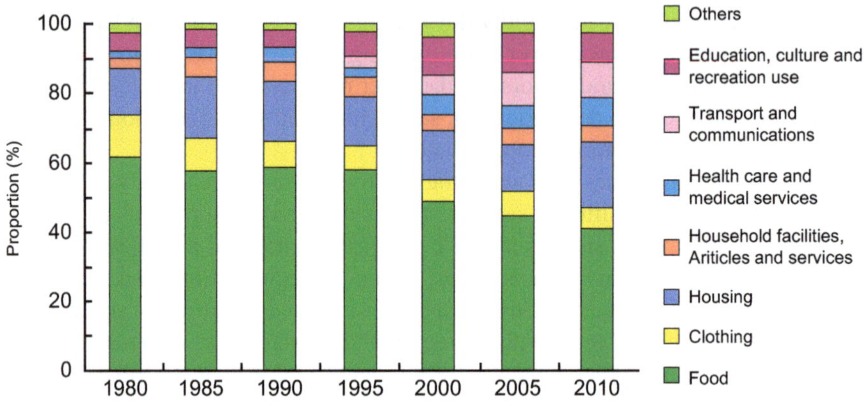

Fig. 4.5 Chinese rural residents' consumption cost structure (1980–2010)

resources per capita are limited and the extent of rural households' land holdings is small. In China, agriculturalists have the right to the use of the land, but do not have the ownership of the land. The land is publicly-owned. Most agriculturalists have their own housing located on or near their farms or grazing areas, and they cultivate grain or vegetables for home consumption near their houses. In the winter or off season or during other slack periods, some of the younger people in the rural labor force migrate to more economically-developed areas to work supporting their families. In some areas, the proportion of migrant workers is quite large, leaving behind mostly the elderly, women and children in the villages.

Chinese farmers' living standards have improved substantially over the past 30 years. Previously, food income was the farmers' main income, and the income in kind accounted for a large proportion of effective income with little use of cash (and no access to credit). Today, farmers and herders income sources are becoming more market-oriented and more diversified and the structure of family income has changed totally. In 2010, national average rural per capita net income reached 5,919 Yuan (for reference this would have been about U.S. $ 1,100, but with a lower cost of living in rural China, the purchasing power of this income would be closer to $ 4,000 in the USA). However, there are regional differences in the increase in farmers' income. For example, in wealthy Guangdong Province, per capita income of farmers reached 7,890 Yuan, with farmers in Gansu province averaging only 3,308 Yuan. Non-farm employment opportunities in the eastern region make farmers' income in the eastern part of China significantly higher than in the central and western regions. In the meantime, rural residents' consumption structure also has changed. Culture, education and entertainment, healthcare, transportation and communications expenses (and opportunities) have rapidly grown. Housing expenses have remained a constant proportion of overall costs, while food expenses represent a decreasing share of overall family expenses (Fig. 4.5). Life in rural areas gradually has entered into an era of electrification and information technology with access to mass media and cellular communications. Nearly every family has a TV set, and most families have washing machines and refrigerators. The rate of adoption of information technology equipment, such as telephones, mobile phones and computers, is becoming more and more rapid. (National Bureau of Statistics of China 2010a).

Although the living standards of farmers have greatly improved, there are still large differences between urban and rural areas. In addition, gaps in education, health care, and employment remain and need to be addressed. In parts of China, the living conditions of farmers and herders still need to be improved greatly. Various types of housing exist in rural China, and some are substandard mud brick houses or caves or in some areas temporary movable shelters. These modes of housing are traditional and use local materials, but are not always commodious, comfortable or easy to keep clean. Running water, indoor plumbing, electricity and drivable roads are not available in some rural areas. In the poorest areas, mostly in western China or mountainous areas, living conditions of some farmers and herders are still very primitive and difficult.

4.2.3 Patterns of Production

As can be seen in Fig. 4.6, there is a demarcation between eastern and western China. This division is due to natural conditions, mostly precipitation and temperature patterns. This dividing line has an impact on the level of socio-economic development and dominant agricultural production modes. Thus, there is a distinct farming area in the east and a grazing (or pastoral) area in the west. The eastern area is also more economically developed, diversified and more densely populated (Xu 2002).

4.2.3.1 The Eastern Agricultural Area

The Eastern agricultural area is located in the monsoonal region; the region spans the temperate, subtropical, and tropical zones from north to south. Annual precipitation of the area is more than 400 mm. With more level terrain and fertile soil, the eastern region is more suitable for farming than the western region. For this reason this area is the main production area for food crops such as wheat, rice, and corn. Thanks to its combination of moderate temperatures and adequate moisture, it is the main production area for various cash crops (such as vegetables, tea, fruits, etc.). This is in contrast to the U.S. which borders the Atlantic Ocean, Gulf of Mexico and pacific Ocean each of which has a beneficial influence on temperature and rainfall, this difference and the influence of three major mountain chains accounts for more widely dispersed areas of agricultural production and the absence of a simple demarcation between pastoral and crop dominated areas as in China.

Wheat grows in a wide range of areas in China (Fig. 4.7). Based on the planting season, wheat can be divided into two types–winter wheat and spring wheat. Winter wheat represents the vast majority of production accounting for 90 % of the total. The Huang-Huai-Hai Plain is the major producing area of winter wheat. Winter wheat mainly is distributed in Henan, Anhui, Hubei and Jiangsu Provinces and there

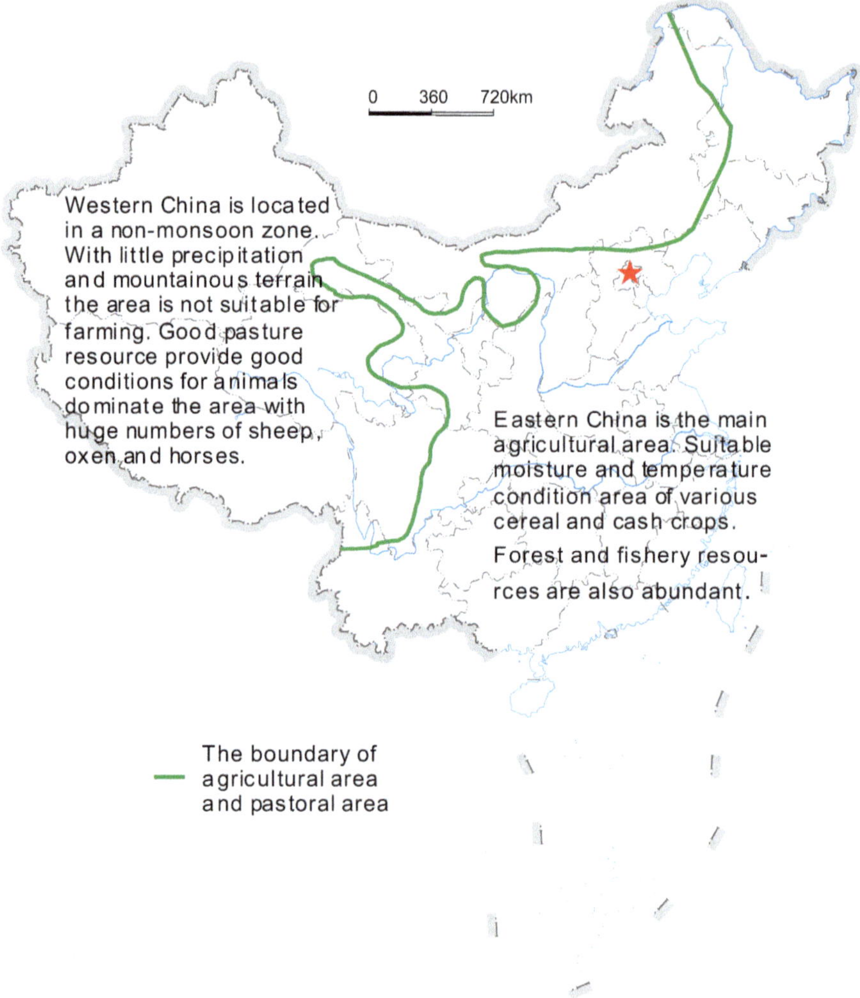

Fig. 4.6 Boundary between eastern agricultural and western pastoral area

is scattered (decentralized) production in the vast areas south of the Qinling-Huaihe Line (Fig. 4.8). Spring wheat is mainly concentrated in the Northeast, Inner Mongolia, some of the oases in Xinjiang Province, the Hexi Corridor of Gansu Province, the Guanzhong Plain in Shanxi Province and other scattered places in areas with cold winter temperatures and adequate soil conditions.

Rice is arguably the most important food crop in China. The Chinese people, especially residents of the south, have the age-old habit of eating rice and foods which use rice as an input (such as rice vinegar, rice flour, rice noodles, etc.). Nowadays, the sown area of rice accounts for about 1/4 of the total acreage of grain crops and the production accounts for about 1/2 of the total national grain output and more

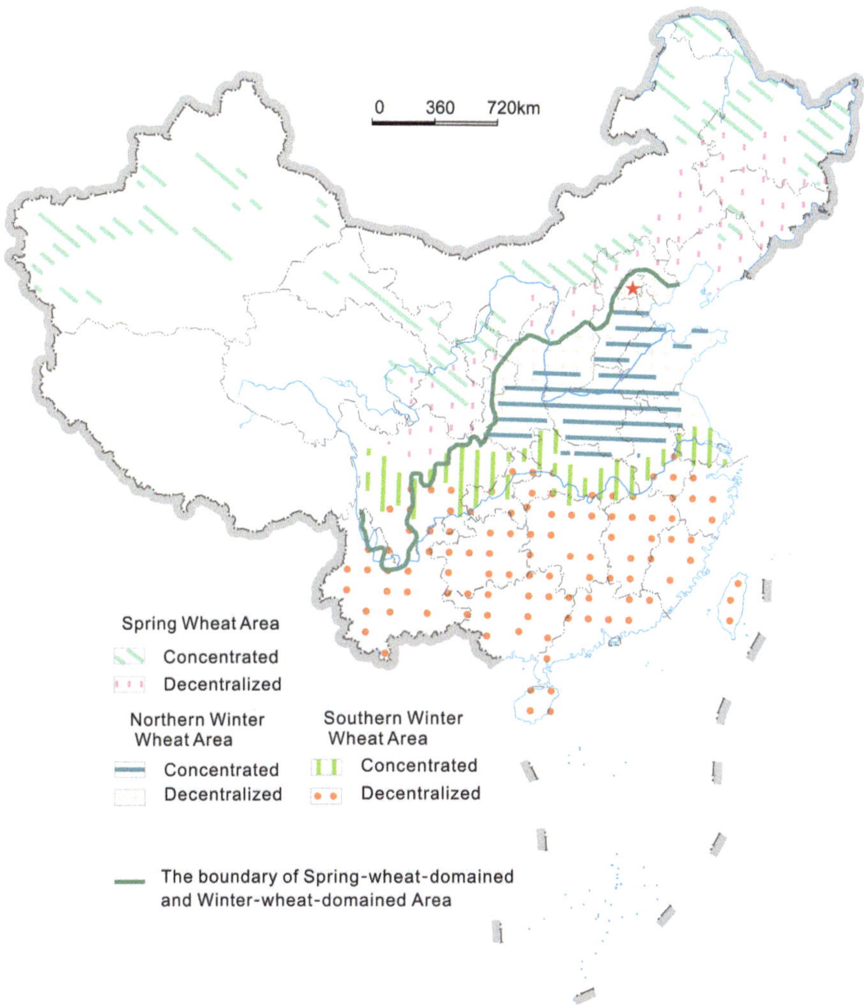

Fig. 4.7 The distribution of wheat in China

than half of the value of grain sold. The distribution of rice is divided by the Qinling-Huaihe Line. The Yangtze River basin flood plain is the main rice growing area in the south, while rice planting in the north is more widely dispersed. Due to the suitable moisture and temperature conditions, Southern China, including Guangxi, Guangdong, Hainan and other provinces, is the main production area for double cropping (producing two crops per year) of rice (Fig. 4.8).

Corn production is widely distributed in China. The corn production regions extend from the northeast plains to the southwest, forming a "Corn Belt" (Fig. 4.9).

The corn belt in China has a different orientation than that of the USA discussed below and in detail in Chap. 10, but both are of vital importance economically.

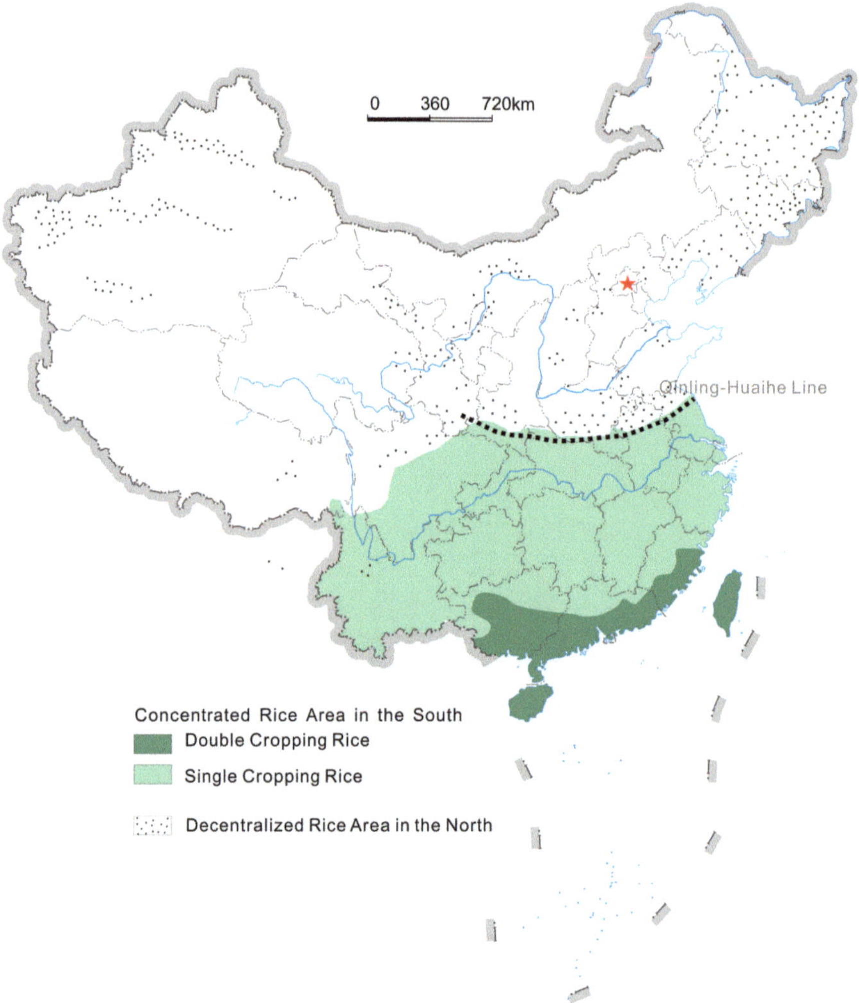

Fig. 4.8 The distribution of rice in China

Heilongjiang, Jilin, Liaoning, Hebei, Shandong, Shanxi, Henan, Sichuan, Guizhou, Yunnan and Guangxi Provinces are the main production areas.

Since the reforms and opening up of 1978, the center of China's food crop production has been moving progressively in a direction from the southeast toward the northwest. Before the 1990s, the focus of the production of food crops was in southeastern China. The southern area has ample water, adequate warmth, and sufficient sunlight and fertile soils. The longer rice growing season and the high yields made it the main grain producing area. With the development of the manufacturing oriented economy in the south, arable land resources have become more valuable and there has been a trend toward use of these scarce resources for freshwater aquaculture, or

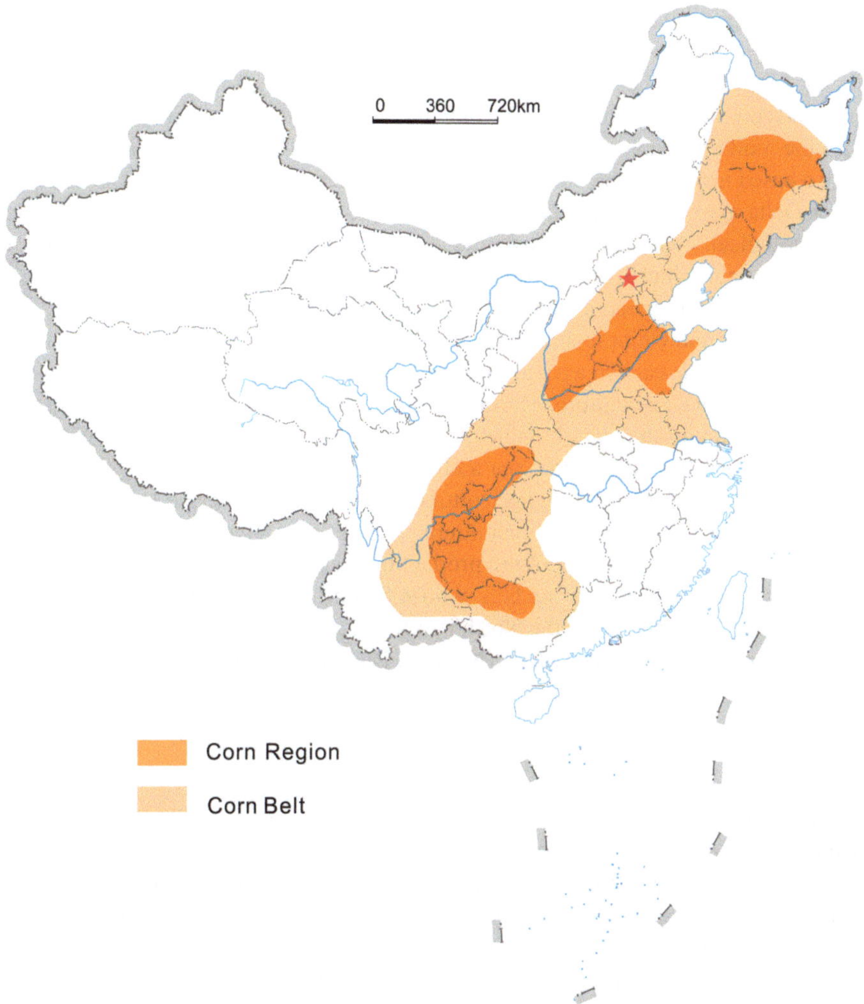

Fig. 4.9 The distribution of corn in China

orchards, vegetable crops and other cash crops (tea, flowers, medicinal herbs, etc.). At the same time, the average area devoted to corn production in the north has expanded gradually. Consequently, the center of food production is gradually shifting to the north and the west.

Eastern agricultural areas include three leading agricultural centers in China: the Northeast Plain, the North China Plain and the Yangtze River region (see Chap. 10). In addition, the southern region is the only area suitable for tropical and subtropical crops such as pineapples, mangoes, tea, etc. It is the most important area for cash crops in China, especially for fruits and sugar cane. It is also the main production base of aquaculture and silkworm rearing for silk manufacture. Farmers in the Pearl

River Delta convert the low-lying land, which is frequently flooded, into ponds in which they can raise fish and shrimp or plant water chestnuts, lotus root and other aquatic crops. Then they pile the sediment and silt which gradually accumulates in the ponds on the embankments, on which they plant mulberries. Silkworms are fed with the mulberry leaves, and the silk produced is used by the textile industry that is also based nearby. These interdependent and synergistic components of the Pearl River Delta agricultural system all form a unique mode of agricultural production with ecologically valuable and sustainable characteristics. With the development of commercial agriculture, the crops grown on the embankments, such as vegetables, fruits, flowers, corn, peanuts, etc. have diversified and their value has increased.

In addition, the Chinese aquaculture industry and fisheries industry are mainly concentrated in the eastern agricultural areas. Coastal areas in this region have well developed marine aquaculture. Marine aquaculture is mainly based on marine fishing. Going from north to south, there are four main fishing grounds–the Bohai Sea, the Yellow Sea, the East China Sea and the South China Sea. China's major rivers all pass through the eastern agricultural area. With the riverine resources, it is feasible to develop freshwater aquaculture. The Yangtze River Delta and the Pearl River Delta are the main aquaculture production areas in China. China has far more widespread aquaculture than the U.S. whose fishing industry is primarily oriented toward wild caught fish from the North Pacific Ocean off Alaska, and to the netting of shrimp in Gulf of Mexico waters (National Bureau of Statistics of China 2010b).

4.2.3.2 The Western Agricultural Areas

The western agricultural area is located in the arid and semi-arid areas of China. Although the soils in this region tend to be infertile, the vast land area of the region provides unique conditions for efficient animal husbandry. In contrast with the concentrated animal feeding operation type of animal husbandry practiced in the eastern agricultural area, livestock grazing and even nomadic herding is dominant in the western agricultural area. Camels and sheep are the predominant species of livestock and the proportions of them in this region are 73 % and 63 % of the national total, respectively. In the alpine pastoral area, yaks and pian-cows (a cross between cattle and yaks) are the dominant species. Wool from sheep and goat wool (cashmere) are produced in this region due to the cold and dry climate and account for about 60 % and 52 % of the national total respectively.

The Inner Mongolia Autonomous Region is in the transition zone from the eastern plains to the Mongolian plateau and it is located between the semi-arid and arid regions. With low rainfall and large variability from year to year, adequate moisture for agriculture is an issue. Annual rainfall decreases from the east to the northwest. Because of the insufficient rainfall, crops in this area are only harvested once a year. However, the vast grasslands provide good conditions to develop a mixed agricultural model combining farming and animal husbandry. The northern part of the region is dominated by pastoral areas, the central part is characterized by farming and pastoral areas, and the southern part is where agricultural areas are most prevalent. Dry land farming dominates the area, with cereals, flax and sugar beets being the main crops grown. In agricultural areas, the cultivated land is extensively farmed with low yields per hectare. Nonetheless, the grassland area in this region ranks first in the country. The eastern area of Inner Mongolia contains a high quality meadow steppe and grass fed animal husbandry occupies an important position here and therefore in China as a whole. Wool and cashmere and dairy foods are the main products (Sun 2003).

The climate of the Northwest portion of China (Xinjiang Province) is arid or drought prone, so that irrigated oasis-type agriculture and grazing herds in the desert are the major agricultural activities. The region possesses rich potential solar energy resources and the temperature conditions are in the most part suitable for agriculture. With intensive solar radiation and low relative humidity, diurnal temperature variations are very large (on the order of 12–16 °C) during the growing season. But the combination of sunlight, temperature, and soil moisture and soil fertility is not ideal for crops. The distribution of agriculture depends on water availability. Water is mainly obtained from the sparse rainfall on the windward slopes and from glacier melt water in streams that run off the mountains. If there is no irrigation, then there is basically no agriculture. There is only a small amount of dry land that is farmed in a few higher elevation areas of the foothills, but the production there is much lower than the irrigated low land (often below sea level) areas. The Hetao Plain, Hexi Corridor and the Yili region are the centers of wheat production in the area. The characteristics of animal husbandry in the desert area are nomadic and seasonal. In the winter, pastoralists graze their herds in the warmer lower elevation plains or basins, and in the summer they move to them to lusher mountain pastures. In addition, south Xinjiang is a major producer of high quality long-staple cotton; therefore this area is one of the three most important cotton producing centers in China. China is in turn the World's leading producer of raw cotton and cotton textiles.

The Qinghai-Tibet area is an important production area for grazing animals and forest resources. High terrain and low temperatures are its main natural characteristics. Most of the region is frigid and only suitable for thick coated grazing animals like sheep, goats and yaks. The southeastern portions of the region which have elevations of less than 4,000 meters above sea level can sustain hardy crops like potatoes and barley while thermophilic crops such as corn and rice can be planted only on the southern edge of the region in the valley bottoms of the Nien-ch'u River, Lhasa River and Chayu River. The main crops in this region are cold-resistant crops such as barley, wheat, peas, potatoes and rape seed (canola), which is cold tolerant when mature. Animal husbandry mainly relies on natural grasslands. Though underdeveloped and conducted on extensive unfenced lands, animal husbandry is still an important sector in this area in sustaining the sparse population, some of whom are semi-nomadic. In this area livestock with alpine characteristics, such as yak, Tibetan sheep, Tibetan horses, and cashmere goats are common.

4.2.4 Major Issues Associated with Chinese Agriculture

In the context of global climate change and global economic integration, China's agriculture is facing some new challenges. At present, China is committed to a targeted process to address these issues, in order to promote sustainable agricultural development.

Issue1: Loss of agricultural land resulting from urban sprawl The expansion of most cities in China as in the U.S. is through the conversion of cultivated land on the margins of existing urban areas. Arable land in China has been reduced from approximately 130 million ha in 1996 to about 122 million ha at the end of 2010. The overlap of high-quality farm lands and urbanized areas has resulted in an obvious contradiction between the protection of farmland and the promotion of urban development, especially in northeastern China and southeastern coastal areas.

To solve this problem, the Chinese government has initiated efforts to improve land management laws and regulations, such as strictly enforcing the system of farmland protection, the introduction of a balanced system of requisition-compensation for farmland and the introduction of a land use control system. Also China has initiated the development of a comprehensive industrial policy that includes preservation of farm land as an objective. In addition, China attaches great importance to the development and reclamation of arable land reserve resources and efforts to improve efficiency of the use of resources. (Zhao and Chen 1999). This is also a major issue in the U.S.

Issue 2: Agricultural inefficiency resulting from small-scale of operators Since 1978, the implementation of the household contract responsibility system caused the development of scattered, small-scale production problems. Due to limited funds available to individual farmers, there are difficulties in purchasing efficient farming machinery, in constructing water conservation facilities or introducing advanced farming techniques. To solve these problems farmers spontaneously created local farmer organizations. Then a large-scale, industrial agriculture management model led by the government was introduced, which has improved land use efficiency and enhanced the market competitiveness of agricultural products. However, the overall level of adoption of economies of scale in agriculture has not greatly improved. In recent years, China has promulgated a series of policies to promote larger scale agriculture and industrial management. For example there are now resources available for the support of large enterprises which can help farmers in the construction of processing and distribution facilities for more efficiency. Also the government has been providing a variety of means to train farmers in modern scientific agricultural practices. This is not an issue in the U.S. where most farms are large scale although in many areas marginal farms of small size have disappeared either reverting to forest as in Appalachia and New England or being merged into larger scale successful farms.

Issue 3: The contamination and degradation of agricultural lands According to incomplete statistics, China's contaminated arable land is as much as 10 million ha which is about 10 % of the total. The pollution sources are various. Factories directly discharge toxic sewage and waste onto cultivated lands. Peasants use large amounts of chemical fertilizers and increasing quantities of synthetic pesticides. Over the long term these can build up in the soil and in aquatic ecosystems. Used mulches build up in agricultural soils. These organic materials can lead to soil compaction, hardening of the soil and the loss of fertility. From these and other processes soil degradation can occur and the productivity of contaminated land therefore declines.

China has made sustained efforts to solve the soil pollution problem in various ways. Public education efforts and outreach to farmers have been aimed at increasing environmental awareness. Laws have been developed and implemented to control pollution of farmland. In addition, soil testing and proper fertilizer use is widely promoted; the application of organic fertilizers has become increasingly common; the areas using green manure and returning straw to the soil have been greatly expanded; the use of appropriate levels of pesticides and use of plastic mulching to conserve water and reduce pests are increasingly implemented by farmers today (Sun 2003).

Issue 4: Agricultural water conservation China's water resources are as scarce as arable land resources; especially in arid areas, there is a sizeable gap between demand and available water. The annual consumption of irrigation water in China is about 4,000 billion m^3, accounting for about 60 % of the country's total water consumption. As a result of adoption of better irrigation methods, more crops are being produced today with only a modest increase in irrigation water consumption. However, waste of agricultural water still exists, mainly due to two factors. First, the utilization coefficient of irrigation water is low; the effective utilization of agricultural water use is only about 43 %, and the channel water conveyance losses accounted for more than 80 % of the irrigation water loss. These losses are due to seepage of water from unlined channels and evaporation of water from canals, reservoirs and other water bodies. Some of the water losses due to seepage can be regained in the form of recharge of groundwater aquifers. Second, the irrigation quota is generally high; in the traditional mode of furrow irrigation, actual irrigation deliveries are double the water actually required. This over irrigation does have the advantage of reducing the likelihood of salinization of the soil since it tends to flush out salts before they can accumulate.

Today agricultural water-saving systems have been constructed to alleviate water shortage and reduce waste. Mostly these technologies have been incorporated into irrigation projects. The methods used include low-pressure pipe irrigation techniques, sprinkler irrigation, micro-irrigation, impermeable membrane channel irrigation and improvements in traditional furrow irrigation methods such as better irrigation scheduling. Like urban sprawl, water conservation and pollution is a major and perhaps the single most important issue facing U.S. agriculture.

4.3 Agriculture in the U.S.

4.3.1 *The Short History of Agriculture in the U.S.*

In contrast to China, a land where agriculture dates back at least six millennia, most of what is now the United States lacked organized large scale agriculture until the last few centuries and in some areas like the "High Plains" (see Chap. 10), it was not until the later part of the 19th century that organized agriculture developed. An important exception to this rule is in parts of the Mississippi Valley including parts of the Midwest "Corn Belt" where Native American societies such as the "Mound Builders" grew crops in areas with rich soils subject to periodic flooding. While the agricultural systems of indigenous tribes were rather primitive, the crops they raised including maize (corn), amaranth, squash, beans, chilies and potatoes were to become staples in many other parts of the world (Silverberg 1968). Not all of these "New World" crops were raised in what was to become the United States, but corn was an important early crop and one that continues to be associated with American agriculture and is currently the most valuable crop grown in the United States (Almanac of American Agriculture 2011).

Europeans first settled in what would become the United States (U.S.) in the tidewater areas of the mid-Atlantic states and New England, later immigrants moved into the coastal plains, the piedmont of the Appalachian Mountains with a few hardy souls like Daniel Boone moving through passes in this major mountain range of the eastern U.S. and into the Ohio and Tennessee River valleys. From there and from the river systems draining into the Great Lakes, settlers moved into the interior regions of the U.S. The interior regions of the U.S. did not see dense settlement except for small trading centers and a few cities such as Saint Louis and New Orleans situated along rivers, until the early years of the 19th century. At that time waves of immigrants from farther east began to push into areas such as Illinois and Iowa. These areas became States in 1818 and 1846 respectively, the date of statehood indicating they had a moderate and stable population. For example, Iowa had 46,000 citizen residents in the Census of 1840. These early settlers were mostly of English, German, Scottish or Irish extraction with most having come from the Northeastern states or Canada and they were mostly small independent farmers who cut down the forests present along river courses and used the logs to construct cabins. These pioneers planted a mixed crop of corn, wheat, and vegetables supplemented with orchards and livestock, particularly pigs and chickens. They found the prairie soils to be dominated by deep mollisols that sustained bumper crops from the outset of agriculture and did not lose fertility rapidly unlike areas with thinner soils further east in the Appalachian Mountains which most settlers had crossed to reach the area (Harding 2009).

In contrast to the Corn Belt of the Midwest (see Chap. 10), the High Plains of Kansas, Oklahoma, Northern Texas, Nebraska and parts of states like Wyoming, Colorado and New Mexico were a harsher environment. Rainfall was less plentiful, native peoples had not built permanent settlements by in large but were nomadic following vast herds of bison and large numbers of elk and antelope and often raiding settled tribes and later European settlements. Tornadoes, hail, grass fires, swarms of locusts, hot dry summers and frequent periods of prolonged drought and harsh winters with unending wind and occasional blizzards all tended to retard settlement. Also surface water was not always available and little stone or wood for building fences or homes was available. Thus many early settlers built homes out of adobe (mud bricks) or sod. However, the prairie soil was, once it was plowed with strong steel plows that began to become available, capable of yielding a good crop of wheat when rains were favorable. The scale of farms and ranches was often vast and reapers pulled by teams of as many as 20 mules were invented to take advantage of the extensive flat acreage of farms. Also innovations such as barbed wire, which allowed exclusion of herds of cattle, better firearms, the advent of railroads, and the invention and perfection of the windmill all made settlement more feasible. Many would argue that the simultaneous near extermination of the bison and the almost complete relocation of the Plains Indian tribes created the favorable conditions for settlement by Europeans of the High Plains by the late 19th century (Hornaday 1889).

Another agriculturally important region especially for specialty crops such as grapes, vegetables, fruits, citrus, cotton and rice is California which has a warm dry climate with numerous rivers bisecting a great Central Valley. California has a different history from other agriculturally important regions, having first been settled in the 17th century by Spanish colonists who raised cattle for export of hides and tallow and they planted vineyards and fruit orchards. In 1848 gold was discovered and a huge influx of population followed as did rapid development of railroads and cattle ranching particularly in the Central Valley. Later in the 19th century a citrus industry developed in Southern California and a fruit raising industry in central and Northern California including in the area of Santa Clara County that is now referred to as Silicon Valley. Besides the Central Valley, the Napa Valley is an important agricultural region noted for vineyards. California has many specialty crops like artichokes, almonds and avocados.

4.3.2 The Development and Characteristics of American Agriculture

4.3.2.1 America's Agricultural Regions

The United States is the world's third largest nation (China is the fourth largest). In such large countries there are many and diverse agricultural regions as well as many areas that lack agriculture entirely. In the USA, the most important agricultural regions include the coastal plain of the Atlantic Ocean and Gulf of Mexico, noted originally for tobacco production and today also for corn, swine and poultry production. The Florida Peninsula (noted for citrus crops that freezing temperatures elsewhere endanger) is another import region. The region with the greatest value of crops is the Central Valley of California; this is because, although relatively small in area, it produces many high value specialty crops such as fruit, nuts and vegetables but also significant amounts of cotton. These regions, though important economically, are more specialized than the two great cereal growing areas of the United States: the Corn Belt of the Midwest and the Great Plains of which the High Plains is a part. Both of these regions are discussed in detail in Chap. 10. Also important is a less well defined region which could be termed the Mississippi Valley, this is an important area for soybeans, as well as corn, and rice and cotton. Many parts of both the High Plains and Corn Belt are inside the drainage basin of the Mississippi River, but this is an area covering more than a third of the U.S. Thus, it ranges in climate from sub-tropical (in Coastal Louisiana) to sub-arctic (at high elevations in Wyoming and Montana). However, the areas in Eastern Arkansas, Northern Mississippi, Missouri and parts of Louisiana that have rich soils and abundant rainfall are the most important agriculturally and are a center of soybean and cotton production. This chapter provides a broad overview of agricultural regions and a geographic portrait of the diversity and distribution of agriculture in America. The High Plains region is most noted for wheat production, the Corn Belt, is as the name implies, famous for corn. Since these cereals are the basis for animal feed, both regions are famous for production of livestock, with the Corn Belt best known for swine and the High Plains best known for cattle production. The map of the conterminous U.S. in Fig. 4.10 shows these two agricultural regions highlighted with the States that contain them labeled. It also shows various other specialized agricultural regions. These delineations are somewhat generalized.

4.3.2.2 Major Agricultural Production Areas in America

North America and the United States of America specifically is blessed with a very diverse range of climatic and physiographic settings that allows many types of crops and animal products to be produced under good and in some cases nearly ideal conditions. Agriculture in the U.S. is generally conducted by large technologically and financially capable corporations and by family farms that have grown and adapted for generations. For example, the Gallo family vineyard in California has over five generations grown to become the largest wine producer in the world. This means that agriculture is conducted on a vast scale with specialization of farming and ranching and animal husbandry into spatially defined areas where only a few crops are raised with large inputs of capital and technology. These major agricultural regions may raise more than one crop or type of crop but many can be characterized by the dominant crop. Thus the Snake River Valley of Southern Idaho is ideally suited due to climate, soil, availability of water and cost of land to growing potatoes. The Wenatchee, Yakima and Chelan Valleys of the Eastern side of the Cascade Mountains in Washington State are ideally suited by climate, elevation, drainage and other factors to produce apples. The Indian Coast region of the Atlantic seaboard of Florida is ideally suited for growing grapefruit and central

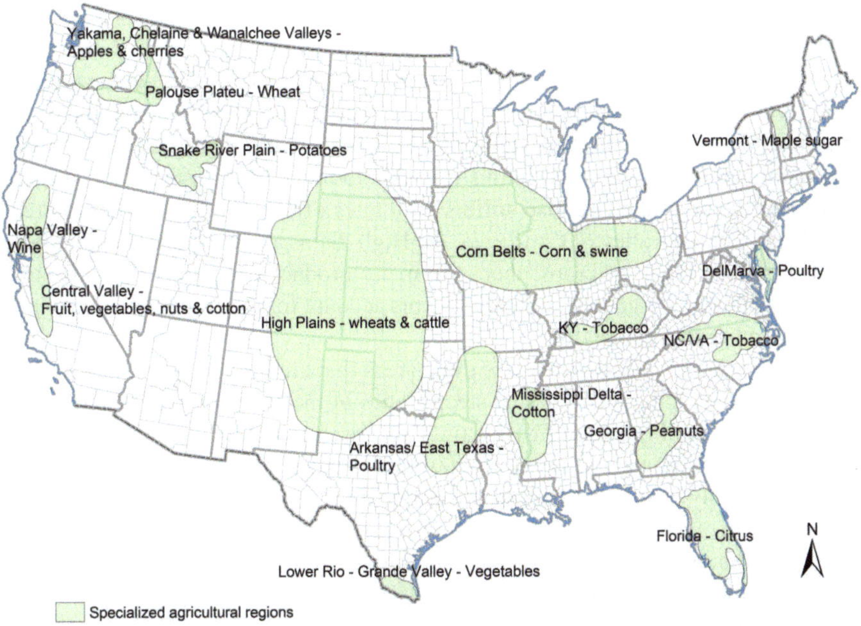

Fig. 4.10 Major specialized agricultural regions of the conterminous U.S. Key areas noted for production of specific crops are delineated and labeled

Florida generally is very well suited, except in some swampy areas, to produce citrus crops largely due to the rarity of freezing weather. This is akin to areas in Fujian being most suitable for tea cultivation. But in the American context the degree of monoculture is greater in many of these regions. Some other crop growing regions are more diverse, so the "Corn Belt" is an area that also has many soybean farms or may grow soybeans in rotation with corn on the same farm. The abundance of corn in the Corn Belt means that swine and dairy cattle production is also common in this region. An area which is both diverse and specialized in terms of agricultural production is the Central Valley of California. This area is blessed with an equitable climate with dry sunny days and abundant runoff from rivers draining the Sierra Nevada Mountains. It is the center of vegetable production in the U.S. but also leads in many types of fruit and nuts. Generally, the quality of the soil and other aspects of the farmland and climate in the Central Valley mean that high value crops are dominant in this region. There are other areas besides the Central Valley with warm sunny weather, rich soils and extensive irrigation that produced bumper crops of fruits and vegetables. These areas include the lower Rio Grande Valley in Texas which is one such area famed for grapefruit but also noted for onions and other vegetables. The Central Valley of California and Rio Grande Valley produce considerable volumes of cotton, but the Mississippi Delta is more famous as a cotton producing region, farther upstream in Arkansas is the largest rice producing region in the U.S., although the delta of the Sacramento River in California also produces rice. The High Plains is the heart of the wheat production area in the U.S., but large scale production occurs farther north in North Dakota and South Dakota. Another famed wheat production region is the Palouse Plateau of Eastern Washington State, an area with some of the highest yields of wheat in the world. This area is dominated by windblown loess type soils. In this it is akin to much of the interior of northern China and the area likewise also suffers high rates of erosion. In addition to areas such as the Central Valley of California, the High Plains and the Corn Belt that raise a fairly diverse range of crops, there are many productive regions that are more highly specialized. In general, climatic factors account for the preponderance of one crop or another in these regions.

In the Hawaiian Islands, there are three areas that are highly specialized. One is the Kona Coast on the well-drained steep slopes of the Mauna Loa Volcano on the West Coast of the big island of Hawaii. This area has rich well drained volcanic soils, a tropical climate and heavy rainfall that is ideal for coffee trees. This is essentially the only area in the U.S. where coffee trees can grow and the coffee that is raised there on small farms is very expensive selling for many times average coffee prices. On the other dryer side of the island in the rain shadow of a volcano is an area where the macadamia nuts are grown. (Kona Coffee Council 2011).

Although Florida is not tropical, but sub-tropical, it has a very temperate climate with abundant rainfall and infrequent freezing weather. This makes it suitable for citrus crops. The citrus region of Florida includes the central portion of the peninsula as well as a band along the coast including the Indian River area noted for grapefruit. Florida also produces oranges for juice, limes and tangerines. The citrus industry in Florida in 2010 produced fruit that was worth over $9 billion and directly employed 76,000 people (Florida Department of Citrus 2011) However, citrus production in Florida is declining with acreage falling from over 800,000 acres in 1970 to 500,000 in 2010. Much of the reduction has been due to conversion of orchards in central Florida into residential areas particularly around Orlando, home of Disney World. Another area that produces grapefruits as well as other vegetables particularly during the winter months is the region of southern Texas along the Rio Grande River. Unlike Florida which receives plentiful rainfall, this region is dry. But the large river that marks the border with Mexico provides irrigation water for the many crops and orchards that are found in the alluvial soils in the lower portion of the Rio Grande Valley which is also an important area for production of winter vegetables.

California is another state that benefits from a temperate (in some areas subtropical, in most Mediterranean) climate. The state is in general dry but the 4,000 m peaks of the Sierra Nevada Mountain range create a great climatic difference between the arid eastern portion of the state and the wetter western portion. However, there is generally not sufficient rainfall to grow the intensively raised crops found in the Central Valley. These crops depend of irrigation derived from the network of rivers that drain the Sierra Nevada. An extensive network of reservoirs and aqueducts connect the dryer southern portions of the State with water sources in the north and east of the State including the Colorado River that is fed by snow melt in the Rocky Mountains as much as 2,000 kilometers to the north east. The dry climate in the Central Valley and other agricultural areas is in fact an advantage in many ways since it reduces pests. The Central Valley is actually composed of two large valleys with an area of about 58,000 km^2 that converge and have a single outlet into San Francisco Bay. They are the Sacramento Valley to the north and the San Joaquin Valley to the south. The value of agricultural production in the million hectares of agricultural land in these valleys totals over $ 20 billion and is about 8 % of the value of total U.S. agricultural production. There are other areas in California like Monterey and Salinas, Southern California and the Napa Valley that add an additional $ 10 billion in production. Overall California is responsible for 14% of national agricultural production. Many of the crops that California is dominant in are high value specialty crops. Thus California in general, and the Central Valley in particular, are leaders in production of almonds, pistachios, olives, onions, peppers, cotton, artichokes, asparagus, spinach, lettuce, strawberries, celery, plums, peaches, pomegranates, oranges and grapes. Also vegetables eaten by ethnic groups are a California specialty, so crops like bokchoy, lemon grass, persimmons, jujube, sapote and bitter melons are raised in California (Umbach 1997). In addition to the Central Valley, California has another famous agricultural valley, the Napa Valley in Northern California noted for vineyards. Just as California is home to several major agricultural regions, it is the location for many specialized micro-climates created by a combination of proximity to the Pacific Ocean and many topographic features. One coastal area in Monterey County near Castroville is ideal for artichokes, another coastal area in Ventura and Santa Barbara counties is ideal for lemons and avocados although avocados in Santa Barbara county have had to compete with housing for millionaires and movie stars and special land use zoning and tax regulations have helped to maintain the avocado industry in towns like Carpenteria in this county. Other California towns are noted for plums, olives, raisins, broccoli and garlic and even in the arid and hot area of eastern California that is near or below sea level dates are produced in Indio while in the far north horse radish production is centered in Tule Lake and blackberries in McCloud. Many of these communities have annual festivals and feature their status as "world capital" of this or that product in their towns' self- identification (see Fig. 4.11).

Other areas in the Western U.S. produce wines including areas in Washington State, but the mountain valleys of the Yakima, Wenatchee and Chelaine areas on the eastern side of the Cascade Mountains in Washington State are most famous for supporting production of apples and sweet cherries. Elevations of orchards range from 250 m to almost 1,000 m in this region. Apple orchards in this area occupy 175,000 acres and employ as many as 45,000 pickers at the peak of the harvest season in the fall. Many apples are put in cold storage and shipped out of the U.S. to world markets (Washington Apple Commission 2010). The migrant pickers generally have to travel long distances to follow the crops that are ready to be picked at different seasons, some may travel from Florida in the South-eastern edge to the U.S. to pick oranges all the way to Washington State in the Northwest corner of the country to pick apples in the Yakima and other valleys. These mountain valleys have cool and dry weather but do not have late freezes which can kill the buds on

Fig. 4.11 Map of California specialty crop centers. The Central Valley produces vegretables, fruit, nuts, cotton and rice also

orchard trees before they can set. Eastern Washington State is also an important area for wheat. In particular, the Palouse Plateau is a major wheat growing region. The rain shadow of the Cascade Mountains produces a dry climate. Also in this region loess soils like those in Northern China yield some of the greatest harvests per hectare of wheat on the planet.

The neighboring State of Idaho is famous for potatoes. In fact the state automobile license plate bears the motto: Idaho—Famous Potatoes. The sandy soils of the Snake River flood plain and dry climate with abundant water available from diversion of rivers running off nearby mountains and a cool climate are ideal for production of potatoes, particularly the russet variety. Recently dairy cattle production has increased in Idaho, partially as a result of dry climate and proximity to crops like alfalfa which is used for cattle feed. Alfalfa does well in alkaline soils in the semi-desert areas.

Farther east and south of Idaho, the High Plains is dominated by wheat and cattle as well as some production of corn and cotton. The surplus of grain generated in the High Plains is often used to feed chickens that have become dominant in the State of Arkansas and neighboring portions of north eastern Texas and eastern Oklahoma. Various factors have led to the development of huge chicken concentrated animal

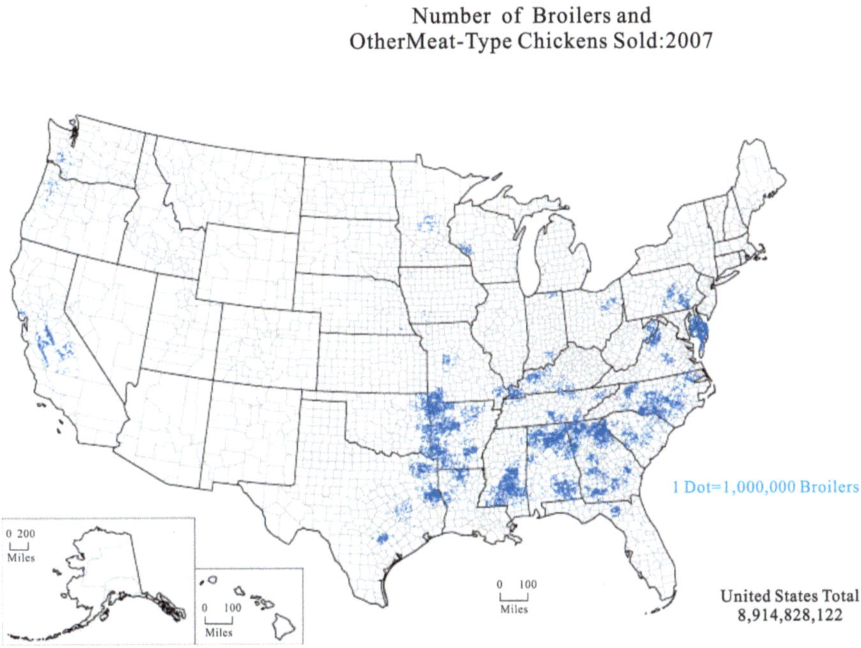

Fig. 4.12 The concentration of chickens (each dot equals 1 million birds) in the south-eastern U.S. is notable, proximity to feed sources and a limited frigid days helps explain this. Note: broilers are chickens raised for meat. (Courtesy: USDA 2007 Census of Agriculture http://www.agcensus.usda.gov/Publications/2007/index.php)

feeding operations (CAFO's) in these states dominated by firms such as Tyson Foods and Pilgrim's Pride and Sanderson Farms. A map of U.S. poultry production for broilers (birds used to produce meat rather than lay eggs) is shown in Fig. 4.12.

Farther east in Arkansas along the Mississippi River is a center of rice production. In the U.S., rice production is limited and although rice is not a staple of the diet of most people in the U.S. it is growing in popularity especially among immigrants. The U.S. is a net importer of rice, especially from the Indian sub-continent and Thailand. Nevertheless, rice is grown in parts of the southern U.S. such as Texas, Louisiana, the Mississippi Delta, as well as the Sacramento Delta of California (Fig. 4.13).

The greatest extent of rice production is in the southeastern portion of Arkansas and the western portion of Mississippi, these areas are termed the Mississippi Delta region. This region is famous for cotton production. It is notable that this region although fertile and productive for agriculture is one of concentrated poverty. Perhaps this is due to a concentration of land ownership. Cotton is a plant that takes a toll on the fertility of soils and the deep alluvial soils of the Mississippi Delta has been able to sustain cotton after other regions that grew cotton switched to other crops.

Further north in the Mississippi River drainage basin in the State of Missouri and neighboring regions of States such as Illinois and Iowa and as for north as

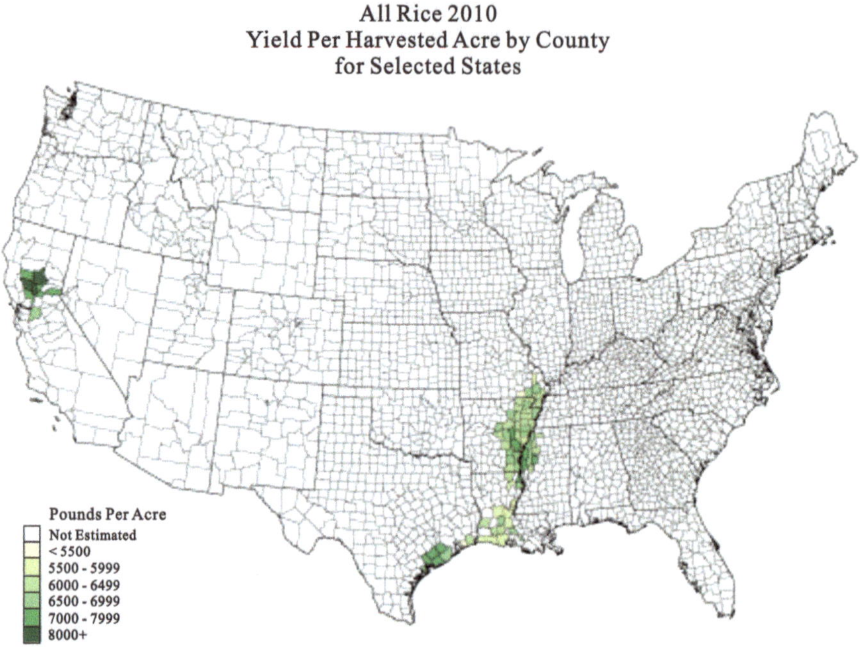

Fig. 4.13 U.S. Rice production in 2010. The three main rice growing areas are the Mississippi Delta region in Arkansas and Mississippi, the coastal regions of Texas and Louisiana and the Sacramento River Delta region in California. (Courtesy: USDA 2007 Census of Agriculture http://www.agcensus.usda.gov/Publications/2007/index.php)

the Dakotas is the center of U.S. soybean production (Fig. 4.14). Soybeans are the second most valuable crop produced in the U.S. Soybeans are frequently grown in rotation with corn and are found in parts of the corn belt but also are grown farther south along the Mississippi River and its tributaries as well as farther north into regions of North and South Dakota. In 2010 a total of 90 million metric t of soybeans were raised on over 30 million ha in the U.S. The value of production in 2010 was $ 37.6 billion. Although production in 2011 is estimated to have declined due to drought, prices for soybeans have increased and estimates of revenues are higher for 2011 at $ 40.2 billion. Of the total U.S. soybean production 45 % was exported in 2011. The top soybean export market was China with exports of 895 million bushels. U.S. soybean production is not increasing as much as corn production and it is expected that Brazil will eventually overtake the U.S. as the world's top soybean producer (Figs. 4.14, 4.15 and 4.16). One of the other crops that farmers in the southern U.S. have diversified into in order to maintain the fertility of their soils is peanuts. Peanuts are legumes that help to fix nitrogen and maintain soil fertility. Peanuts are raised in many areas of the south.

The arid climate of the border area of Western Texas and Eastern New Mexico along with irrigation water diverted from the Rio Grande River is another center

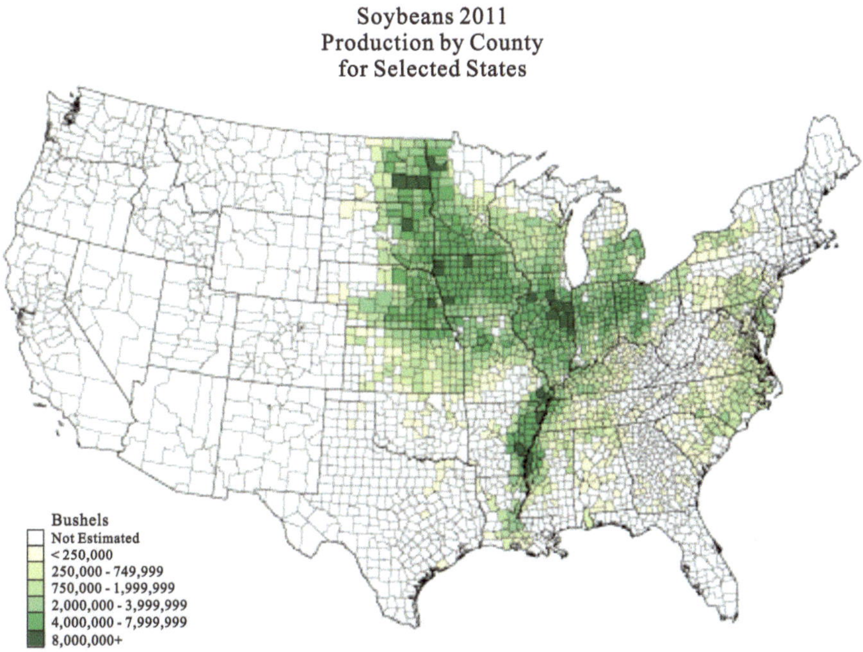

Fig. 4.14 Soybean production is centered in the Mississippi River basin. Production is shown in bushels, there are 27.2 kg of soybeans in a bushel, the most productive counties yield over 200 million metric t a year. (Courtesy: USDA 2007 Census of Agriculture http://www.agcensus.usda.gov/Publications/2007/index.ph)

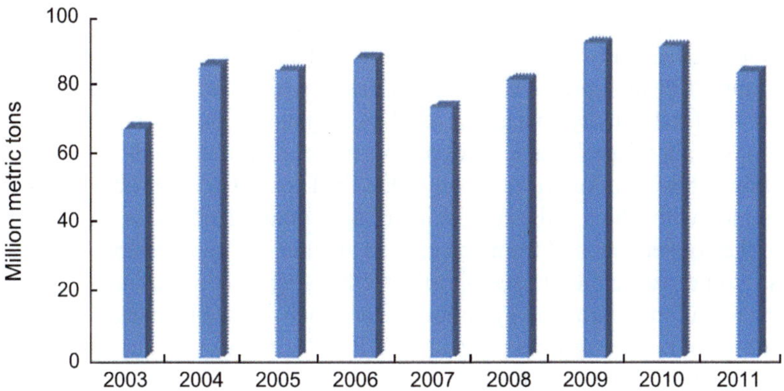

Fig. 4.15 U.S. soybean production. The record production in 2009/2010 declined in 2011/2012, but prices are higher and second only to corn. (Courtesy: USDA)

of peanut production. Peanuts are susceptible to mold, in particular aflotoxin, and much of the peanut production in this area goes to poultry feed. Thus, a dry climate is ideal for bulk handling and storage of this crop.

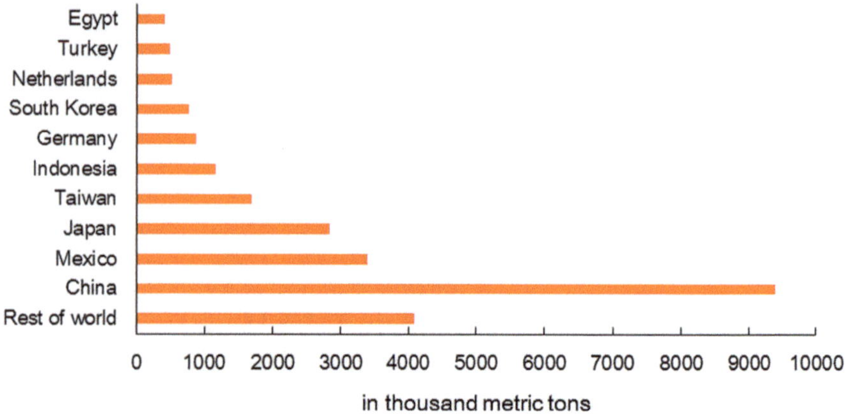

Fig. 4.16 The U.S. is the World's top soybean exporter. Most U.S. exports go to Asia with China the largest single importer of U.S. soybeans. Courtesy: U.S. Department of Agriculture National Agricultural Statistical Service 2012. http://www.nass.usda.gov/Charts_and_Maps/A_to_Z/in-soybeans.asp)

4.3.3 Major Issues Associated with U.S. Agriculture

There are many issues that impact agriculture in the U.S. These issues include most notably sustainable agriculture which in turn involves issues of soil and water conservation and minimizing negative externalities associated with agriculture such as water pollution, groundwater overdraft and odor which is most frequently associated with large-scale CAFOs (Environmental Health Sciences Research Center 2011). Also of importance and currently generating great controversy is the question of genetically modified (GM) foods and animals. Another important issue is the impact of food products on human health which traditionally had related to famine and food scarcity on a global level, but in the U.S. now ironically may be related to overabundance or at least excessive intake of certain foods associated with increasing obesity in American society. One final issue that is a problem in the U.S. in common with China is the conversion of agricultural lands. This includes the issue of the conversion land from one crop to another, the conversion of agricultural lands into non-agricultural uses (mostly for housing or commercial development) and the retention of farmland by various means such as use of zoning, tax incentives and the conservation reserve program (CRP) which is also related to sustainable agriculture.

Issue 1: Sustainable agriculture Sustainable agriculture is a set of practices designed to produce crops and animal products from the land while not, over the long run, diminishing the ability of the land to sustain that production. A related issue is avoiding negative externalities associated with production. Traditionally, this has involved minimizing soil erosion due both to water and wind. Many meth-

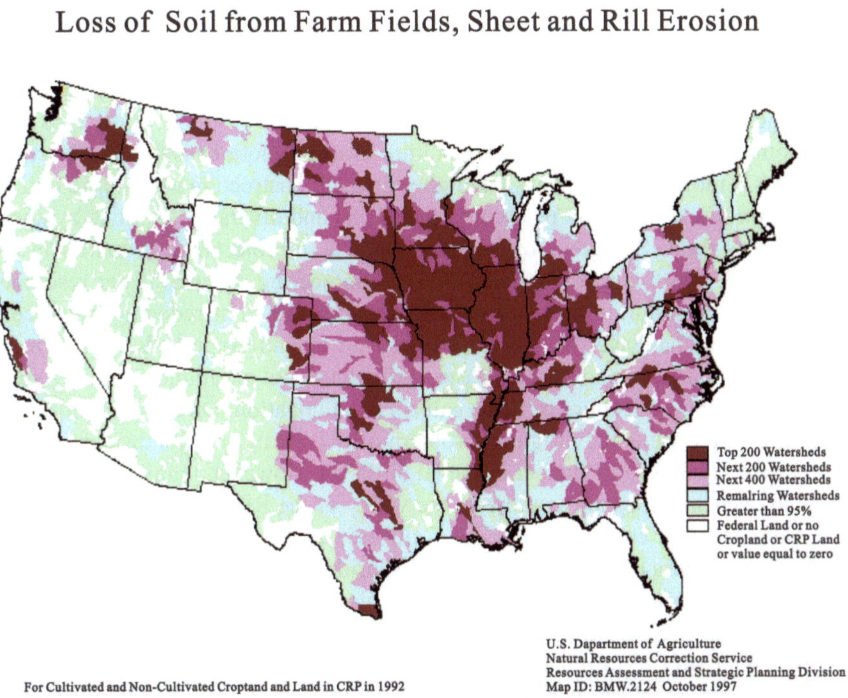

Fig. 4.17 In the areas with the highest water related erosion up to 2 metric tons of topsoil can be lost per hectare per year. (Courtesy: U.S. Department of Agriculture Resource Assessment and Planning Division, Report on Conservation Reserve Program Lands 1987)

ods for this have been employed some of which are discussed in more detail in Chap. 10. However, in general, soil erosion can be reduced by using contour plowing, using strip cropping, maintaining buffer strips of natural vegetation along water courses, using gabions and check dams and geo-textiles on rills, ditches and gullies and by a variety of other methods to keep topsoil from being washed away. Wind erosion, which is a greater problem in the arid and semi-arid western U.S., can be controlled by conservation tillage, planting windrows (rows of tall trees) along the edges of fields and by maintaining grasslands in vulnerable areas. Grasslands can be maintained by reducing over-grazing, planting perennial grasses, through use of fire to promote healthy grasslands and/or by enrolling lands in the Grasslands Reserve Program (GRP) a program sponsored by the U.S. Department of Agriculture (USDA) which pays farmers and ranchers to maintain sensitive grasslands. Programs to control soil erosion were introduced by the Soil Conservation Service (now the Natural Resources Conservation Service or NRCS) in the 1930's and are implemented by numerous local soil and water conservation districts in agricultural areas throughout the U.S.; these work cooperatively with farmers and the USDA to reduce the erosion of top soils (Helms 1991). A map of the agricultural areas most impacted by water induced erosion is shown in Fig. 4.17. This map is delineated by watersheds. Wind erosion is a greater factor farther west, but water borne erosion impacts lands with the greatest fertility and thus causes the greatest losses to agricultural productivity. The Corn Belt, the Palouse Plateau region, the Mississippi Delta region and certain parts of the High Plains are the areas most affected by water induced erosion.

Another aspect of sustainable agriculture is to promote conservation of water resources. Traditionally, this has taken the form of construction of water retention, flood control and irrigation reservoirs, aqueducts and diversion facilities by agencies such as the U.S. Bureau of Reclamation. These projects are primarily in the western U.S. More recently, structural civil engineering oriented water resources management has shifted to conservation measures aimed at reducing water consumption. This has taken the form of improved irrigation methods, use of more drought tolerant crops and in some cases taking lands out of production in dry years or discontinuing agriculture of lands subject to salinization. Directly related to agricultural water use is the on-going problem of over-draft of groundwater aquifers such as the Ogallala and Dakota in the High Plains. However, many other aquifers are also over-drafted, such as those in the valley of the Rio Grande River in Texas and New Mexico, coastal aquifers in California and parts of the Central Valley in California, parts of South Carolina and Florida (Floridian Aquifer) and several aquifers in Arizona (Salt River Valley) and Colorado (San Luis Valley). Related to this issue is the increase in salinization of soils, which is a particular issue in the rich Central Valley of California and in Arizona, but this has been reduced through use of buried tile (now plastic) drains. Another related issue is water pollution from agriculture; this primarily involves issues of non-point source pollution from pesticide residues in water, particularly of shallow aquifers and the issue of eutrophication of surface water bodies due to run-off of nutrients from fertilizers and animal wastes. Fertilizers and concentrated animal wastes in run-off causes eutrophication. This process is due to an over-abundance of limiting nutrients such as available nitrogen and phosphorus that cause growth of algal blooms that subsequently die and whose decomposition uses up available oxygen in aquatic environments, particularly in shallow or stagnant lakes and canals that in turn results in conditions in which fish cannot survive. Controlling water pollution involves many strategies, which include reducing erosion, using less toxic pesticides and implementing integrated pest management which uses natural predators like spiders, beneficial insects and selected use of targeted pesticides. Many of the most toxic, and persistent pesticides such as those chlorinated hydrocarbons with a tendency to bio-accumulate have been banned from use in agriculture in the U.S. starting in the 1970's; these banned chemicals include DDT and lindane and the soil fumigant DBCP. Also water pollution can be controlled by timing and better control of pesticide applications, which in the U.S. are frequently applied by "crop dusting" aircraft that now use geographic information systems and global positing systems to precisely control pesticide application and avoid over-spraying or spraying near water.

Finally, sustainable agriculture can be extended to include the conservation of habitat and diversity. This is partially achieved through the Conservation Reserve Program (CRP) and Wetlands Reserve Program (WRP) and Grasslands Reserve Program (GRP) administered by the Natural Resources Conservation Service (NRCS 2010). These programs pay farmers an annual fee (rental payments) for agreeing

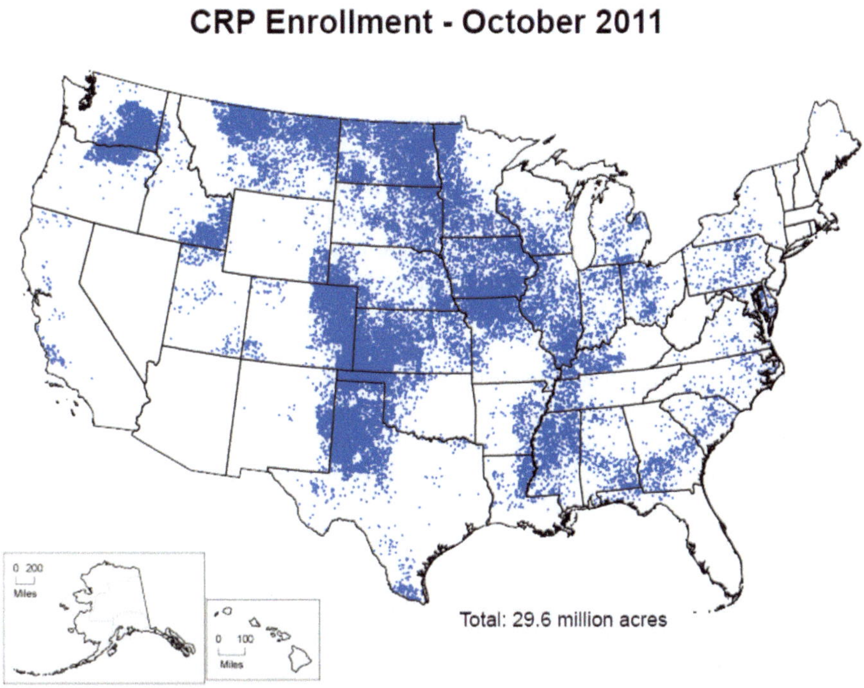

Fig. 4.18 The Conservation Reserve Program pays farmers not to farm on lands vulnerable to erosion and substitute sustainable natural vegetation for habitat and erosion control. Each dot equals 405 ha of protected lands. (Courtesy: U.S. Department of Agriculture, Farm Service Administration 2011 http://www.fsa.usda.gov/Internet/FSA_File/crpenrolldotden1211.pdf)

over a number of years not to farm on lands vulnerable to erosion or that have special habitat characteristics, such as being wetland or grassland areas. The programs are an adaptation of a program started in the 1930's to support agricultural prices by curtailing production of crops grown in surplus of the then low demand for them. However, it now takes the form of a land, water and habitat conservation oriented approach. In 2011, 29 million acres of farm lands were enrolled in the CRP program, the WRP had 2.3 million acres of wetland areas under protection and the GRP had 335,000 acres of grasslands enrolled in it: for the extent of the CRP See Fig. 4.18.

There are many other aspects of sustainable agriculture in the U.S., but erosion control, water conservation, water pollution minimization, and preservation of sensitive habitat are among the more important aspects of the concept. It is an idea originally associated with soil and water conservation that has been embraced in the U.S. starting in the 1930's and is becoming increasingly important in the last two decades. The supporters of the concept are also starting to embrace the issue abatement of the effects of global climate change, both in terms of assessment of how climate change may impact agricultural productivity, crop selection and management but also in terms of global carbon balance. This link between sustainable land management and the global carbon cycle is particularly important in the area of silvaculture and forestry. In the U.S. context, forestry is usually a separate discussion from agriculture, but many forest lands, particularly those in the south are on private lands and are operated as tree farms and are located on land that was at one time devoted to crops such as cotton.

Issue 2: CAFO's and GM foods Large scale concentrated animal feeding operations (CAFO's) have a range of negative externalities associated with them. Swine CAFO's and cattle feedlot operations are discussed in detail in Chap. 10, but all large concentrations of animals including aquaculture and poultry CAFO's generate concentrated animal wastes that contribute to surface water pollution. Another issue of concern is the build-up of nitrite in groundwater that can occur near facilities such as cattle feed lots. Other nitrogenous wastes can cause eutrophication of surface waters.

Scientific farming on an industrial basis not only has developed the modern CAFO, but also has made changes in animal feeds, animal waste management and is responsible for the creation of genetically modified (GM) plants and animals. Genetic engineering related innovations have been among the most controversial in agriculture. The issue of genetically modified (GM) crops and animals is one that is complex. At present many genetically modified seeds are used in U.S. agriculture. Mostly these are seeds that grow crops such as soybeans that are resistant to herbicides that ordinarily would kill both the crops and the weeds that typically infest them. The genetically modified seeds may have other attributes such as insect resistance, higher yield, plant disease resistance or drought or cold tolerance. There is some consumer resistance to consuming genetically modified foods and there are real concerns that alteration of physiological processes in a plant may confer undesirable characteristics on the food made from it. For example, the food created from the GM crop might become the source of allergic reactions in sensitive persons. Conversely, genetic modification opens the promise of crops that are naturally resistant to pests with a concomitant reduction in the use of toxic pesticides, which are themselves a source of human and ecological health concerns. Also soil fertility and water conservation might be achieved with development of better nitrogen fixation or better drought resistance or ground cover characteristics, etc. in crops (National Research Council 2010).

The use of GM seeds is increasing rapidly. In 2010, 81 % of all soybeans, 64 % of cotton, 29 % of corn and 23 % of canola globally were grown from plants which originated from GM seeds and in the U.S. the use of GM seeds is even higher (see Fig. 4.19). The most common modification in GM seeds is herbicide tolerance (HT), where plants are given a gene that allows farmers to spray them with the weed killer glyphosate, without harming them. Sixty-one percent of GM crops carry this gene. The other commonly seen trait is the addition of a gene from the soil bacteria Bacillus thuringiensis that renders plants noxious to insects (called Bt). About 17 % of GM crops carry this Bt trait. About 22 % of GM crops contained both genes. The addition of multiple modifications raises concerns, as it adds to the complexity of the compounds that the plants produce, some of which may end up in the human diet. GM crops have been commercially planted in the U.S. since 1996. A total of 29 countries worldwide now plant GM crops, with the United States planting the

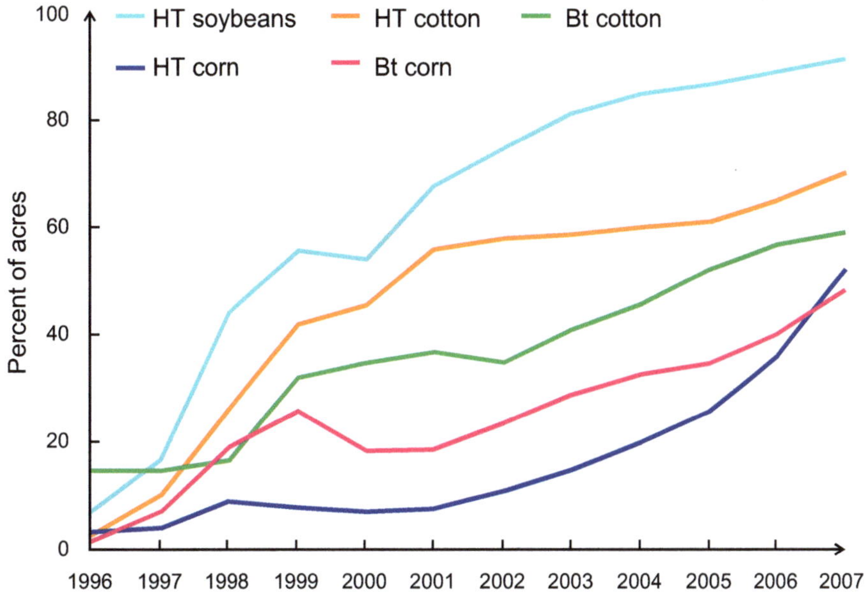

Fig. 4.19 Herbicide tolerant (HT) soybeans are near 100% of acreage, with rapid increases in corn and cotton acreage using GM seeds that have either insect resistant (Bt) or herbicide tolerant (HT) traits. (U.S. Department of Agriculture Economic Research Service 2012. http://www.ers.usda.gov/data-products/adoption-of-genetically-engineered-crops-in-the-us.aspx)

most, at 165 million acres in 2010. (International Service for the Acquisition of Agri-biotech Applications 2011).

Issue 3: Obesity and healthy eating Food has long been something that was in short supply. Famine has impacted many parts of the world. China has suffered from many famines in its long history. The early European settlers in America had poor diets during their first few years, but soon the abundance derived from the largely virgin soils of the New World produced an abundance of food. At many times in U.S. history, the major issue was finding markets for this bounty. The food production in the U.S. has reflected consumer tastes, but agri-business companies have also tried to shift those tastes. The development of fast food restaurants and instant gratification of the desire for sugar and fats has produced a fundamentally unhealthy diet among most Americans. In 2012 the majority of Americans adults were overweight and about a third were obese. The most profound change in typical American physiognomy in recent years has been the growth in the waistlines of children and young adults (see Fig. 4.20). This is a result of many trends. Cultural and spatial factors are also important since childhood obesity is much higher is some parts of the U.S. such as the south as shown in Fig. 4.21. One is a more sedentary life style than in past years. Another important cause is a shift in food production and consumption toward meats (chicken and pork in particular) but also an increase in intake of high fructose corn syrup (HFCS) and refined cane sugar. High

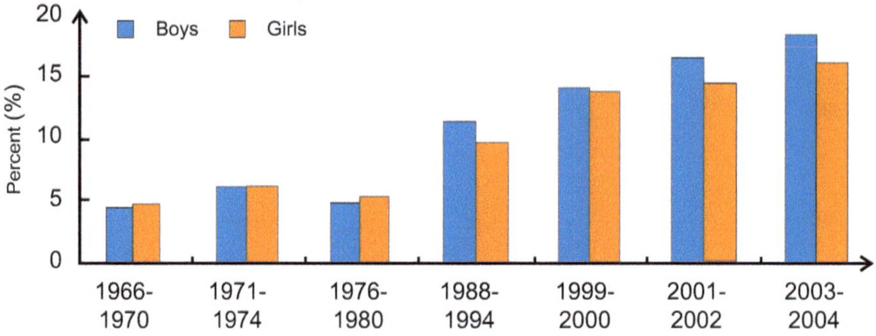

Fig. 4.20 Between 1966 and 2004 the percentage of overweight children has quadrupled. (U.S. Department of Health and Human Services Assistant Secretary for Planning and Evaluation 2010, http://aspe.hhs.gov/health/reports/child_obesity/)

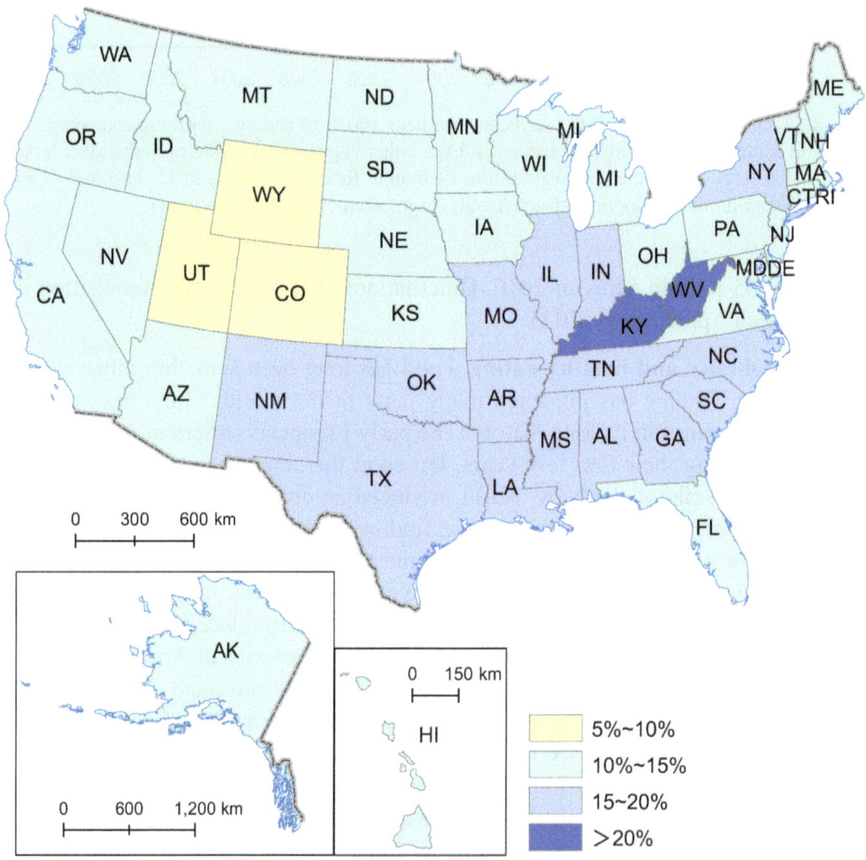

Fig. 4.21 Obesity in children is higher in the South and highest in the Appalachian region in Kentucky and West Virginia. It is lowest in the West in states of Wyoming, Colorado and Utah. (National Conference of State Legislatures 2010, http://www.ncsl.org/Portals/1/oldsite/programs/health/obesitymap.jpg)

fructose corn syrup in particular has been singled out as source of obesity (Bray et al. 2004). It is the primary sweetener used in beverages and represents about 16% of the caloric intake of the average American. Americans consume far more sweetened beverages than in the past, but also before the 1970's those beverages were sweetened with cane sugar. Nutritionists have developed evidence that fructose is particularly likely to produce obesity and consequently diabetes, heart disease, stroke and certain cancers like breast and colon neoplasms. But other foods are associated with heart disease such as foods high in cholesterol and saturated fats. The agricultural industry would argue that changes in production and food composition are driven by consumer demand and economic factors and in fact the costs of some products like HFCS sweetened beverages per calorie are low compared to fresh fruits and vegetables. Many unprocessed fruits and vegetables, particularly organic ones, are quite costly as a source of calories. Part of the reason is that in many areas of the U.S., particularly inner cities, there is a shortage of markets that sell a wide range of fresh fruits and vegetables or organic and unprocessed foods. This in turn has led to several important societal trends. There is an effort by many groups including the Federal Government to try to address availability of healthy foods. One trend is the growth of organic farms, farmers markets, and urban and community gardens and the development of an integrated organic foods industry. U.S. sales of organic foods and beverages have grown from $ 1 billion in 1990 to $ 26.7 billion in 2010. Sales in 2010 represented a 7.7% growth rate over 2009 sales. Experiencing the highest growth in sales among organic products during 2010 were organic fruits and vegetables, with sales up 11.8% over 2009 sales (Organic Trade Association 2011). Another response has been to try to educate the public in nutrition and the health consequences of food choices. The entire food industry has responded in various ways to these trends, such as developing many more diet and low calorie foods and particularly beverages, changing the offerings at fast food restaurants to include salads, fresh fruit, oatmeal, fruit juices and milk instead of just hamburgers, french fries and soda sweetened with high fructose corn syrup. These may be primarily cosmetic changes, such as the effort to rename high fructose corn syrup as "corn sugar" on product packaging. They may involve greater changes in perceptions than in the realities of food production and consumption (Hsu 2012). Nevertheless, industry has made some major adaptations, such as the growth in offerings of organic products, the increase in production of "diet" and "lean" and "whole" foods and more offerings and emphasis on those foods like salmon and tuna fish that contain fish oils that have omega 3 fatty acids that are linked to heart health. Also many farms now manage their animals with less use of pens and more use of pastures and use less antibiotics and/or hormones (Rao 2012).

Issue 4: Conversion of agricultural land Conversion of agricultural land to other uses is an on-going issue in the U.S. as in China. Historically, there was a pattern of conversion of forested land and grassland to cropland. Later some of the more marginal cropland reverted to grazing land and forests. In the 20th century many areas of wetlands were drained and/or filled, and this has resulted in the loss of perhaps half of the historic acreage of wetlands in the U.S (Environmental Protection Agency 2011). Crops raised in the U.S. have changed over time. Thus, areas in the

South that once grew a mixture of food crops for subsistence such as corn, sweet potatoes, and swine might have switched to indigo as a cash crop when the plant was a valuable source of dyes in the 18th century, then to cotton after the perfection of the cotton gin in the early 19th century, then to peanuts in the early 20th century as the fertility of the soil was exhausted by lack of soil conservation and over-production of cotton. After the Second World War, with growing use of synthetic fertilizers and pesticides, the same land might have been used to raise soybeans and now it may be used to grow corn due to high current prices produced by demand related to production of fuel grade ethanol. The conversion of land from raising one crop to another still maintains the land in a condition fit for agriculture and in some cases, as in leaving the land fallow in rotation or raising legumes, it may increase the fertility of the land. Conversion of land from agriculture to forest (or grassland or wetland) also preserves the ability of the land to raise crops at a future date, as well as having other benefits such as helping to remove carbon dioxide from the atmosphere, and reducing the potential for soil erosion. Conversion of land from agriculture and/or forestry type uses to residential, commercial or industrial uses represents a qualitative shift and is an ongoing process in the U.S. The sprawl of cities facilitated by the automobile and the low density suburban single family home centered development pattern has characterized land use in the U.S. over the period from the end of the Second World War to the present time. In particular, suburban sprawl has characterized the years after 1980 through 2008, especially in the Western U.S. and has resulted in a profound shift in land use patterns. Many areas of once productive farmlands have been converted to residential housing and related land uses. In most of the years since 1970 between 1 million and 2 million acres of agricultural land has been lost annually in the U.S. to development. However, most of the land that was converted from agricultural uses is classified as grazing land or pasture, not prime crop land. Between 2002 and 2007, 4,080,300 acres of crop lands were converted to other developed uses. Between 1982 and 2007, 41,324,800 acres of rural land (i.e., cropland, pasture, range, land formerly enrolled in the CRP, forest and other rural land) were converted to other developed uses. During the 25-year span from 1982 to 2007, every state lost prime farmland. States with the biggest losses of agricultural land included Texas (1.5 million acres), Ohio (796,000 acres), North Carolina (766,000 acres), California (616,000 acres) and Georgia (566,000 acres). Between 2002 and 2007, 7,491,300 acres of rural lands were converted to developed uses. This is an average annual conversion rate of 1,498,200 acres (American Farmland Trust 2011).

The conversion of farmland to industrial uses is much less common today, although it certainly took place in the past century, particularly in States like Michigan and Ohio. Some areas of rich farmland have more recently been converted to a mix of housing and light industry, for example in California, where the fruit orchards of the Santa Clara Valley have become the technology parks and housing of the "Silicon Valley' and in Southern California where orange groves have been replaced by technology firms and the entertainment industry. However, most of the lost acreage has been supplanted by housing and retail establishments. Many types of retail firms rely on large "big box" stores located on the edge of existing communities.

Thus one frequently sees a farmer's field on the outside of a medium sized town which happens to be located on the edge of a major road or highway transformed into, for instance, a new Wal-Mart store, or a car dealership or home improvement center. Frequently, this was prime farmland, since the retailers want large flat areas for their stores, parking lots and associated smaller retailers, restaurants and gas stations located in the "pad" and these characteristics may be associated with better farm land. Also such retail establishments can afford to pay enough to purchase prime crop land, something that other buyers may not be able to do. But retailers are not the largest user of formerly agricultural acreage, the most common reason for land use changes is associated with suburban residential developments. In many areas premium prices are paid for larger lots in gated or otherwise planned subdivisions which often offer amenities such as golf courses.

Overall, the amount of cropland in the U.S. has not declined sharply over the last few decades, what has changed more greatly is the amount of grazing land which has declined. Most of the lost pasture and other grazing land are being converted into residential uses; some of the more fertile pastures have been converted into cropland replacing other croplands that have been converted to residential uses. What has increased in the U.S. is the area occupied by housing and the area occupied by forests. However, in specific areas where cropland is the major land use, urbanization has had a significant impact on cropland. Thus for example, the recent growth of the suburbs of Chicago, Minneapolis, Cincinnati and Saint Louis has all been in areas that were formerly farm lands.

4.4 Comparison of the Agricultural and Food Production in China and the U.S.

The agricultural foundation of China and the United States differs in resource endowments and historical development. With population tied to the land, less arable land resources and a long history of development, China has developed a famous tradition of precision farming. Driven by business concerns from its early years American agriculture flourishes because of its rich natural resources. Today, the total area of cultivated land in China is less than the United States with a decreasing trend, while the arable land in the United States changes little in relative terms. Thus, the gap in the area and quality of the cultivated land between the two countries is widening. Besides, China has nearly 30 % of the population engaged in agricultural production, while in the United States those engaged in agricultural production account for only about 2 % of the national population. There is a huge disparity between the two countries on the amount of labor involved in agriculture. Due to the huge differences in resource endowments, China and the United States have different focuses in the agricultural technology development process. With fewer people and more land, agricultural technology development in America tends to emphasize labor savings, while in China, the contradiction between space needed for people and land need for farming guide technology development which tends to favor conservation of land oriented technologies and methods.

In addition, there are great differences in agricultural technology and service support system related to agricultural development between the two countries. The development of agricultural productivity in the United States is built on top of the development of the most advanced technology. As early as the 1980s, the United States began to adopt biotechnology had adopted the widespread use of computer information technologies including GIS and GPS for agricultural development. The U.S. government provides the basis for agricultural education, training, research and conversion of agricultural technology. With a relatively low level of agricultural support, China has an urgent need to increase the investment in agricultural science and technology. For a long time, agricultural development in both China and the United States has benefited from the protection and support of the government, such as financial assistance, and commodity price protection, but the methods of financial support differ. The developed system in America providing support and services in agricultural research credit and machinery is what China can learn from.

Due to the limited funding and technical support and considerable population, China mainly has developed labor-intensive agriculture. Characterized by lack of alternative funding and technology and inefficient agricultural production, Chinese agriculture is in stark contrast to the capital and technology-intensive agriculture in America. The United States has the world's largest commercial agricultural production.

The differences and comparative advantages of the agricultural production in America and China, makes for a good basis of complementary agricultural trade. Both of them are the world's most important agricultural production and trade powers. With the agricultural trade dependence gradually increasing, the two countries are essential agricultural trade partners with each other. Agricultural production and trade cooperation of the two countries is conducive to the promotion of bilateral trade and is a win-win development.

4.5 Conclusions

China is a country with a very ancient tradition of agriculture and a population of more than 1.3 billion people; 51.3 % were living in rural areas in 2010. These are China's most fundamental national characteristics. Thus agriculture is of primary importance in China and defines its physical, environmental, economic and cultural characteristics to a large degree. This also makes Chinese agriculture critical to the world economy and food price stability. Because much of China's land is not suitable for agriculture, intensive and efficient agriculture must be practiced on the scarce lands that are suitable. Ultimately 7 % of the world's land must help feed 22 % of the world's population. China's farmers are meeting this challenge with the help of scientists, engineers and educators. During the past 30 years, Chinese agriculture has rapidly developed, with increased agricultural production. Agricultural production conditions continue to improve, and farmers' living standards continue to advance. However, as a developing country China has issues with investment in agricultural related infrastructure, research and environmental protection. China has become a large agricultural export country and played an important role in the world food market's supply and demand. However, due to natural conditions and development history, there are still large differences in living standards between urban and rural areas and between the eastern and western parts of China.

America has a much shorter history of agriculture than China, but in that brief time its farmers, scientists, engineers and educators have been responsible for many historic innovations in agricultural production such as invention of the cotton gin (1794), mechanical reaper (1831), steel plow (1837), grain elevator (1842), barbed wire (1874), and milking machine (1879) all the way through development of many important agricultural chemicals such as glyphosate herbicides (1970) and more recently development of GM seeds and foods. Agriculture in the U.S. involves a declining and now surprisingly small proportion of a much smaller population than in China. Thus of about 318 million people in the U.S. in 2010 less than 2 % were involved directly in agriculture and only 960,000 depend on farming, ranching or fishing for their livelihoods. Nevertheless, these few agriculturists are able to produce a vast surplus of agricultural commodities, in excess of domestic demands in most cases. Thus the U.S. while the world's third most important agricultural producer, after China and India, is the world's leading exporter of many types of food such as cereals, and meat. Agriculture in the U.S. faces a number of important challenges most notably loss of high quality farm lands. A network of land grant funded state agricultural research universities have pioneered the development of methods for sustainable agriculture. The question remains however, can the U.S., China and the other countries like India and Brazil work internally and cooperate internationally to expand production while minimizing externalities like pollution and negative social consequences like loss of the viability of small farms. The long term survival of the billions of people on earth depends on the effective leadership efforts in these spheres in both the People's Republic of China and the United States of America.

References

Almanac of American Agriculture. (2011). Production of major crops. http://almanacofamericanagriculture.com/About_Us.html. Accessed 25 Aug 2013.

American Farmland Trust. (2011). Natural resources inventory reveals loss of farmlands. Every State has lost Prime Farmland. Washington DC: American Farmland Trust.

Bray, G., Nielsen, S., & Popkin, B. (2004). Consumption of high fructose corn syrup in beverages may play a role in the epidemic of obesity. *The American Journal of Clinical Nutrition, 79*(4), 537–543.

Environmental Health Sciences Research Center. (2011). Environmental impacts of industrialized livestock production. University of Iowa. http://www.ehsrc.uiowa.edu/cafo_impacts.html. Accessed 25 Aug 2013.

Environmental Protection Agency. (2011). Wetlands status and trends. http://water.epa.gov/type/wetlands/vital_status.cfm. Accessed 25 Aug 2013.

Florida Department of Citrus. (2011). Florida's citrus industry. http://floridajuice.com/. Accessed 25 Aug 2013.

Harding, S. (2009). Colonial history: Settlement patterns. http://www.articlesbase.com/history-articles/introduction-to-colonial-american-history-settlement-patterns-3457642.html. Accessed 25 Aug 2013.

Helms, D. (1991). Two centuries of soil conservation. *OAH History Magazine, 5*(3), 24–28.

Hornaday, W. (1889). The extermination of the American bison. Washington D.C.: Reprinted by the Smithsonian Institution.

Hsu, T. (2012). FDA rejects bid to rename high fructose corn syrup "Corn Sugar". Los Angeles Times Newspaper. http://www.latimes.com/business/money/la-fi-mo-fda-corn-sugar-20120531,0,790458. Accessed 23 May 2012.

International Service for the Acquisition of Agri-biotech Applications. (2011). Global status of commercialized biotech/gm crops. Ithaca. http://www.isaaa.org/itemdescription.asp?ItemType=BRIEFS&Control=IB043-201. Accessed 25 Aug 2013

Kona Coffee Council. (2011). Kona coffee. http://www.kona-coffee-council.com/. Accessed 25 Aug 2013

Liu, S. J. (1996). Chinese geography. Beijing: Higher Education Press.

Lu, C. (1984). Agro-geography of China. Beijing: Agriculture Press.

National Bureau of Statistics of China. (2011). China statistical yearbook 2011. Beijing: China Statistics Press.

National Bureau of Statistics of China. (2008). China compendium of statistics 1949–2008. Beijing: China Statistics Press.

National Bureau of Statistics of China. (2010a). China rural statistical yearbook 2010. Beijing: China Statistics Press.

National Bureau of Statistics of China. (2010b). Compendium of statistics of the second agricultural census. Beijing: China Statistics Press.

National Research Council. (2010). *Impact of genetically engineered crops on farm sustain-ability in the United States*. Washington DC: National Academy Press.

Natural Resources Conservation Service (NRCS). (2010). The Conservation reserve program. http://www.fsa.usda.gov/FSA/webapp?area=home&subject. Accessed 25 Aug 2013.

Organic Trade Association. (2011). Survey of organic foods industry statistics and projected growth. http://www.ota.com/organic/mt/business.html. Accessed 25 Aug 2013.

Rao, A. (2012). Burger king to go cage free by 2017. *Slate Magazine*. http://slatest.slate.com/posts/2012/04/25/burger_king_to_go_cage_free.html. Accessed 25 Aug 2013.

Silverberg, R. (1968). The mound builders of ancient America. New York: New York Geographic Society.

Sun, J. T. (2003). Agrogeography of China. Inner Mongolia Inner-Mongolian University Press.

Travis, W. (2007). *New geographies of the American West: Land use and the changing patterns of place*. Washington DC: Island Press.

Umbach, K. (1997). A Statistical Tour of California's Great Central Valley. California Research Bureau. On-line at: http://www.library.ca.gov/CRB/97/09/

USDA, Economic Research Service. (2007). Summary of statistics from the 2007 census of agriculture. USDA, Washington D.C. http://usda.ers.gov. Accessed 25 Aug 2013.

Washington Apple Commission. (2010). Washington apples. http://www.bestapples.com/facts/facts_washington.aspx. Accessed 25 Aug 2013.

Xu, J. M. (2002). Economic geography of China. Jinan: Jinan Press.

Zhao, J., & Chen, C. K. (1999). Chinese geography. Beijing: Higher Education Press.

Zhou, L. S. (2000). *Agro-geography of China*. Beijing: Science Press.

Chapter 5
Economic Geography

Hongmian Gong and Huasheng Zhu

5.1 Introduction

China and the U.S. are in different stages of economic development and display very different economic geographies. The U.S. experienced industrialization and urbanization in the late nineteenth century and early twentieth century as farmers moved to the cities to meet the labor demand of the rapidly growing manufacturing. By mid-twentieth century, the employment in the tertiary (services) sector in the U.S. has forever passed that in the secondary (industrial) sector (Fig. 5.1). Based on the experiences of the U.S. and other developed countries, western theories (Clark 1940) have suggested that as economies develop, labor markets shift from the primary sector to the secondary sector and then to the tertiary sector. This domestic labor transition was deepened as the U.S. was integrated into the globalization process in the late twentieth century. As a result of the international division of labor, manufacturing jobs moved out of the U.S. to developing countries such as China and Mexico while the U.S. increasingly relies on exporting advanced service to the world market for its economic growth. Today, the U.S. is one of the few countries that has reached the post-industrial stage of economic development, in which services is the most important economic sector in both income and employment (Table 5.1). As of 2010, the services sector contributes 76.8 % of the gross domestic product (GDP), much higher than the industrial sector (22.1 %) and the agricultural sector (1.1 %) (The Richest 2012). Similarly, the services sector employs about 83 % of the labor force, while that percentage is around 15 % for the secondary (industrial) sector and 2 % for the primary (agricultural) sector (de Blij and Muller 2006).

China began its industrialization much later. It was not until 1970 (Gong 2002) that the contribution of the industrial sector in GDP has forever passed that of the

H. Gong (✉)
Department of Geography, Hunter College of the City University of New York, New York, USA
e-mail: gong@hunter.cuny.edu

H. Zhu
School of Geography, Beijing Normal University, Beijing, China
e-mail: zhuhs@bnu.edu.cn

R. Hartmann, J. Wang (eds.), *A Comparative Geography of China and the U.S.*,
GeoJournal Library 109, DOI 10.1007/978-94-017-8792-5_5,

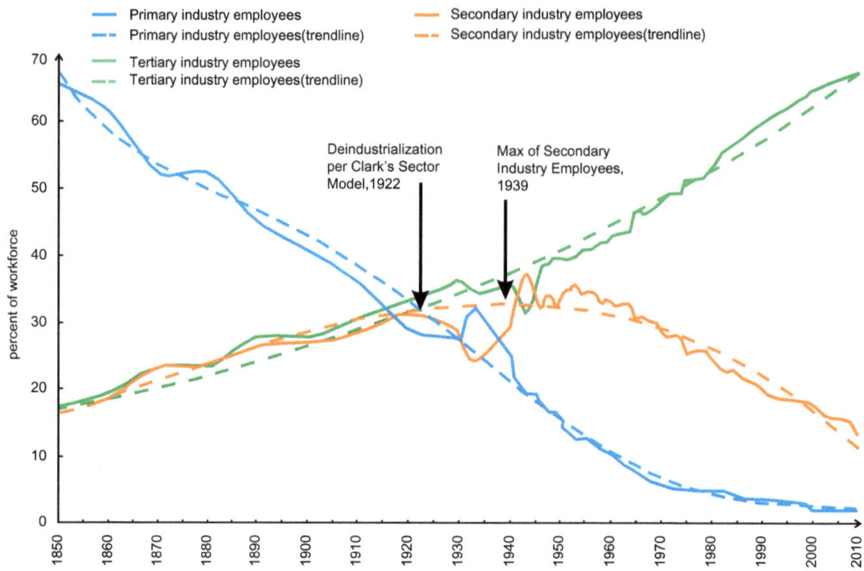

Fig. 5.1 Sector model for the U.S. economy, 1850–2009. (Kossik 2011)

Table 5.1 GDP and employment structures of China and the U.S. (The Richest 2012; CSSB 2009a; de Blij and Muller 2006)

Structures	Countries	Primary/agricultural sector (%)	Secondary/ industrial sector (%)	Tertiary/services sector(%)
GDP	China	10.2	46.9	43.0
	United States	1.1	22.1	76.8
Employment	China	40	27	33
	United States	2	15	83

agricultural sector to become the leading sector. As of 2010, the industrial sector contributes 46.9% of the GDP, slightly higher than the 43% of the services sector (The Richest 2012). China is still in the industrialization stage and has not yet reached the post-industrial stage of economic development. In the globalization process and the international division of labor, China is at the receiving end of manufacturing that is moving out of developed countries such as the U.S. and Japan. China has become the largest manufacturer in the world (Mitra-Thakur 2011).

In terms of employment, 40% of China's labor force was employed by the primary sector in 2008, 27% by the secondary sector, and 33% by the tertiary sector (CSSB 2009a). The percentage of the tertiary sector in the total labor force was lower than that of the secondary sector in the 1970s and 1980s, but passed it since 1994 (Fig. 5.2) as China became more industrialized and labor productivity in the manufacturing sector increased. Overall, China followed a similar path of sectoral

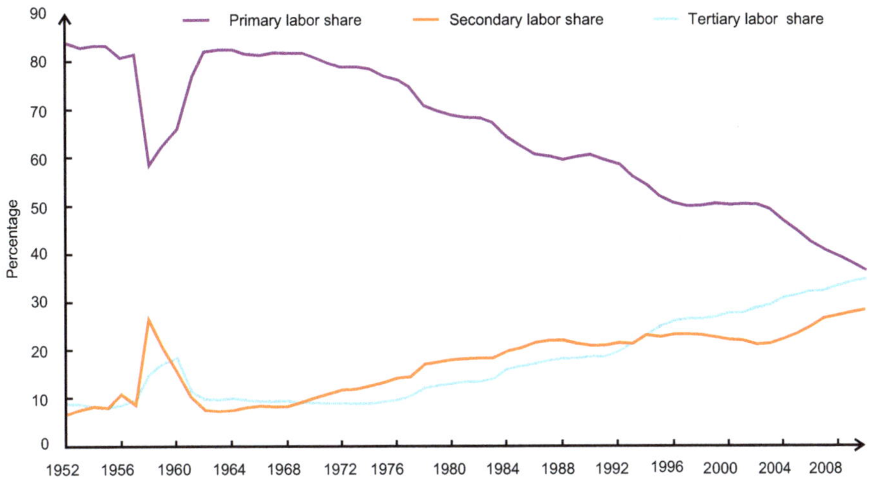

Fig. 5.2 Change in employment structure in China, 1952–2010. (CSSB 2011)

labor shift in developed countries from the primary sector to the secondary sector and then to tertiary sector (Gong 2002). In the future, the percentage of the primary sector in the total labor force is expected to continue to decline and eventually become the lowest among the three sectors, while that of the tertiary sector in the total labor force will become the highest.

Since China and the U.S. are in different stages of economic development, in addition to their different physical geographies as discussed in a previous chapter, the economic geographies of China and the U.S. display different spatial patterns. For example, the economic activities in the U.S. are almost entirely concentrated in urban areas; while a good portion of those in China (10.2 % of GDP and 40 % of labor force) are primary (agricultural) activities that widely spread across the rural areas. This chapter is devoted to discuss the three economic sectors and their geographies in the two countries. The primary sector typically includes agriculture and mining, economic activities that make direct use of natural resources. Although agriculture is the major component of the primary sector and these two are often used interchangeably, agriculture has been discussed in the last chapter. The discussion of the primary sector in the next section will therefore focus on mining (including quarrying and extraction of oil and gas) of natural resources.

5.2 Natural Resources and Mining

Both the U.S. and China have large territories and abundant natural resources. While both countries take advantage of their natural resources through mining, they import raw materials from the world markets to sustain their economy development. As the

U.S. has reached the post-industrial stage, its import of raw materials is mostly oil and gas to keep up with its high per capita energy consumption. China, on the other hand, imports raw materials necessary for its production of manufactured goods.

China has discovered and extracted a multitude of metal ores and non-metallic minerals to provide the raw materials for domestic industrial production. China has large coal reserves. Raw coal production reached 3.24 billion tons in 2010, accounting for 40 % of the total coal production worldwide (CSSB 2011). Most of the coalfields are found in the central interior and western areas of the country. The three provinces of Shanxi, Inner Mongolia, and Shanxi account for two thirds of the known coal reserves in China. By contrast, Eastern China where most of the industrial production takes place has only 5 % of the coal reserves. This imbalance has resulted in constant energy shortage in Eastern China and large amount of coals being shipped eastward (in Chinese Pinyin, the "Xi Mei Dong Yun" policy).

China is not only the leading producer but also by far the largest consumer of coal. In 2010, China consumed 47.5 % of the coal produced worldwide. While the U.S. is another major consumer of coal—the U.S., Europe, and India combined is responsible for nearly 30 % of the growing world coal consumption—it is a coal exporter, with large coal reserves available in Appalachia and beneath the Great Plains. In recent years, the U.S. has started to export coal to China where energy needs have considerably risen along with a growing economy and greater output of manufactured goods.

China's coal powered economy is aided by other energy sources such as oil and natural gas. In order to reduce its coal dependency, China has made significant progress in developing renewable energy sources, most importantly solar and wind energies. The proportion of clean energy resources has increased from 9.9 % in 2001 to 13 % in 2011 (CSSB 2012) and further growth can be expected. In the United States efforts in generating more energy from renewable sources have been made as well but change has come very slowly; the share of clean energy in total energy consumption has surpassed 5 % but has not yet reached 10 %. The U.S. economy remains largely dependent on the uses of fossil fuels, with 36 % of its energy coming from petroleum, 26 % from natural gas, and 20 % from coal in 2011 (U.S. Energy Information Administration 2011).

The production and consumption trends of oil and natural gas in China and the U.S. emphasize again the growing importance of a globalizing world economy. The U.S. has been an importer of oil and natural gas for several decades now—with more than 50 % of the oil and natural gas consumed imported from the neighboring countries of Canada and Mexico, from the Middle East, Venezuela and more recently from African countries, whereas China has become an oil importer not until the mid-1990s. However, the tremendous needs of a rapidly growing economy have resulted in a considerable increase of oil and natural gas imports over the past ten years (Ying 2012), mostly from countries in the Middle East, Angola in Africa, Kazakhstan in central Asia, Russia, and Venezuela.

Importing more and more oil and natural gas is a trend that China shares with the U.S. In both countries, domestic oil and natural gas production has become a high priority. New technologies in extraction, most notably hydraulic fracturing ("fracking"), have enabled the U.S. to significantly increase its domestic production in states such as Colorado, North Dakota, and Pennsylvania, and the goal of reducing imports considerably below the 50 % mark in the near future have found a more realistic basis. In China, major efforts in increasing domestic production are made as well. Crude oil and natural gas resources are mainly concentrated in the lacustrine petroleum basins of Northeastern China, Northern China around the Bohai Sea, Sichuan Province, and Northwestern China. Xinjiang, in particular, has the potential of developing into a national base for oil and gas production and processing. China also has large crude oil and natural gas reserves on its continental shelves. The East China Sea holds the largest reserve in China, followed by the South China Sea, the Bohai Sea, and the Yellow Sea. Exploitation efforts of these offshore reserves have not come without controversy as neighboring Asian countries have increasing energy needs as well.

The increasing use of fossil fuels, with nearly 70 % of its energy coming from coal, has brought about high level of manufacturing carbon emissions in China. According to the report of UK Tyndall Climate Change Research Center (Zhang 2012), China ranked the top in the world as it produced 28 % of the global carbon emissions in 2011, followed by the U.S (16 %). However, China's carbon emissions per capita are lower than that of the U.S. Chinese central government promised to reduce carbon emission per unit of GDP by at least 40 % by 2020 from the 2005 level (Wang and Feng 2011). For different reasons, both China and the U.S. opted for voluntary measures in addressing national and international concerns regarding their respective 'carbon foot print'.

The availability and use of metals for industrial production in China have also a global dimension. Although endowed with significant resources, China imports some of these raw materials for the high quality production of manufactured goods. For instance, China is rich in iron ore but the average grade is relatively low and the proportion of high-grade ore is less than 5 %. Consequently, the import of iron ore increased six times in China between 2002 and 2011, mainly from Brazil and Australia. As to lead, zinc, and copper ore reserves, China has sufficient domestic resources, though.

China is especially endowed with rare earth that is useful for the production of specialized high-tech goods. China has 23 % of the world's rare earth reserves and accounts for more than half of the global production. However, overexploitation not only makes rare earth resources decrease in value quickly, but may also have serious environmental impacts. In recent years, China's dominance in the global rare earth market has been weakened as other countries made efforts to explore rare earth resources. Some countries in the global economy launched new exploitation efforts while other countries that were used to depend on China for rare earth have managed to employ substitutive technologies (Wei and Pan 2012). Responses to China's once prominent role in rare earth extraction have also been implemented by U.S. companies inside and outside of the United States, which again shows the forces of the global market in action.

In general, mining of natural resources has been growing in both the U.S. and China to meet their demands for consumption and production. Table 5.2 shows the

Table 5.2 Employment and growth in mining in the U.S. and China. (Calculated from U.S. Census Bureau (2007a) and CSSB (2005, 2009b))

U.S.				China			
NACIS	Industry	2007	2002–2007 Annual growth (%)	Code	Industry	2008	2004–2008 Annual growth (%)
21	Mining, quarrying, and oil and gas extraction	703,129	8.2	B	Mining	9,908,000	2.8

employment and its annual growth rate in mining (including quarrying as well as oil and gas extraction in China's statistics) in the two countries. The employment data come from their latest economic censuses (2007 for the U.S. and 2008 for China). The average annual growth rates are calculated from the last two economic censuses (2002 and 2007 for the U.S. and 2004 and 2008 for China). The annual growth rate in the U.S. was 8.2% between 2002 and 2007, almost three times of that in China, as the U.S. encouraged domestic production (mostly in Texas and Louisiana, see Fig. 5.3) to reduce the import of oil and gas. On the contrary, China invested heavily in overseas, such as Africa, South America, and Australia, for its growing quest for energy and raw materials (Stevenson 2010). As a result, China's oil import rose while domestic oil production declined, for example, in May 2012 (China Daily 2012).

5.3 Industrial Sector

While the economies of the U.S. and China are at different stages of development, they have become increasingly integrated into the rapidly growing global economic system over the past twenty-five years. This applies most evidently to the production and distribution of manufactured goods. The industrial sectors of both countries are intricately and tightly connected to the global economy. This section will discuss the growth and geographies of the industrial sectors in the two countries, with a focus on China being integrated into the global production network.

5.3.1 Industrial Structure and Growth

The industrial sector, also called the secondary sector, includes utilities, construction, and manufacturing industries. Because the U.S. has passed the industrial stage, the employment in the industrial sector has been experiencing decline as a whole and in manufacturing in particular (Table 5.3). Manufacturing jobs in the U.S. have been outsourced to developing countries, such as Mexico and China, to take advantage of the cheap labor and land there to reduce the production costs and increase

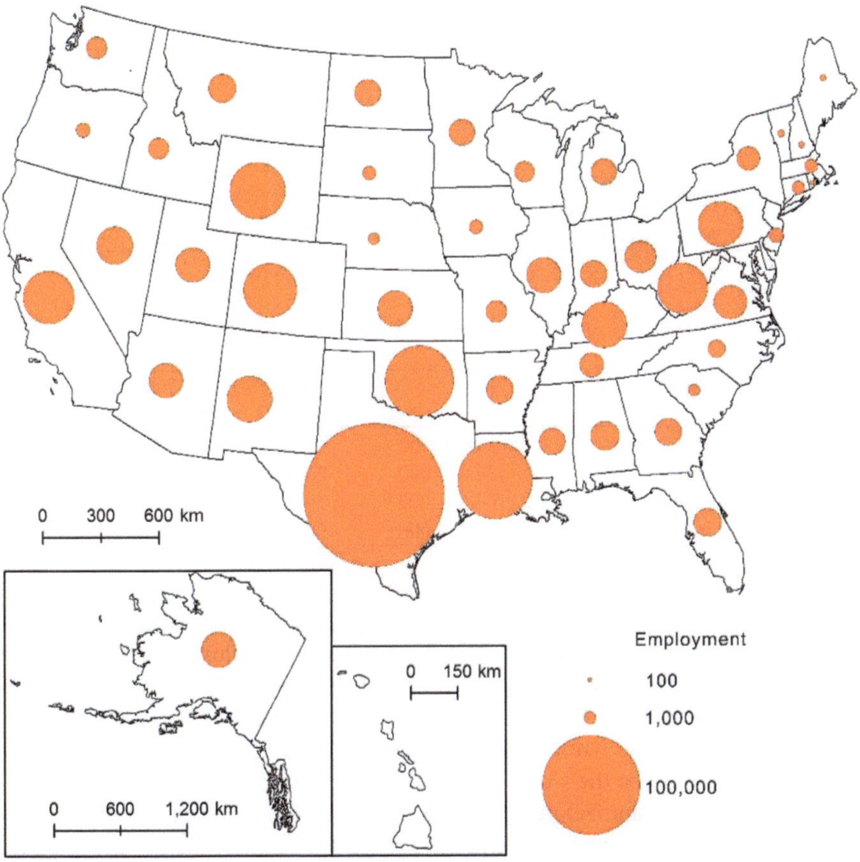

Fig. 5.3 Employment in mining, quarrying, and oil and gas extraction in the U.S. by State, 2007

the profits. Although the employment in manufacturing declined in the U.S, the revenues from manufacturing increased from $ 3,915 trillion in 2002 to $ 5,339 trillion in 2007 (U.S. Census Bureau 2007a), an annual increase of 6.4 %. The labor in the U.S. has been shifting from labor-intensive manufacturing to high-tech manufacturing and high-order services to capture higher value-added and profits.

Employment in construction in the U.S. experienced minor growth (Table 5.3) between 2002 and 2007, due to the housing market boom before the financial meltdown in 2008. As it is much more difficult to outsource construction work to overseas than manufacturing, the U.S. has used many foreign workers in construction, many of them are illegal immigrants from Mexico and other countries in Latin America. When the housing market crashed in 2008, the number of construction workers in the U.S. dropped from 7.4 million in 2007 to 5.4 million in 2010 (U.S. Census Bureau 2010), an annual decline of 11.1 %!

Table 5.3 Employment and Growth in the Industrial Sector in the United States. (Calculated from U.S. Census Bureau (2007a))

NACIS	Industries	2007	Percent (%)	2002–07 Annual growth (%)
22	Utilities	632,432	3.0	−1.0
23	Construction	7,399,047	34.6	0.6
31–33	Manufacturing	13,333,390	62.1	−1.9
22–33	Total in industrial sector	94,116,804	100	−1.1

Table 5.4 Employment and growth in the industrial sector in the China. (Calculated from CSSB (2005, 2009b))

Code	Industries	2008	Percent (%)	2004–2008 Annual growth (%)
D	Utilities	4,046,000	2.7	2.6
E	Construction	39,077,000	26.5	8.8
C	Manufacturing	104,331,000	70.8	5.6
C–E	Total in industrial sector	147,454,000	100.0	6.3

China is booming in the industrial sector, especially in construction and manufacturing (Table 5.4). Manufacturing is the predominant industry in the industrial sector, accounting for 70.8 % of the employment in China's industrial sector, higher than the 62.1 % in the U.S. (Table 5.3). Manufacturing is mainly responsible for China being called the world's factory floor (Hennock 2002) and for China surpassing Germany to be the world's top exporting country in 2009 (Mufson 2010). The proportion of manufactured goods in the total exported goods of the country increased dramatically from 49.4 % in 1985 to 94.7 % in 2011. Meanwhile, the proportion of manufactured goods in the total imports declined from 87.5 to 65.3 % in the same period (CSSB 2012).

Compared to that in the U.S., manufacturing in China is primarily labor-intensive, although technology and capital-intensive products have been steadily increasing. The proportion of labor-intensive products to the export commodities in China was 62.2 % in 1991, but declined to 39.8 % in 2010 (Fig. 5.4). Technology and capital intensive products grew quickly; it percentage in manufacturing export increased from 15.3 % in 1991 to 55.0 % in 2010. The export of high-tech products reaches 31.2 % of all exports in 2010, nearly 3 percentage points higher than that in 2005 (Fig. 5.5). While China has clearly been integrated into the global production network, technology and capital-intensive products (such as chemicals and related products) are still less competitive than labor-intensive ones (such as light textile industrial products) in terms of trade competition index (Table 5.5). Evidence shows that China focuses on some low-skill-labor-based manufacturing and is still at the low-end of the value-added chain (Xing and Detert 2011).

In terms of gross industrial output value in 2010, the top ten sectors in China's manufacturing are transportation equipments manufacturing, communication equipments, computers, and other electronic equipments manufacturing, ferrous metals

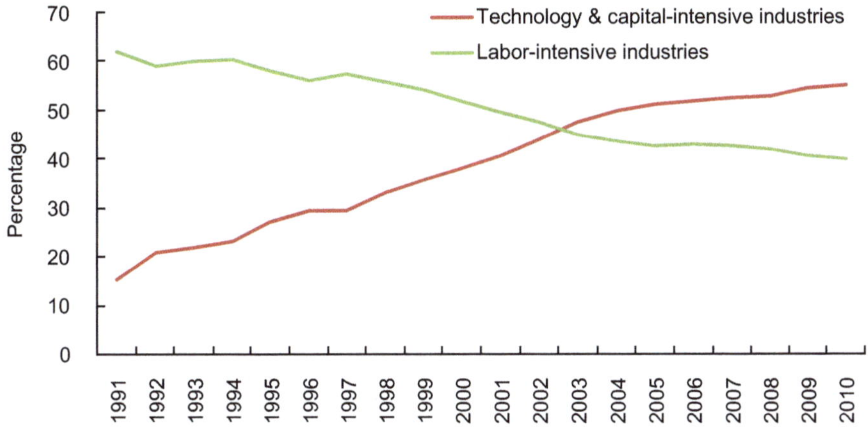

Fig. 5.4 Export structure of China's manufactured goods, 1991–2010. (CSSB 2011)

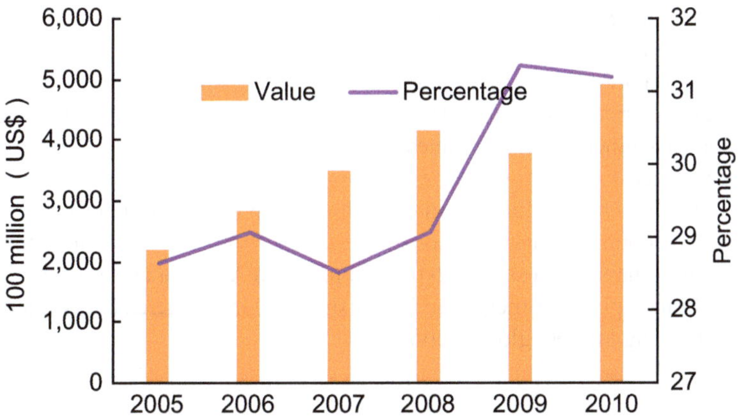

Fig. 5.5 Percentage of high-tech products in China's total export, 2006–2011 (CSSB 2006–2011)

Table 5.5 Trade Competitive (TC) Index of China's Products. (CSSB 2007–2011)

Year	Chemicals and related products	Machinery and transport equipment	Light textile industrial products, rubber products, minerals and metallurgical products
2006	−0.32	0.12	0.34
2007	−0.28	0.17	0.36
2008	−0.20	0.21	0.42
2009	−0.29	0.18	0.26
2010	−0.26	0.17	0.31

TC (Xi − Mi)/(Xi + Mi), where Xi represents the export trade and Mi represents the import trade of the product i (Grubel and Lloyd 1975). TC ranges between −1 and 1. The higher the value of the TC index, the more competitive the product

smelting and pressing, raw chemical materials and chemical products manufacturing, electrical machinery and equipment manufacturing, general purpose machinery manufacturing, agricultural products and byproducts processing, non-metallic mineral product manufacturing, petroleum processing, coking, and nuclear fuel processing, and textile industry. These sectors combined accounts for 67 % of the manufacturing, indicating that the heavy industry overweighs the light industry. The heavy industry occupies a dominant position in the industry sector in China, with a high percentage (70 %) in terms of value added (CSSB 2011).

Inwards foreign direct investment makes great contribution to the speedy growth of Chinese manufacturing. The amount of overseas investment actually utilized on manufacturing increased from $ 25.84 billion in 2000 (CSSB 2001) to $ 52.10 billion in 2011 (CSSB 2012). However, its proportion in the total inwards foreign investment decreased from 63.4 to 44.9 % in the same period. During this time, manufacturing did not grow as fast as services. Moreover, multinational corporations shifted investment from China to Southeast Asia since the mid-2000s as the labor cost in China started to increase. Furthermore, reindustrialization in some advanced economies caused some foreign manufacturers to close their factories in China and other developing countries and to focus on economic activities at home. General Electric and Caterpillar are two cases in point (Zeng 2012).

A growing number of Chinese manufacturing companies have been speeded up their pace of internationalization through going abroad since the late 1990s. A large proportion of Chinese overseas investment in manufacturing concentrates in Africa and Southeast Asia. However, a few large companies, such as Haier in home appliance manufacturing, Lenovo in electronic computer manufacturing, ZTE and Huawei in telecommunication equipment manufacturing, and Geely in automobile manufacturing, have successfully expanded to North America and Europe through setting up branch plants or merger and acquisition. China is increasingly integrating its manufacturing into the global economy.

5.3.2 Geography of Manufacturing

In general, manufacturing in China concentrates in the area along the Yangtze River and the coast, especially in the Pearl River Delta, the Yangtze River Delta, and the Bohai Rim. In terms of employment, five provinces in the three areas mentioned above, Guangdong, Jiangsu, Zhejiang, Shandong, and Liaoning, account for more than half of the country. In addition, Hubei, Sichuan, Henan, and Jilin are also important manufacturing provinces in China (Fig. 5.6). In contrast, Tibet, Xinjiang, Gansu, Inner Mongolia, Ningxia, Qinghai, Yunnan, and Guizhou in western China account for less than one percent of China's manufacturing employment.

To a great extent, the spatial distribution in manufacturing is a result of its physical geography, historical development, and government policies in China. Historically, manufacturing was concentrated in the Northeast, a by-product of Japanese occupation in this area to extract the rich natural resources, and in a few major cities along the coast. When the People's Republic of China was established in 1949, the

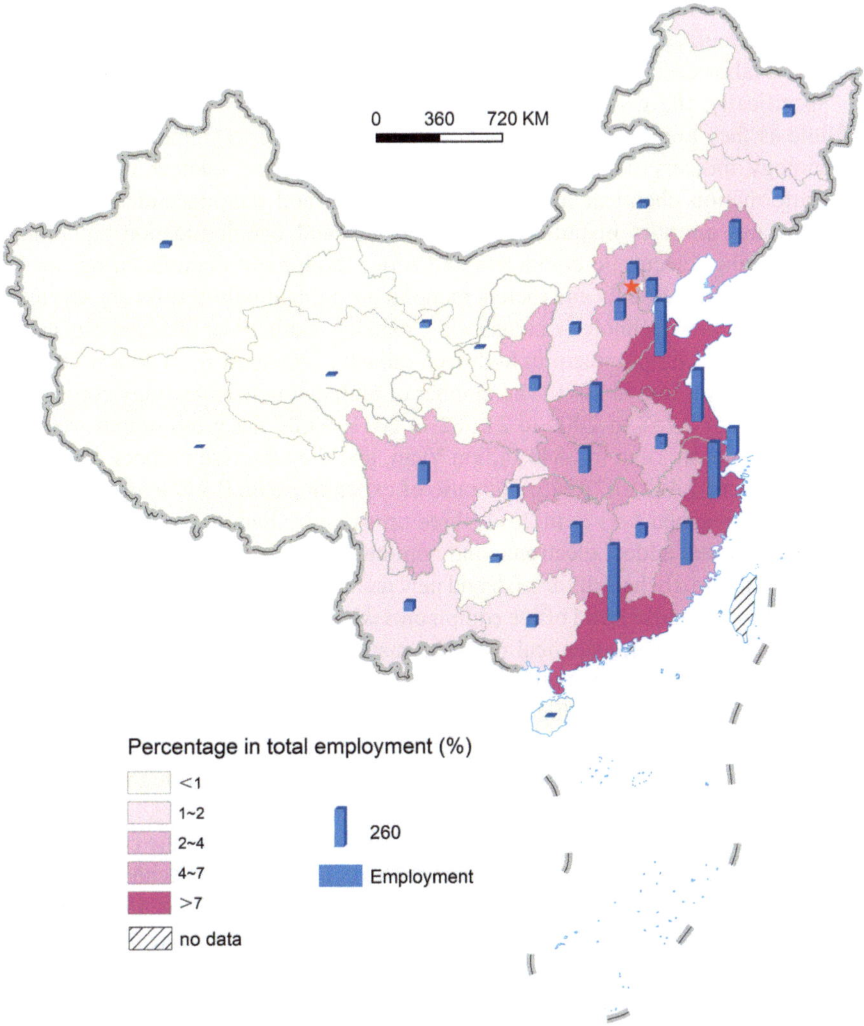

Fig. 5.6 Distribution of manufacturing employment in China by province. (CSSB 2011)

Chinese government adopted a regional balanced development strategy to change the excessive concentration of manufacturing on the East. The gap between the inland and the coastal areas shrank significantly. Since the open-door policy and economic reforms in the late 1970s, the strategic status of the eastern coastal belt and the Yangtze River has been highlighted by the Chinese government to increase the productivity in manufacturing. The two areas constituted a "T"-style spatial structure of the national economy, and the agglomeration economies of manufacturing were further intensified. Since 2000, China has successively implemented strategies to encourage economic development in Western and Central China and to revitalize old Northeastern industrial bases, to reach a more balanced regional development. Under such circumstances, the inland areas have more opportunities to attract manufacturing companies at home and from abroad to take advantage of the lower labor costs and improved infrastructure.

Specifically, the geography of high and new technology industries is worthy of note as they are the future of manufacturing in China. In China, high and new technology industry consists of eight sub-sectors, namely nuclear fuel processing, information chemical manufacturing, medical and pharmaceutical products, aviation and aircrafts manufacturing, electronic and communication equipment manufacturing, electronic computers and office equipment manufacturing, medical treatment instruments and meters manufacturing, and public software services. The high and new technology industry in China has made great progress especially since the 1990s. Employment in high-tech industry increased from 4.48 million in 1995 to 114.7 million in 2011. The number of high and new technology companies increased from 9,758 in 2000 to 21,682 in 2011, while the gross output went up from 1041 billion Yuan to 8843 billion Yuan, and the effective number of patents increased from 1,443 to 77,725. The ratio of expenditure on R&D to prime operating revenue reached 1.41 % in 2011, three times more than that in 1995 (0.46 %). In terms of value added, electronic and communication equipment manufacturing accounted for nearly half of the high and new technology industry in 2011, followed by electronic computers and office equipments whose percentage was close to 24 %. The medical and pharmaceutical products manufacturing ranked the third with a proportion of 16.8 % (CSSB 2012). In 2008, China ranked first in the world for its high technology manufacturing exports of $ 381 billion, accounting for 19.11 % of the world, followed by the U.S. ($ 231 billion). Meanwhile, China ranked fourth in the world for its high proportion of manufacturing exports in high technology products (28.66 %) in the same year (World Bank 2010).

High and new technology companies in China agglomerate in the coastal areas. Figure 5.7 shows the location of 105 national-level high and new technology industrial development zones (NHNIDZs) in 2012. Of the 105 NHNIDZs, 45 of them are in the Eastern area, 13 in the Northeast area, 23 in the Central area, and 24 in the Western area. Seven of the eight NHNIDZs with more than 10 billion Yuan of gross output in 2010 lie in the Eastern China while only one is in the western province of Shaanxi.

Similar to China, manufacturing in the U.S. was traditionally concentrated in the Northeast, the so-called manufacturing belt. However, the geography of manufacturing in the U.S. changed significantly during deindustrialization in the late twentieth century. While many manufacturing jobs moved out of the Northeast of the U.S., domestically, they moved from the manufacturing belt (also called Snowbelt) to the Sunbelt areas, i.e. the South and the West.

In Fig. 5.8, the size of the pies represents the magnitude of employment in the industrial sector while the blue, green, and red colors in each pie represent the percentages of total industrial employment in manufacturing, construction, and utilities in each state. In the Sunbelt, California in the West and Texas in the South have the highest employment in the industrial sector (Fig. 5.8) and in manufacturing. In terms of percentage, Midwest states such as Wisconsin and Indiana have the higher proportions of manufacturing employment in the industrial sector.

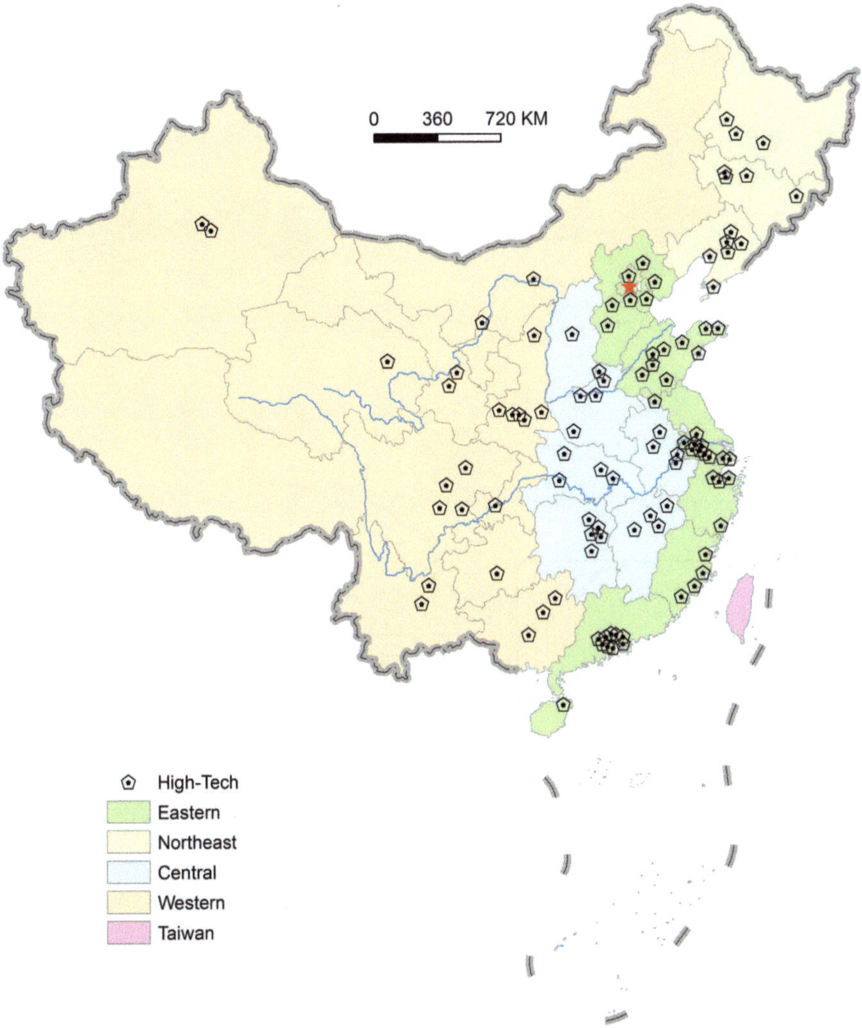

Fig. 5.7 Distribution of national-level high- and new-tech industrial development parks in China

5.4 Services Sector

5.4.1 Structure and Growth of Services

The services sector was called the tertiary sector in the past, but is subdivided into tertiary, quaternary, and quinary services when the services sector expanded to become the predominant economic activities in developed countries such as the U.S. (Table 5.6). Tertiary services in the U.S. today include low-order services such as

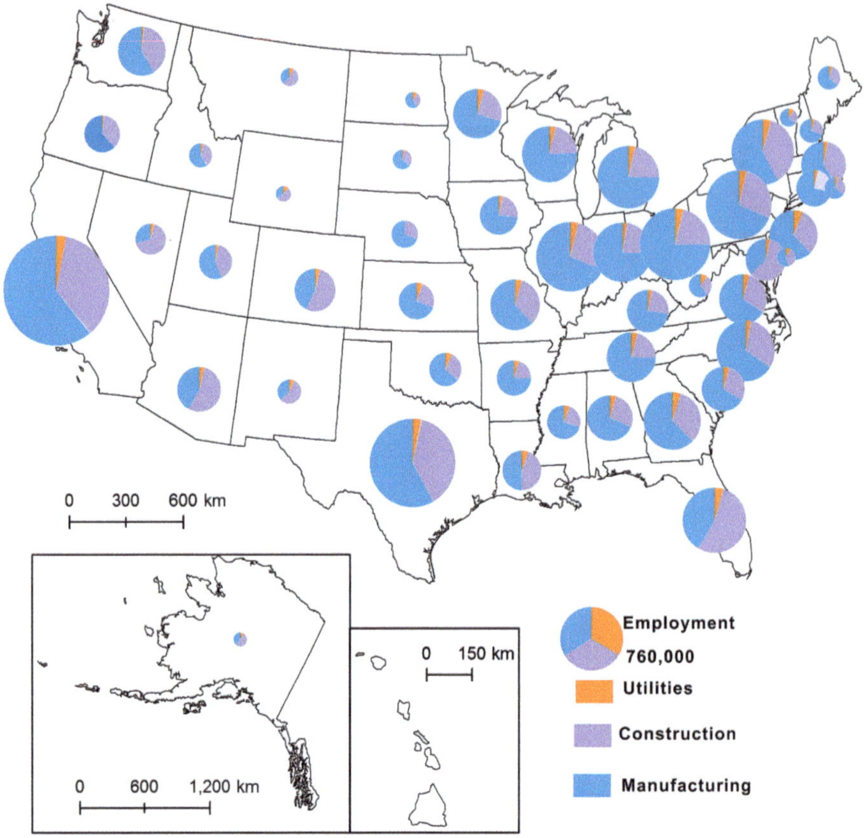

Fig. 5.8 Industrial employment and structure in the U.S. by state, 2007

retail, wholesale, transportation, and warehousing. Quaternary services, also known as producer services, primarily cater to manufacturing producers or other services and include intermediate services such as business and professional services, finance and insurance, and real estate services. Quinary services typically include health, education, arts, entertainment, and recreation services. Quaternary and quinary services are also called high-order services or knowledge-based services, the demand for which tends to increase as an economy develops and per capita income rises. The services sector in China is much less developed and is often reported together in government statistics as the tertiary sector. For comparison purpose, it is subdivided and reported in Table 5.7 in this chapter by using data from the first two national economic censuses in China.

Of the services sector, a lower percentage of the labor force is employed in low-order tertiary services in the U.S. (23.3 %) than in China (27.8 %) and a higher percentage in quaternary or producer services in the U.S. (29.3 %) than in China (24.6 %). These are indications that the services sector is more advanced

Table 5.6 Services employment and growth in the United States. (Calculated from U.S. Census Bureau (2002, 2007a, b) and de Blij and Muller (2006))

NACIS	Industries	2007	Of services Total (%)	Of U.S. total (%)	2002–2007 annual growth (%)
42–49, 72	Tertiary services	26,341,579	23.30	19.3	2.00
51–56	Quaternary services	33,179,368	29.30	24.3	1.50
521–523	Finance	4,184,028	3.70	3.1	0.14
5613	Employment services	5,189,787	4.58	3.8	4.49
61–81[a]	Quinary services	53,695,857	47.40	39.4	2.00
62	Health care and social assistance	16,859,513	14.89	12.4	2.29
72	Accom-modation and food services	11,587,814	10.24	8.5	1.25
42–81[a]	Services[a] total	113,214,804	100.00	83.0	1.3

[a] For compatibility with China, number of full-time equivalent government employees in the U.S. is added to the Economic Census data for calculation

Table 5.7 Services employment and growth in China. (Calculated from CSSB (2005, 2009b))

Code	Industries	2008	Of services total (%)	Of China total (%)	2004–2008 annual growth (%)
F, H, O	Tertiary services	31,680,000	27.8	9.2	8.1
G, J–N	Quaternary services	27,996,000	24.6	8.1	9.2
J	Finance	4,870,000	4.3	1.4	6.8
P–S, I	Quinary services	54,124,000	47.6	15.7	4.3
Q	Health, social security and social welfare	6,804,000	6.0	2.0	5.5
I	Accom-modation and food services	5,858,000	5.2	1.7	8.1
F–S	Services total	113,800,000	100.0	33.0	6.5

in the U.S. than in China. However, the average annual growth rates for tertiary, quaternary/producer services, and quinary services are much higher in China than in the U.S., indicating the momentum and potential of growth in China's services sector. In fact, the average annual growth rate for the services sector (6.5 %, Table 5.7) in China is higher than that for the industrial sector

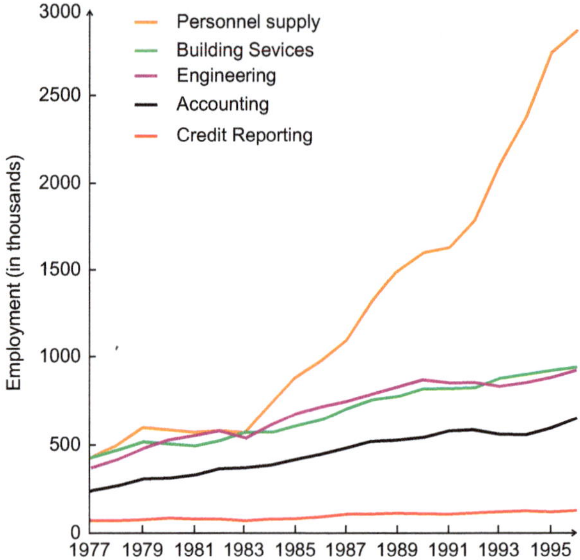

Fig. 5.9 Growth of employment services in the U.S., 1977–1996. (Gong 2001)

(6.3 %, Table 5.4) and for mining, quarrying, and oil and gas extraction (2.8 %, Table 5.2). More inward foreign investment in China was invested in the services sector ($ 58 billion, 50 % of the total) than the industrial sector ($ 52 billion, 45 %) in 2011 (CSSB 2012).

As the statistics for the subsectors in services are not always directly comparable in the U.S. and China, the discussions below and in the next section may focus on only a few subsectors. Employment services that provide placement services for employers or those seeking employment, for example, is worthy of special attention. Not listed separately in China's statistics, employment services experienced tremendous growth in the late twentieth century and in the 2000s in the U.S. Figure 5.9 shows the employment increase in five producer services in the U.S. in the late twentieth century. Employment services, called personnel supply services at that time under the old Standard Industrial Classification (SIC), grew much faster than other fast-growing producer services. Between 2002 and 2007, employment services still increased at an average annual growth of 4.5 %, more than three times of that (1.3 %) for the services sector as a whole.

Within the quinary services, health services are becoming increasingly important in the U.S. economy as income rises. Health services accounts for 14.9 % of the services employment in the U.S. (Table 5.6), much higher than the 6.0 % in China (Table 5.7). Of the total employment in the country, this gap is even bigger for the two countries. The 12.4 % in the U.S. is more than six times of the 2.0 % in China. Although China is currently in a stage of developing producer services to serve its industrialization, as income in China continues to rise in the next few decades, health services will no doubt play a much bigger role in China's economy.

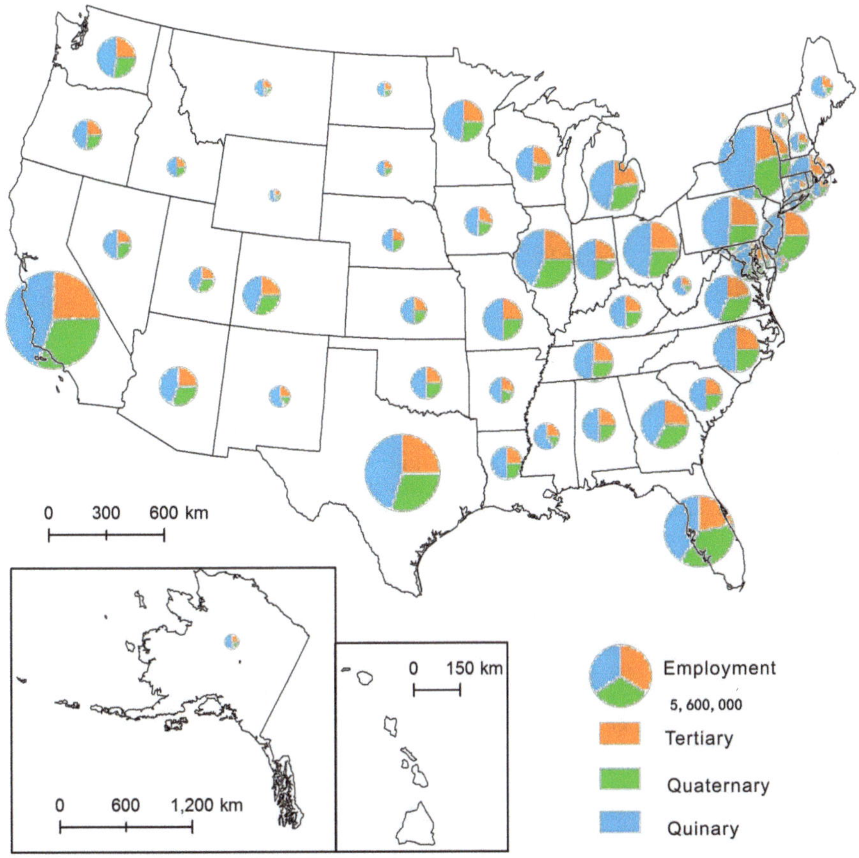

Fig. 5.10 Employment and structure of services in the U.S. by State, 2007

5.4.2 Geography of Services

States in the U.S. have a good balance of tertiary, quaternary, and quinary services in the services sector (Fig. 5.10). The top four states in services employment are California, Texas, New York, and Florida. New York metropolitan area has gone through the economic structuring from manufacturing to services and ranks top in many advanced services, including business and professional services (Fig. 5.11) and financial services (Gong and Keenan 2012). It is the home of the New York Stock Exchange and one of the top three international financial centers in the world. Together with London and Tokyo, New York City is well known as a global city. It has always been the most visited city in the U.S. (Tourism to be 2008), followed by Los Angeles in California.

In China, coastal areas play an important role in the development of the services sector. Guangdong, where services employment reached 22.75 million in 2010, was

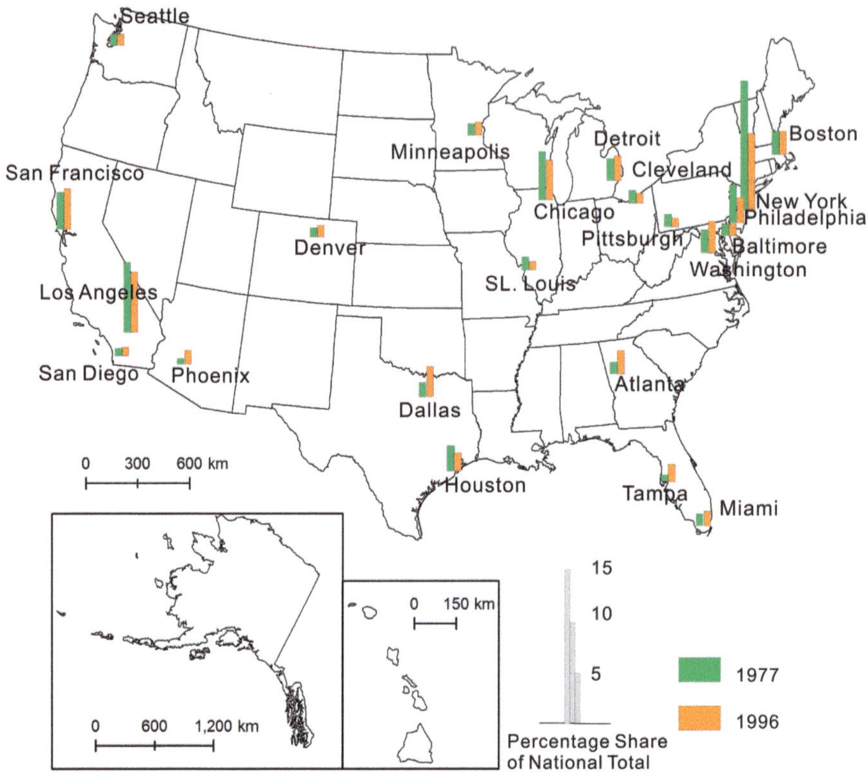

Fig. 5.11 Share of large metropolitan areas in the U.S. employment in business and professional services. (Gong 2001)

in the first place. Jiangsu, Shandong, and Zhejiang each had a services employment more than 10 million. In the Northeast, the services employment of 9.52 million in Liaoning province is the highest. Henan, Hunan, Hubei, and Anhui were relatively higher in the central region, while Sichuan was the most outstanding in the western region (Fig. 5.12).

While finance accounts for 3.7% of the services employment in the U.S. (Table 5.6), its percentage in China is higher, at 4.3% (Table 5.7). Since the late 1970s, China's financial service has grown significantly. In terms of the output value in 2010, Beijing, Shanghai, Guangdong, Zhejiang, and Jiangsu were the top five financial centers at the provincial level. The only two stock exchanges in Mainland China are located in Shanghai and Shenzhen (in Guangdong). With the development of Pudong New District in Shanghai since the early 1990s, Lujiazui Finance and Trade Zone has been developing rapidly. Building Shanghai into an international financial center by 2020 has been a development strategy of China. Shenzhen has also strengthened its position in the financial services of China based on the cooperation with its neighboring city—Hong Kong. As the state capital, Beijing is attractive to the headquarters of financial institutes at home and abroad.

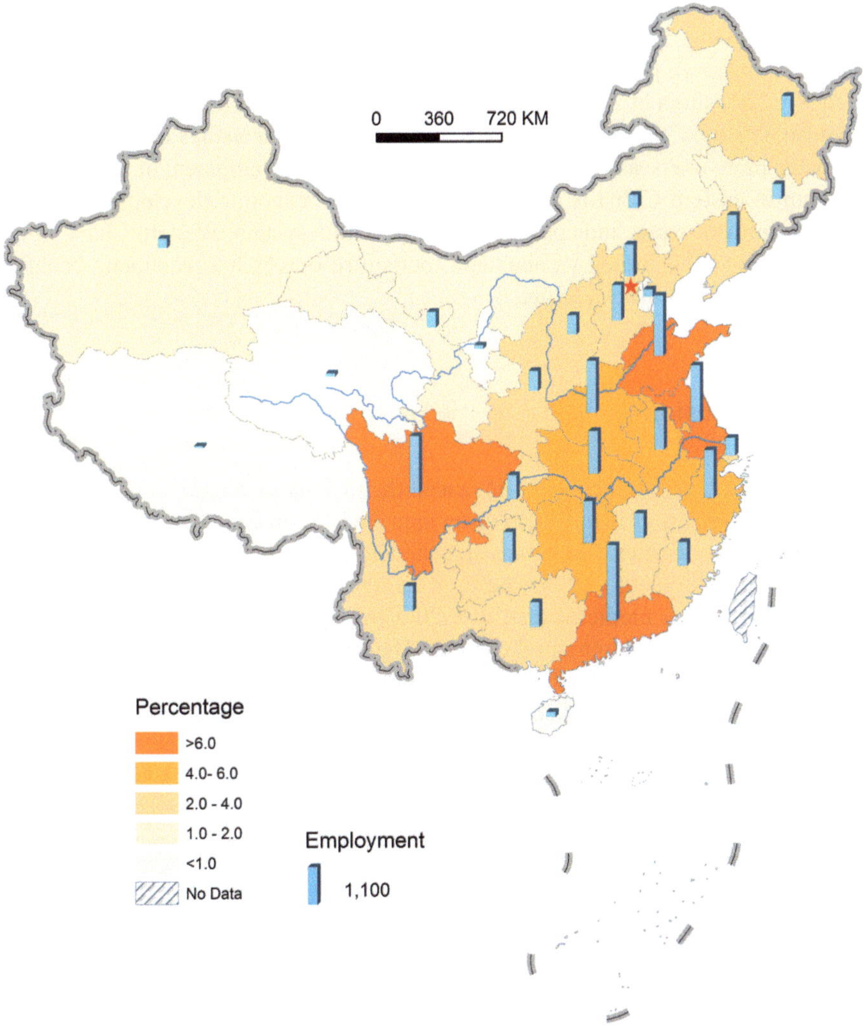

Fig. 5.12 Employment in the services sector and as a percentage of total employment by province in China, 2010. (CSSB 2011)

China's tourism industry has also been developing rapidly since the late 1970s. By the end of 2009, there were 20,399 travel agencies and 14,237 star-rated hotels in China (CSSB 2011). Related to tourism, accommodation and food services in China grew at 8.1 % annually between 2004 and 2008 (Table 5.7). Domestic and international tourism reached 2.1 billion person-times in 2010, with about 1.3 trillion Yuan of domestic tourism income and 45.8 billion U.S. dollars of foreign currency revenue from international tourism. As more and more visitors coming from overseas, China has advanced its rank in the world's tourism from the seventh in 1995 to the third in 2010 and in foreign currency revenue from the 14th to the 4th (Editorial Board of the Yearbook of World Economy 2012).

The eastern areas of China are the main tourism destinations for overseas visitors. Guangdong, Shanghai, Zhejiang, Jiangsu, Beijing, Fujian, and Shandong are the most attractive places. Guangdong, in particular, ranked the top with 31.4 million foreign visitors in 2010, followed by Shanghai (7.3 million). In western China, Yunnan, Guangxi, and Shaanxi have built up their famous names in the world and attracted more overseas visitors in recent years. Rapid development of international tourism in eastern China is attributed to the rapid economic development, better tourism infrastructure, and convenient transportation system. Western China, however, has not fully turned its abundant tourism resources into economic benefits because of its poor infrastructure.

5.5 Conclusions

Although the U.S. is much more advanced than China in economic development, China's economy is growing much faster than the U.S., in fact, the rest of the world, in the past few decades. China was still the sixth largest economy in the world in 2004 (China Daily 2005), but has become the second largest economy in 2010 (The Richest 2012). It is predicted that by 2020, China will have overtaken the U.S. to become the largest economy in the world (The Richest 2012). On such a fast track of economic development, China will certainly benefit from the experience of the U.S. in an advanced stage of economic development and from a comparison study of these two largest economies in the world.

References

China Daily. (2005). China's 2004 GPD ranks 6th in world economy. http://www.chinadaily.com.cn/english/doc/2005-12/20/content_504977.htm. Accessed 20 Dec 2011.

China Daily. (2012). China's oil production slightly down in May. http://www.chinadaily.com.cn/china/2012-06/24/content_15519525.htm. Accessed 20 Dec 2012.

China State Statistical Bureau (CSSB). (2001). *China statistical yearbook, 2001*. Beijing: China State Statistical Press.

China State Statistical Bureau (CSSB). (2005). First National Economic Census Major Data Bulletin, No. 1. http://www.stats.gov.cn/zgjjpc/cgfb/t20051206_402294807.htm. Accessed 31 Oct 2011.

China State Statistical Bureau (CSSB). (2006). *China statistical yearbook, 2006*. Beijing: China State Statistical Press.

China State Statistical Bureau (CSSB). (2007). *China statistical yearbook, 2007*. Beijing: China State Statistical Press.

China State Statistical Bureau (CSSB). (2009a). *China statistical yearbook, 2009*. Beijing: China State Statistical Press.

China State Statistical Bureau (CSSB). (2009b). Second national economic census major data bulletin, no. 1. http://www.stats.gov.cn/tjfx/fxbg/t20091225_402610155.htm. Accessed 31 Oct. 2011.

China State Statistical Bureau (CSSB). (2011). *China statistical yearbook, 2011*. Beijing: China State Statistical Press.

China State Statistical Bureau (CSSB). (2012). *China statistical yearbook, 2012*. Beijing: China State Statistical Press.
Clark, C. (1940). *The conditions of economic progress*. London: Macmillan.
de Blij, H. J., & Muller, P. O. (2006). *Geography: Realms, regions, and concepts* (12th ed.). Hoboken: Wiley.
Editorial Board of the Yearbook of World Economy. (2012). *The yearbook of world economy, 2011–2012*. Beijing: China Social Sciences Press.
Gong, H. (2001). A hierarchical change model of business and professional services in the United States. *Urban Geography*, *22*(4), 340–359.
Gong, H. (2002). Growth of tertiary sector in China's large cities. *Asian Geographer*, *21*(1–2), 1–16.
Gong, H, & Keenan, K. (2012). The impact of 9/11 on the geography of financial services in New York: A few years later. *The Professional Geographer*, *64*(3), 370–388.
Grubel, G. H, & Lloyd, J. P. (1975). *Intra-industry trade: The theory and measurement of international trade in differentiated products*. New York: Wiley.
Hennock, M. (2002). China: The world's factory floor. http://news.bbc.co.uk/2/hi/business/2415241.stm. Accessed 10 Dec 2012.
Kossik, J. (2011). Clark's sector model for US Economy 1850–2009. http://www.63alfred.com/whomakesit/clarksmodel.htm. Accessed 20 Jan 2012.
Mitra-Thakur, S. (2011). China beats the US to become world's largest manufacturer. http://eandt.theiet.org/news/2011/mar/china-manufacturing.cfm. Accessed 20 Dec 2011.
Mufson, S. (2010). China surpasses Germany as world's top exporter. http://www.washingtonpost.com/wpdyn/content/article/2010/01/10/AR2010011002647.html. Accessed 20 Dec 2012.
Stevenson, J. (2010). China seeks global sources of raw materials. http://www.voanews.com/content/china-seeks-global-sources-of-raw-materials-108556309/166646.html. Accessed 20 Dec 2012.
The Richest. (2012). The world's largest economies 2012. http://www.therichest.org/world/worlds-largest-economies/. Accessed 1 Aug 2012.
Tourism to be. (2008). Top 10—The most visited US cities in 2007. http://tourismtobe.wordpress.com/2008/03/14/top-10-the-most-visited-us-cities-in-2007/. Accessed 20 Dec 2012.
U.S. Census Bureau. (2002). 2002 Census of governments. http://www.census.gov/govs/cog/historical_data_2002.html. Accessed 20 Dec 2011.
U.S. Census Bureau. (2007a). 2007 economic census and surveys. http://factfinder.census.gov/servlet/DatasetMainPageServlet?_program=ECN&_submenuId=datasets_4. Accessed 20 Dec 2011.
U.S. Census Bureau. (2007b). 2007 Census of governments. http://www2.census.gov/govs/apes/emp_compendium.xls. Accessed 20 Dec 2011.
U.S. Census Bureau. (2010). 2010 County Business Patterns. http://censtats.census.gov/cgi-bin/cbpnaic/cbpdetl.pl. Accessed 20 Dec 2011.
U.S. Energy Information Administration. (2011). Annual energy review 2011. http://www.eia.gov/totalenergy/data/annual/pecss_diagram.cfm. Accessed 20 Dec 2012.
Wang, F., & Feng, G. (2011). Determinants of low-carbon Development of China's economy and their functions to carbon emission reduction. *Economic Issues in China*, *3*, 62–69.
Wei, L., & Pan, A. (2012). Research on sustainable development of rare earth resource in China. *Soft Science*, *26*(11), 27–32.
World Bank. (2010). *World development indicators 2010*. Beijing: China Financial & Economic Publishing House.
Xing, Y., & Detert, N. (2011). How the iPhone widens the United State trade deficit with the People's Republic of China (in Chinese). *Finance Research*, *2011*(3), 198–206.
Ying, D. (20 November 2012). Fu helps Sinopec navigate overseas markets. *China Daily*, 13.
Zeng, X. (2012). Made in China: Loss of manufacturing, collect the market. http://www.infzm.com/content/81435. Accessed 20 Dec 2012.
Zhang, J. (2012). The study says China's carbon emissions was the first in the world, per capita was less than the United States. http://news.qq.com/a/20121203/001420.htm. Accessed 21 Feb 2013.

Chapter 6
International Trade Issues and Status for China and the U.S.

Gregory Veeck and Yuejing Ge

6.1 Trade Makes the World Go 'Round'

Taken collectively, the importance of domestic economic activities in China and the United States far exceeds the monetary benefits accrued from international trade. This is common ground for our nations, but uncommon for the nations of the world at large. Overall, the United Nations estimates that global merchandise trade will grow at an average of 7.6 % over the years from 2011–2013 (United Nations 2012). This is only an estimate and tracking international trade volumes in the new millennium is similar to riding a "roller coaster"! The growth of world exports in percentage terms for 2010 was the highest on record going back to 1950 while 2009, just a year earlier, experienced one of the greatest drops in global trade levels in history. In 2010, world merchandise exports were up 22 % over 2009, rising from US$ 12.5 to 15.2 trillion in a single year, while world exports of commercial services rose 8 %, from US$ 3.4 to 3.7 trillion (WTO 2012). All nations are impacted by these fluctuations as the current fiscal crisis within the EU clearly shows, but for those nations that have disproportionately large manufacturing sectors dependent on international exports such as China, global recession can be especially daunting. Further, as supply chains grow in complexity, those nations that supply raw materials and/or product components to China for processing and assembly are also negatively impacted by fluctuations in the global trade in manufactured goods (Dunaway 2009).

Fortunately, for two such land extensive and populous nations as China and the U.S., the importance of the domestic market for manufactured goods and services will probably always be paramount with respect to GDP and economic stability. However, the positive and negative effects of international trade—the topic for this chapter—for both nations is indisputable, and cannot be overlooked in any assess-

G. Veeck (✉)
Department of Geography, Western Michigan University, Kalamazoo, USA
e-mail: gregory.veeck@wmich.edu

Y. Ge
School of Geography, Beijing Normal University, Beijing, China
e-mail: geyj@bnu.edu.cn

R. Hartmann, J. Wang (eds.), *A Comparative Geography of China and the U.S.*,
GeoJournal Library 109, DOI 10.1007/978-94-017-8792-5_6,

ment of the economic geographies of these two great nations. International trade not only generates revenue and employment opportunities, it also aids all nations in myriad additional ways including elevated international status, technology transfers, improvements to human capital and domestic product development (Mikic 1998). The clear benefit of international trade comes as no surprise to economic geographers. What might surprise readers is how, increasingly, national statistics related to the proportional share of international trade, especially for merchandise, for our two nations appear to be converging. While wealthier nations such as the U.S., Japan, and the EU nations have service-sector domestic economies, services still account for a minority share of international trade. This chapter, then, largely concentrates on the "lion's share" of international trade; that is trade in manufactured goods (Miroudot et al. 2010). In 1993, China's share of total international trade (imports and exports) was estimated by the WTO to be approximately 5.5 % of the world's total, while the U.S. claimed 28.5 % for the same year. However, by 2009, the U.S. share had declined to 21.6 %, of total merchandise trade while China's share had increased to 18.0 %. Since then, the relative shares achieved by China from 2009 to 2011 of total imports and exports have remained around 10–11 % for both vis-à-vis global totals (United Nations 2012). Broken down by merchandise imports and exports, the pattern is all the more striking. China's arrival as a major trading nation is immediately apparent (Ministry of Commerce 2010). In 1993, China and the United States accounted for 2.5 and 12.6 % of world merchandise exports respectively. As of 2011, these figures had shifted dramatically with China accounting for 9.9 % of merchandise exports, while the U.S. share declined to 8.7 % (WTO 2012; United Nations 2012). The case is similar for imports. In 1993, the United States imported 15.9 % of all merchandise traded internationally, while China imported only 2.7 %. As with exports, these shares shifted significantly by 2010 when the U.S. share of global merchandise imports declined to 12.9 % while China's imports rose to 8.1 % of the world total (WTO 2012).

Indisputably, trade is an essential component in China's remarkable rise after the late 1970s (Yao and Luo 2010). The shifts in the above numbers reflect mercurial changes in the global trading system; changes all the more remarkable considering that total global merchandise trade (imports and exports) grew from $ 7.46 trillion US$ in 1993 to US$ 18.2 trillion in 2011, surpassing the previous peak of US$ 16.2 trillion from 2008 (Hsu 2012). In short, since the mid-1990s, the global trade "pie" has almost tripled in size, while the combined share of the "trade pie" for China and the U.S., just for merchandise now accounts for almost 40 % of total trade. Trade, then, is essential for both nations, and the many nations within supply chains of Chinese and American firms (Gao and Cai 2008). The high level of trade between the two nations is certainly one of the most important aspects of the complex relationship that has developed between the two states (Francois 2010). This is not to say that international trade and trade issues are viewed in similar fashion by the citizens of the two nations. In the following sections, we hope to illustrate some basic similarities and differences in these perceptions.

Table 6.1 China's foreign trade: 1990–2010 (in billion US$). (China State Statistics Bureau 2011; MOFCOM 2011)

Year	GDP	Foreign trade	Exports	Imports	Balance	Export/GDP
1990	390.28	125.24	67.79	57.45	10.35	17.37
1991	409.28	146.50	78.81	67.69	11.12	19.26
1992	409.17	183.83	94.04	89.79	4.26	22.98
1993	488.22	218.30	102.74	115.56	-12.82	21.04
1994	559.22	268.82	137.41	131.41	5.99	24.57
1995	727.98	323.86	167.18	156.68	10.50	22.96
1996	856.08	332.88	171.65	161.23	10.42	20.05
1997	952.68	377.36	207.29	170.07	37.22	21.76
1998	1,019.46	374.35	207.61	166.74	40.88	20.36
1999	1,083.28	417.83	221.13	196.70	24.43	20.41
2000	1,198.47	540.30	279.30	260.99	18.31	23.30
2001	1,324.82	581.55	299.00	282.55	16.45	22.57
2002	1,453.82	706.27	364.50	341.27	23.73	25.07
2003	1,640.97	952.29	484.63	467.66	16.97	29.53
2004	1,931.64	1,288.26	655.43	632.83	22.60	33.93
2005	2,236.62	1,579.00	835.85	743.15	92.70	37.37
2006	2,658.41	1,952.13	1,060.37	891.76	168.61	39.89
2007	3,383.82	2,425.59	1,340.34	1,085.25	255.09	39.61
2008	4,329.24	2,866.13	1,575.05	1,291.07	283.96	36.38
2009	4,990.52	4,822.91	2,207.53	1,201.61	1,005.92	37.55
2010	5,879.06	5,720.81	2,953.99	1,557.75	1,396.24	59.19

6.1.1 International Trade and China

6.1.1.1 The Growth of Chinese Trade 1990–2010

Over the past few decades, the expansion of international trade has served as one of the important "engines of growth" for China and the Chinese people. Many Chinese families, especially in new manufacturing regions such as those for high tech or pharmaceuticals have seen their real wages grow at amazing rates due to international trade in these products. As noted in many other chapters, GDP in China has grown at nearly 10 % per year since reform and opening-up began in December of 1978, and international trade is an important part of this miracle. Exports in 1980 were valued at 18.27 billion $ which was only 0.9 % of total world exports, ranking the nation 26th in the world. Since China's entry to the WTO in 2001, international trade has grown at an ever quickening pace. Foreign trade in 2010 reached 2.97 trillion $ (China State Statistics Bureau 2011). Due, in part, to the 2008–2009 global financial crisis, China has replaced Germany to become first with respect to the total value of commodity exports. For the past decade, the annual growth of commodity exports exceeded 15–20 %, more than double the growth rate of GDP (Table 6.1 and Figs. 6.1 and 6.2). Imports in 2008 were 1.29 trillion $ accounting for 6.7 % of world imports for that year good for third among all nations. In 2010, international

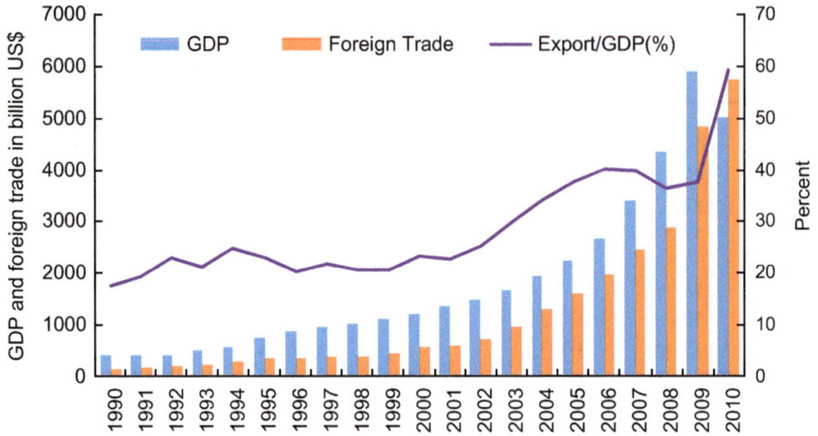

Fig. 6.1 GDP, Trade, and export as a share of GDP for China: 1990–2010

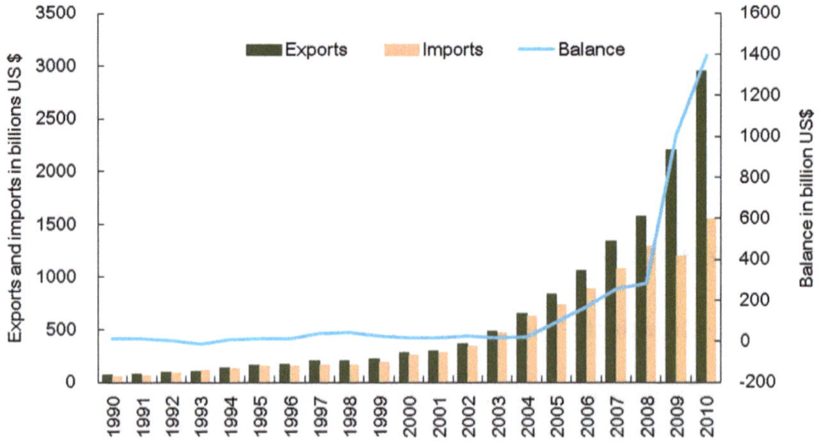

Fig. 6.2 Export, import, and trade balance for China: 1990–2010. (China State Statistics Bureau 2009; 2011)

trade accounted for approximately 1/3 of China's GDP. Most Chinese people, then, recognize the benefits of the explosion of international trade that has occurred since the December 1978 reforms were initiated to themselves and their nation.

6.1.1.2 The Geographical Distribution of Principal Trading Partners

At the continental scale, the principal destinations for Chinese goods are nations/districts within Asia, Europe, and North America. Trade with Asian nations/districts reached $ 1.57 trillion in 2010, accounting for 52.69 % of total trade volume while European trade was over $ 573 billion or 19.27 % of the total for the same year.

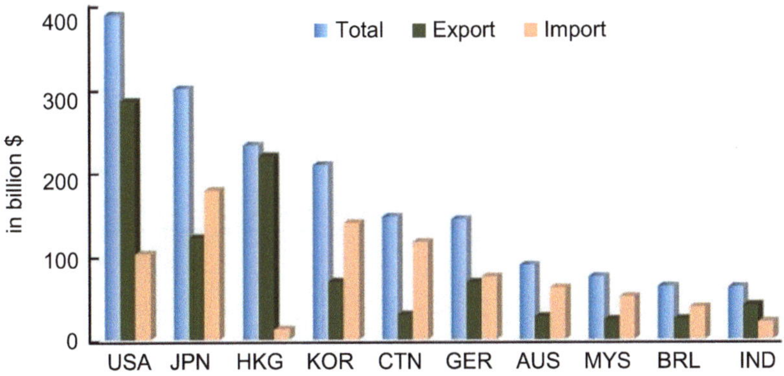

Fig. 6.3 Major partners and their foreign trade volume, 2008. (MOFCOM 2011)

North America accounted for $ 423 billion in trade (14.22 % of total). Other regions including Latin America, Africa, and Oceania combined for almost $ 410 billion in 2010 or 13.82 % of the China's total trade (imports/exports) volume (Table 6.2; Fig. 6.3).

Based on official statistics, the top five trading "blocks" or trading partners for mainland China in 2010 include the EU, the U.S., Japan, Hong Kong (China), and the ASEAN nations taken collectively. Of these nations or trade blocks, mainland China has a trade surplus with Hong Kong, the U.S. and the EU and a trade deficit with the ASEAN group and Japan. Looking only at exports, the main partners do not change: the U.S., the EU, Hong Kong (China), Japan, and ASEAN. With respect to imports, the top five import partners of mainland China include Japan, the EU, the Republic of Korea, ASEAN and Chinese Taiwan. Based on individual tariff regions, the top five trading partners for mainland China include the U.S., Japan, Hong Kong, the Republic of Korea, and Chinese Taiwan (Fig. 6.3). Mainland China has a trade surplus with Hong Kong and the U.S. but deficits with Japan, Korea and Chinese Taiwan. By nation, the top five export partners of mainland China are U.S., Hong Kong, Japan, Korea and Germany, and the top five import partners of mainland China are Japan, Korea, Chinese Taiwan, U.S., and Germany.

6.1.2 International Trade and the United States

6.1.2.1 The Benefits of Trade to the U.S. Economy

The official population of the United States in mid- 2012 was 314.3 million representing less than 5 % of the world population, yet the civilian workforce of approximately 155–165 million which is still increasing generated more than 20 % of the total global value of goods and services (Bureau of Economic Analysis 2010; Fig. 6.4). Of course, as noted in the introduction, much of this wealth in U.S. was

Table 6.2 Major trading partners of mainland China in 2010 (in billion US$, % share of total)

Country (region)	Total	Exports	Imports	Share of total (%)	Share of exports (%)	Share of imports (%)
Total	2,973.99832	1,577.75432	1,396.24401	100.00	100.00	100.00
Asia	1,566.91107	731.95484	834.95623	52.69	46.39	59.80
Africa	127.04602	59.95405	67.09196	4.27	3.80	4.81
Europe	573.05809	355.18797	217.87012	19.27	22.51	15.60
Latin America	183.63967	91.79803	91.84164	6.17	5.82	6.58
North America	422.91973	305.84271	117.07703	14.22	19.38	8.39
Oceanic and Pacific Islands	99.03469	33.01671	66.01798	3.33	2.09	4.73
Others	1.38905	0	1.38905	0.05	0.00	0.10
Major economics						
United States	385.38529	283.28655	102.09873	12.96	17.96	7.31
Japan	297.77959	121.04349	176.73610	10.01	7.67	12.66
Hong Kong, China	230.56247	218.30205	12.26042	7.75	13.84	0.88
Korea Rep.	207.11512	68.76626	138.34885	6.96	4.36	9.91
Taiwan, China	145.41314	29.67449	115.73865	4.89	1.88	8.29
Germany	142.30840	68.04718	74.26122	4.79	4.31	5.32
Australia	88.34232	27.22026	61.12205	2.97	1.73	4.38
Malaysia	74.24884	23.80204	50.44680	2.50	1.51	3.61
Brazil	62.58587	24.46050	38.12538	2.10	1.55	2.73
India	61.76120	40.91496	20.84625	2.08	2.59	1.49
Singapore	57.07598	32.34723	24.72875	1.92	2.05	1.77
Netherlands	56.18320	49.70423	6.47897	1.89	3.15	0.46
Russia	55.53311	29.61207	25.92104	1.87	1.88	1.86
Thailand	52.93702	19.74108	33.19594	1.78	1.25	2.38
United Kingdom	50.07223	38.76704	11.30519	1.68	2.46	0.81
Italy	45.14624	31.13944	14.00680	1.52	1.97	1.00
France	44.75684	27.65139	17.10546	1.50	1.75	1.23
Saudi Arabia	43.19549	10.36644	32.82905	1.45	0.66	2.35
Indonesia	42.75028	21.95357	20.79672	1.44	1.39	1.49
Canada	37.13988	22.21613	14.92375	1.25	1.41	1.07

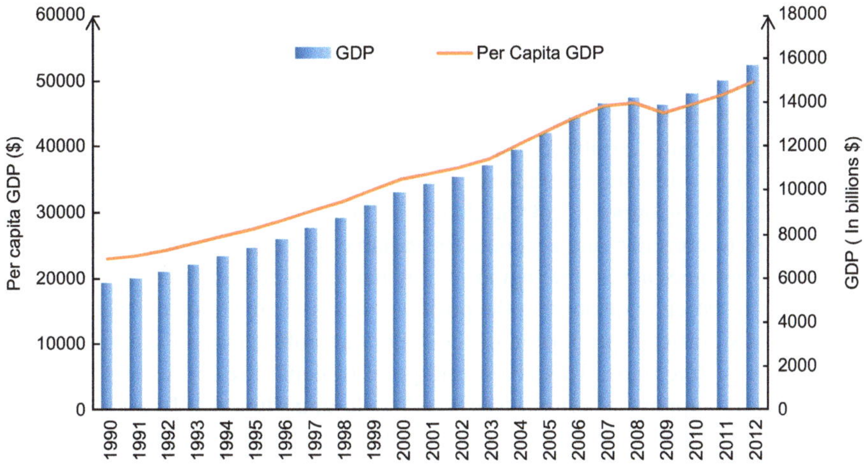

Fig. 6.4 GDP and per capita GDP for the United States from 1990 to 2010. (Source: http://www.usgovernmentrevenue.com/us_gdp_history#copypaste, http://data.un.org/Data.aspx?d=SNAAMA&f=grID%3A101%3BcurrID%3AUSD%3BpcFlag%3A1)

produced through domestic manufacturing and retail and most importantly through the provision of services and the licensing of intellectual property at home and abroad. Trade in raw materials, manufactured goods and agricultural products remains quite important, particularly in terms of employment in major industrial areas and for the maintenance of the farm sector.

Reading newspapers or listening to television commentators in the U.S., it is clear that many Americans do not always support policies and agreements that promote international trade. Particularly in the years since the most recent recession began in 2008, many issues related to international trade have been hotly debated throughout America. Indeed, trade-related issues such as the shift of manufacturing jobs "off-shore" were major points of debate in the fall 2012 presidential election. Domestic economic problems in the U.S. always seem to spur public debates on the benefits of "free trade" and it seems as if those espousing "protectionist" views at the present time have never been as vocal or as visible as in 2013.

Proponents of this "protectionist" view expect that tighter controls on trade will result in a national economy with lower levels of unemployment, more "steady" paychecks, and promising futures for industrial workers. Concerns ranging from the "outsourcing" of jobs to foreign nations, declining real wages, unprecedented high rates of unemployment, and declining tax revenues have fueled these protectionist views, especially in the old "Rust Belt" and Southern manufacturing regions that have lost the greatest number of jobs and fear losing more. This "geography" of opinion regarding "free trade" is an important component in understanding U.S. domestic politics.

Most professional economists in the U.S., however, agree that global trade and global trading institutions such as the GATT, WTO, and NAFTA have been essential for long-term American prosperity. Of course, globalization has certainly resulted in important labor shifts that have caused many firms to close their doors, but the benefits of increases in international trade outweigh the problems. Ultimately, however, for the United States, the globalization of trade is rooted in local processes and local places and represents a "geographic" phenomenon where locational advantages or disadvantages create variable conditions whereby some U.S. regions have gained from growth in international trade while other regions have, in real terms, lost a great deal. This "geography" of the domestic "winners and losers" in any nation is important to consider, and accounts for variable degrees of political support for "free trade" policies across the regions of any nation, including the U.S.

Despite research underscoring the positive effects of free trade on both income and GNP, a significant proportion of the American public remain skeptical about the benefits of free trade. This is a useful point of departure for readers in China who typically support all efforts to expand trade opportunities. In a recent (November 2009) PEW survey, a large portion of "the general public report negative opinions about the specific impacts of free trade agreements on jobs, economic growth and wages" (PEW Research 2009). 53 % of respondents to the PEW survey—a leading U.S. NGO (non-governmental organization) "think-tank"–believe free trade agreements lead to job losses. 53 % felt that international trade agreements lead to lower wages. Finally, 42 % of those surveyed associated free trade agreements with slower economic growth. When people were simply asked if they supported or did not support free trade, a significant portion of the population still remain opposed to "free trade" (32 % in 2009, but down from 43 % of respondents in 2008, perhaps due to the fall 2008 collapse as the survey is conducted each year in November). Somewhat counter-intuitively, of the persons surveyed by the annual PEW Research Survey on National Issues—a more general survey on many topics of national interest–43 % agreed with the more general statement "Free trade agreements are good for the country" (Pew Research 2009). It seems that Americans, like the citizens of most nations, are often of mixed opinions regarding trade, and poll results are often shaped by "current events" not sound reasoning. When U.S.-based trade specialists and foreign policy experts are asked the same questions, 88 % support free trade and international trade organizations. These differences in views between citizens and experts are very important in the U.S. and account for the "mixed messages" that U.S. politicians and government experts send to the citizens and governments of other nations.

Still, evidence from economic research favors free trade in the long run. According to the Peterson Institute for International Economics, "American real incomes are 9 % higher than they would be over time as a result of trade liberalizing efforts since the Second World War. In terms of the U.S. economy in 2008, that 9 % represented $ 1.3 trillion in additional American income" (Nanto and Donnelly 2009). Trade not only increases GDP and total wages by keeping America's factories, farms, and offices working, but contributes in many less observable ways as well. Producing products for export where the nation has comparative advantage helps focus investment capital, rewards innovation, lowers production costs and raises per worker productivity, while increasing tax revenues, wages, profits and total GDP (Fig. 6.5).

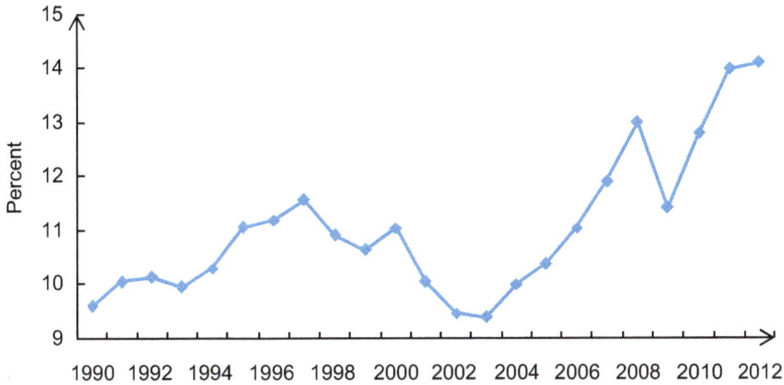

Fig. 6.5 U.S. Exports of goods and services as a percentage of GDP: 1990–2010. (http://www.census.gov/foreign-trade/statistics/historical/, accessed May 1, 2013)

Of course, access by any of the world's consumers to both domestic and imported goods and services adds consumer options while restrain rising prices through increased competition. Meeting the demands of highly competitive markets spurs innovation and provides consumers with the most choices and the best possible products. Indisputably, free trade is essential to the U.S. economy, just as it is for China. The only difference is how citizens in each nation view the issue. Trade and growth in GDP are closely related. Despite steady and significant population growth in the U.S. for the 22 year period from 1990 to 2011, the Pearson's product moment correlations between total export volume, and GDP and per capita GDP are 0.957 (p=0.0001) and 0.956 (p=0.0001) respectively.

Consider that three quarters of world purchasing power and almost 95 % of world consumers can only be accessed by U.S. manufacturers and farmers through international trade in goods and services. Population in the U.S. will continue to increase, but the domestic demand increase will not be sufficiently to maintain per capita incomes and productivity without international trade. Hufbauer (2008) estimates that elimination of remaining global trade barriers would further increase the gains Americans already enjoy from trade by another 50 %. Also similar to China, a considerable portion of economic growth in the U.S. is associated with international trade. Between 2005 and 2008, exports rose by 43 %, accounting for 47 % of overall GDP expansion. Although this tailed off in 2009, sustained growth in 2010 and 2011 continue the upward trend. In 2008, U.S. exports totaled $ 1.8 trillion or a record high 13 % of U.S. GDP. In fact, during the past decade, the share of U.S. GDP accounted for by exports rose from 9 % in 2001 to 13.0 % in 2010. In spite of a significant drop in value in 2009 due to the global recession, the general upward trend is that exports will account for an increasing share of U.S. GDP over time (Fig. 6.5). Better enforcement of trade regulations, greater trade "transparency", the reduction of tariffs around the world, and the inclusion of effective regulations for the control of trade in IP will allow this share to increase significantly in the future.

Table 6.3 Top ten trading partners of the United States by total trade: 2011. (United States Census Bureau 2012; Wikipedia 2010)

	Total merchandise trade ($ billions)	Exports (year-to-date)	Imports (year-to-date)	Total trade (year-to-date)	Trade balance 2011	Percent of total trade (%)
2011	Total, all countries (regions)	1,480.7	2,207.0	3,687.6	−726.3	100.00
Rank	Total, top 15 countries (regions)	1,015.8	1,596.1	2,611.9	−580.3	70.80
1	Canada	280.9	316.5	597.4	−35.6	16.20
2	Mainland China	103.9	399.3	503.2	−295.4	13.60
3	Mexico	197.5	263.1	460.6	−65.6	12.50
4	Japan	66.2	128.8	195.0	−62.6	5.30
5	Germany	49.1	98.4	147.5	−49.3	4.00
6	United Kingdom	56.0	51.2	107.1	4.8	2.90
7	Korea, South	43.5	56.6	100.1	−13.1	2.70
8	Brazil	42.9	31.4	74.3	11.5	2.00
9	France	27.8	40.0	67.8	−12.2	1.80
10	Taiwan, China	25.9	41.3	67.2	−15.4	1.80

6.1.2.2 Major Trading Partners of the United States

The United States trades with more than 200 nations and territories (U.S. Census Bureau 2012). In reality, however, only a few nations command the "lion's share" of total trade with the country, and of course have a commensurate share of disputes and trade conflicts. In 2011, the top 15 countries accounted for 70.8 % of total trade, while the top 10 accounted for 62.89 % (Table 6.3). Chinesed Taiwan overtook The Netherlands as the tenth largest trading partner of the U.S by the end of August, 2009 and has remained in this position. On the basis of total trade, the largest U.S. partners at the end of 2011 include Canada, mainland China, Mexico, Japan, Germany, the United Kingdom, South Korea, Brazil, France, and Chinese Taiwan.

From the perspective of trade deficits, a slightly different picture emerges. Actually, the bulk of the U.S. trade deficit is generated by trade with five nations: China, Japan, Mexico, Canada, and Germany. Trade with the oil exporting countries driven by recent increases in energy imports and higher costs adds the nations of Nigeria, Venezuela, and Saudi Arabia to this group. These latter countries had more balanced trade histories with the U.S., and given the surprising boon in gasoline and fuel exports from the United States from 2011–2013, there is evidence that at least in the short term, balanced trade with these oil-rich nations will remain the norm for some time to come as more fossil fuels are mined and processed domestically.

Table 6.4 China export-import volume grouped by trade type in 2010 (billion $, %). (National Bureau of Statistics of China 2011)

Year	General trade		Processing trade		Other trade	
	Export	Import	Export	Import	Export	Import
1985	23.73	37.27	3.32	4.27	0.30	0.70
1990	35.46	26.20	25.42	18.76	1.21	8.39
1995	71.37	43.37	73.70	58.37	3.71	30.34
2000	105.18	100.08	137.65	92.56	6.37	32.46
2005	315.06	279.63	416.47	274.01	30.42	106.31
2006	416.20	333.07	510.36	321.47	42.38	136.92
2007	538.46	428.61	617.56	368.48	61.76	158.86
2008	662.86	572.09	675.11	378.38	92.72	182.09
2009	529.83	533.87	587.03	323.31	90.90	154.00
2010	720.73	767.98	740.52	418.77	153.40	196.00

6.2 The Structure of International Trade

6.2.1 *The Structure of China's International Trade: Export-oriented Processed Goods*

Since the 1990s, processed exports from China have increased significantly, but declined slightly in proportional terms as overall trade has expanded so dramatically. Processed exports accounted for 55 % of total exports in 2000, this gradually declined to 46 % by 2008, which is just about where it is in 2010 and 2011. At the same time, processed products included 30.28 % of all imports, with a value of $ 418 billion in 2010. Mechanical or electronic products represent an increasingly important trade category, products included in these two categories account for about 55 % for both exports and imports.

According to SITC (Standard International Trade Classification), primary goods are mainly import-oriented in China, while manufacturing products are export-oriented. In 2010, the export value of all primary goods was $ 81.68 billion which represented only 5.18 % of total exports by value. On the other hand, the value of exported manufactured products was $ 1.49 trillion—the other 94.82 % of total exports by value. Turning to imports, machinery and transportation equipment were the leaders in terms of greatest value by commodity group. Imports of raw materials such as fuels, minerals and ores and grains were valued at $ 433.85 billion in 2010, which was 31.07 % of the total value of imports (Table 6.4, 6.5, and 6.6). As noted earlier, the majority of imported goods in 2010 (68.9 %) were classified as manufactured goods valued at $ 962.39 billon. A full 57 % of the share of manufactured goods included machinery and transport equipment.

Imports related to services still play a minor role in China's trade. China's import and export of material goods accounted for 90 % of total trade while services were only 10 % of value in 2008 which was only half of this percentage for the U.S. In

Table 6.5 Major import categories for China in 2008. (MOFCOM 2009)

Import Categories	Amount in billion U.S. $	Percent of total China import (%)
Machinery and transport equipment	441.92	39.00
Mineral fuel and lubricants	169.11	14.92
Non consumption materials	167.21	14.76
Chemicals	119.20	10.52
The materials of the finished product	107.16	9.46
Miscellaneous goods	97.62	8.62
Foods	14.05	1.24
Plant and animal oil and grease wax	10.49	0.93
Undefined others products	4.42	0.39
Drinks and tobacco	1.92	0.17

Table 6.6 Major export categories for China in 2008. (MOFCOM 2009)

Export categories	Amount in billion U.S. $	Percent of total China export (%)
Machinery and transport equipment	673.33	47.13
Miscellaneous goods	334.61	23.42
The materials of the finished product	261.74	18.32
Chemicals	79.31	5.55
Foods	32.76	2.29
Mineral fuel and lubricants	31.64	2.21
Non consumption materials	11.35	0.79
Undefined others products	1.72	0.12
Drinks and tobacco	1.53	0.11
Plant and animal oil and grease wax	0.57	0.04

2008, the ratio of services to goods was 1:7.1 (exports 1:8.3, imports 1:6.1), far below the global average ratio of 1:4.

6.2.2 *The Structure of U.S. International Trade*

With higher wages rates, an aging workforce in industrial manufacturing (including higher healthcare costs and costly retirement packages), higher fixed and variable costs, higher tax rates, and [often] stricter environmental controls, the United States is losing, or has lost, comparative advantages in manufacturing in many industries to firms in many developing nations. Many of the products where the U.S. has lost comparative advantages are imported, not only from China but from many other nations as well including Japan, South Korea, Mexico, Brazil, India, Malaysia, and even the EU nations such as Germany and Italy. The latter two are nations with exceptional reputations for the highest quality of industrial machinery and technology.

U.S. imports from year to year are typically no more or less volatile than those of other industrial nations, representing a mix of raw materials and finished goods.

Table 6.7 Major import categories for the United States in 2009. (Workman 2010)

Import categories	Amount in billion U.S. $	Percent of total U.S. import (%)
Crude oil	188.5	12.1
Medicinal product dental and pharmaceutical preparations	81.4	5.3
Passenger cars	53.2	3.4
Other household goods (example: clocks, kitchen products)	47.3	3.0
Computer accessories	43.9	2.8
Automotive parts and accessories	47.3	3.0
Cotton apparel and household goods	43.9	2.8
Computers	40.8	2.6
Telecommunications equipment	37.3	2.4
Video equipment	36.1	2.3

Table 6.8 Major export categories for the United States in 2009. (Workman 2010)

Export categories	Amount in billion U.S. $	Percent of total U.S. export (%)
Civilian aircraft including parts	74.7	7.1
Medicinal, dental and pharmaceutical preparations	46.1	4.4
Semiconductors	37.5	26.0
Other industrial machines	30.9	2.9
Automotive parts and accessories	30.0	2.8
Telecommunications equipment	28.7	2.7
Passenger cars	27.5	2.6
Medicinal equipment	26.9	2.5
Electric apparatus	26.1	2.5
Plastic materials	25.5	2.4

Despite a recent change in fuels, over the *long term,* America's top imports range from crude petroleum and other raw materials to high-tech electronics. Crude oil deliveries in 2009 took up the highest percentage of U.S. imports—a situation that was of great domestic concern. This worry about energy in the 1980s and 1990s spurred growth in alternative energy investments and research—in part funded by the government. Raw materials and semi-processed products for medicines were second, while new and used passenger cars were third in 2009 (Table 6.7). In 2011, however, processed petroleum-based fuels (jet fuel, gasoline) were the NUMBER 1 export of the United States by value. Traded products can change quickly due to changing economic, social and political conditions.

Turning briefly to exports, there is a similar diversity of products and services. For products, civilian aircraft continue to lead U.S. exports. Other high-tech products such as medicines and semiconductors are also important (Table 6.8). There is optimism in the United States that as services are more effectively included in future WTO agreements, markets in this field will expand. Issues related to intellectual property rights and trade in agricultural products have not really been dealt

Table 6.9 Top ten states in U.S. exports for total merchandise trade: 2009. (Trade Stats Express 2010)

	U.S. export values ($ 1,000)	Share in %
U.S. total	1,056,931,976	100.0
Texas	163,046,235	15.4
California	120,142,220	11.4
New York	57,320,623	5.4
Washington	51,739,397	4.9
Florida	46,919,556	4.4
Illinois	41,513,559	3.9
Ohio	34,083,697	3.2
Louisiana	32,714,797	3.1
Michigan	32,553,939	3.1
Pennsylvania	28,253,146	2.7
% Share top 10 states		57.6

with adequately in existing WTO agreements, and underscore the importance of these agreements to the U.S. economy.

6.2.3 Location and International Trade

There is also a "geography" of U.S. exports that should be at least mentioned in this brief chapter—especially as most contributors and readers are geographers. Of course, larger states such as Texas, California and New York have largest volumes of exports, but the coastal location of most top exporting states should also be recognized. The "top ten" exporting states account for 57.6% of total merchandise exports (Table 6.9), and service exports are even more concentrated. The Midwestern states on the list including Ohio, Illinois and Michigan have access to international markets via low-cost waterways such as the Ohio-Mississippi River System that connects the region to the Gulf of Mexico, and the Great Lakes-St. Lawrence Seaway with access to the Atlantic.

Historically, since Sino-U.S. diplomatic relations were formally reestablished in 1978, international trade between China and the U.S. has experienced a transition from initial confrontation to neutral contact, to strategic collaboration, and now, largely, to constructive cooperation. As noted earlier, bilateral trade between China and the United States has played an outstanding role in both national economies. Since 2003, China has become the second largest market for U.S products, and the U.S. has become the largest market for China during the last 10 years. A Gallup poll conducted in January of 2012 found that a majority of Americans felt that China was "the leading economic power in the world today" (53% of those polled and up 1% from 2011 when the same question was asked). 33% of Americans felt the United States was the "leading economic power", and Japan was third with 7% of the vote (Jones 2012) (Gallup is a well-known opinion poll company in the U.S.). This perception underscores the importance of Sino-U.S. trade, not only to these nations, but to the global economy, but to U.S. domestic perceptions of the greater world (Jones 2012).

The lessons of the recession that began in mid-2008 underscores how inter-related the Chinese and U.S. economies have become, especially in the past two decades (Francois 2010). As American consumers tightened their belts and started saving once the recession was wide-spread, the savings rate for working American's rose from an all-time low of 1.5 % in the first quarter of 2008 to 5.4 % just over a year later. Given that many of the consumer products that China exports represent highly "elastic" purchases, workers in factories throughout China suffered even as citizens in the U.S. tried to curb their appetites and save for their future. Similarly, as Chinese investors and the Chinese government sought sound and stable investments, U.S. treasury bills and other dollar assets remain attractive, despite the rhetoric to the contrary. China has $ 2.3 trillion in "official" reserves, and Dunaway, of the Council on Foreign relations, a prestigious "think tank", estimates that 70 % is held in U.S. dollar assets, mostly U.S. government securities. Further FDI flows into, and from, both nations are at record levels. The economic interests of these two nations are closely related, and further, the dramatic increases in trade we have seen over the past several decades have been, as noted earlier, exceptionally good for both nations.

Indeed, to date, economic and trade contacts between the two countries largely represent a win-win situation. Of course, as the level of trade increases, so will typically minor disputes about particular products and services. This is to be expected. In current dollars, the China's bilateral trade surplus was $ 226 billion in 2009, up from $ 161 billion only 5 years earlier, representing 36.9 % of the nation's total trade deficit in 2009, if down somewhat again in 2012. The high-water mark of 2008 is a dramatic increase from 2004, when China's share represented only 22 % of the total trade imbalance. However, it is important to notice that the total U.S. deficit with the world is down by $ 121 billion from 2004 to 2009 as well (Francois 2010). While exports to the U.S. are high, China is far less dependent on the U.S. market than Canada or Mexico, and this dependence (like Canada's) is trending downward over time as China develops more trading partners throughout the world.

Further, it should come as no surprise that Chinese products are welcomed by virtually all U.S. consumers—they cost less than domestic equivalents and are of increasing good quality. Trade with China helps reduce household expenditures in many areas for all families in the United States by around five hundred dollars every year.

Sino-U.S. trade has steadily increased in volume, but the trade imbalance has also increased, and as noted earlier, this "gap" has also grown to be a source of concern in the domestic politics of the United States (Table 6.10). Even how to calculate the trade "surplus" or "deficit" is controversial. While beyond the scope of this paper, an example using Chinese data is instructive. Based on Chinese data, China's trade balance with to the U.S. recorded a $ 1.41 billion deficit in 1990 (of no concern to U.S. voters!), but registered a $ 170.86 billion surplus in 2008, 26 times greater than the surplus in 1993 when the trade balance first shifted to China's favor. If based on U.S. data, the situation seems even more uneven. Official American data (and government officials) indicated that China's trade surplus began in 1983, actually 10 years earlier than the official Chinese statistics reaching $ 10.41 billion in

Table 6.10 Trade between China and the U.S. with ratios of dependency (0.1 billion $, %). (National Bureau of Statistics 1990–2009; MOFCOM 2009; BEA 2010; http://www.census.gov; http://www.commerce.gov)

Year	Dependency ratio of China to the U.S.				Dependency ratio of the U.S. to China			
	Exp. − Imp.	Exp. + Imp.	Export	Import	Exp. − Imp.	Exp. + Imp	Export	Import
1990	−14.1	3.02	1.33	1.69	−104.1	0.35	0.08	0.26
1991	−18.2	3.47	1.51	1.96	−126.9	0.42	0.10	0.32
1992	−3.1	3.58	1.76	1.82	−182.6	0.52	0.12	0.41
1993	62.7	4.51	2.77	1.74	−227.7	0.61	0.13	0.47
1994	74.9	6.34	3.84	2.50	−294.9	0.68	0.13	0.55
1995	85.9	5.61	3.39	2.21	−338.1	0.77	0.16	0.62
1996	105.3	5.00	3.12	1.89	−395.2	0.81	0.15	0.66
1997	164.0	5.14	3.43	1.71	−497.0	0.91	0.15	0.75
1998	210.2	5.39	3.73	1.66	−569.0	0.98	0.16	0.81
1999	224.7	5.67	3.87	1.80	−686.7	1.02	0.14	0.88
2000	297.4	6.21	4.35	1.87	−838.1	1.18	0.16	1.02
2001	280.8	6.07	4.10	1.98	−830.5	1.20	0.19	1.01
2002	427.2	6.68	4.81	1.87	−1,031.2	1.41	0.21	1.20
2003	586.1	7.70	5.64	2.06	−1,240.0	1.65	0.26	1.39
2004	802.7	8.78	6.47	2.31	−1,619.8	1.97	0.30	1.68
2005	1,141.7	9.46	7.28	2.18	−2,015.5	2.29	0.34	1.95
2006	1,442.6	9.88	7.65	2.23	−2,325.8	2.56	0.41	2.15
2007	1,633.2	8.93	6.88	2.05	−2,562.0	2.75	0.46	2.28
2008	1,708.6	7.71	5.83	1.88	−2,680.4	2.82	0.48	2.34

1990 and over $ 270 billion in 2008. The latter value is almost $ 100 billion higher than that reflected in official Chinese statistics. The different numbers and statistical findings are due to different accounting methods—issues too complex to review in this limited space, but such "accounting" disagreements are not unusual among major trading partners. Different estimates of commodity value moving as imports and exports cannot help but lead to trade frictions between the two countries. On the other hand, according to Chinese statistics, although the absolute surplus has grown dramatically, for the last few years, the growth rate of China's trade surplus is slowing. That is the annual growth rate of China's trade surplus for the 4 years from 2005 to 2008 was 42.2, 26.4, 13.2 and 4.6% respectively.

Trade and investment issues between any two nations are always more complicated than they might first appear. According to a recent Ministry of Commerce (PRC) report, by the end of 2009, the United States had set up nearly 60,000 commercial or manufacturing enterprises such as off-shore factories, fast food restaurants, consulting firms, etc. in China since diplomatic relations were reestablished in 1978. Of these, more than 30,000 enterprises remain in operation. The enterprises involve 29 different manufacturing categories and over 100 types of different service provision firms (imports/exports, banking, insurance, product design, marketing, etc.). Based on a preliminary survey by the Ministry of Commerce, these China-based enterprises realized $ 150 billion in sales in 2009. This level of FDI was actually doubled the value of U.S. exports to China ($ 70 billion) for the same period. From the Chinese perspective, it could be argued that if the revenues of China-based U.S. firms were included in the calculations then, by and large, Sino-U.S. trade is balanced and the Chinese side actually faces a slight deficit. Of course, the U.S. Department of Commerce does not agree with this method of accounting, but it is clear that there is room for far more than one or two opinions on the matter.

For major trading partners of any nation, more conflicts arise over particular issues and particular products as trade grows increasingly important and diversified. As the importance of bilateral trade increases for any two given nations, these conflicts become symbolically more important and more politically charged. Trade relations between China and the United States must be evaluated in this context. Another contentious issue related to Sino-U.S. trade in the past several years has been the issue of China's currency valuation as it related to access to markets and the relative price of goods. Again, this is a complex issue and is beyond the scope of this chapter, but it is important to note that many foreign economists find themselves arguing that while greater revaluation of the Yuan would certainly have some effect on the U.S.-China trade, it would NOT have the effects that many U.S. politicians suggest.

For major trading partners of any nation, more conflicts arise over particular issues and particular products as trade grows increasingly important and diversified. As the importance of bilateral trade increases for any two given nations, these conflicts become symbolically more important and more politically charged. Trade relations between China and the United States must be evaluated in this context. Another contentious issue related to Sino-U.S. trade in the past several years has been the issue of China's currency valuation as it related to access to markets and the relative price of goods. Again, this is a complex issue and is beyond the scope of this chapter, but it is important to note that many foreign economists find themselves arguing that while greater revaluation of the Yuan would certainly have some effect on the U.S.-China trade, it would NOT have the effects that many U.S. politicians suggest.

6.3 Conclusions

6.3.1 From the Chinese Perspective

In the contemporary global trading system, both China and the United States are major players and their interaction adds up to a remarkable 500 billion $ of bilateral trade in 2012. This high level of exchange will continue into the future. Since the period of reform and opening up, China has achieved rapid growth in international trade both with respect to gross value and the number of trading partners. During the past two decades, China has maintained a favorable trade balance, especially with the United States. However, compared to the United States, China still faces major challenges including those related to raising total economic output, minimizing the trade dependency ratio, upgrading the nation's trade structure and improving the spatial distribution of international trading manufacturers and firms. Strategically, trade is a win-win opportunity associated with expanded mutual cooperation between China and the United States. There are some important issues related to this trade that readers should recognize.

First, in the past 20 years, considering the total value of trade as well as the trade dependency ratio, China has grown rapidly, but China's trade dependency ratio is over twice of that of the U.S. The nation's bilateral trade dependency ratio is far higher than the United States' ratio with China. China's economic growth is still driven to some extent by foreign trade.

Secondly, considering the structure of bilateral trade, Sino-U.S. trade is complementary. China has comparative advantage in all labor-intensive products whether in goods or services. In recent years, thanks to the development of the iron and steel industry, China's capital intensive products are growing increasingly competitive on international markets, but "high-tech" and knowledge- intensive industries, particularly computers and financial services industry are still in the early stages of development and lag behind those of Japan, the United States and many EU nations.

Finally, from a geographical perspective, there are important differences. Although the two countries share several major trading partners from the Asia-Pacific regions and the European Union, the nations—from the Chinese perspective—face different situations. The border trading partners of the United States including Canada and Mexico are members of NAFTA. They are also complementary in trade. The U.S. is usually the first and most important partner in the complementary and profitable exchanges with these nations, and usually has a trade surplus with these nations as well. In contrast, the dominant exporting industries in neighboring Japan, Korea, Chinese Taiwan, and the ASEAN countries are almost all the same as those of China. In a sense then, China faces competition rather than cooperation with these countries that often results in trade deficits for China with these nations. Developing greater trade cooperation with these economies, while building up the entire East-Asian economic community, will prove essential for China in the future.

6.3.2 From the U.S. Perspective

Past history as well as present conditions in both nations would be radically different in the absence of the (relatively) free international trade that exists at the present time—and especially after the World Trade Organization (WTO) was initiated under the Marrakech Agreement in 1995 (Garg 2004). Trade with China still offers excellent opportunities for countless U.S. firms now and in the future. Further, Sino-U.S. trade represents a vital component of the global trading system. As supply chains grow longer and incorporate an ever-growing number of nations, perturbations in Sino-U.S. trade will have far reaching effects on dozens of other nations as well.

Free and unfettered trade, in the long run, typically benefits more persons than it hurts, but as geographers we must recognize that old industrial regions in BOTH nations have faced particular challenges as products from many nations flow into the country. High rates of unemployment are challenging to all nations, but especially to nations such as the United States that typically have limited experience with high unemployment or underemployment. For this reason, and also due to the sheer size of the deficit, trade with China has become enmeshed in the domestic politics of the country. It will remain there for some time.

Still, and most importantly, for many reasons noted in other chapters of this book, it is becoming increasingly clear that China and the United States are destined to have a special relationship in this new century, and—like marriage for better or worse—trade, despite the disputes that invariably accompany trade, will form one of the enduring ties that will bind these two great nations. Trade and intellectual property issues are always complex, and the United States has a long history of difficult negotiations in these areas with many close allies. Good faith negotiations and increasing transparency will prove essential for their resolution, but solving these problems is in the best interests of citizens in both nations, and indeed for consumers throughout the world.

References

Bureau of Economic Analysis (BEA). (2010). US international trade in goods and services, CB10-65, BEA 10-20, FT-900 (10-03), May 12. http://www.bea.gov/newsreleases/international/trade/tradnewsrelease.htm. Accessed 6 June 2012.

China State Statistics Bureau. (2009). The China statistical yearbook (1990–2009). http://www.stats.gov.CN/.

China State Statistics Bureau. (2011). *China statistical yearbook 2011*. Beijing: China Statistics Press.

Dunaway, S. (2009). The U.S.-China economic relationship: Separating facts from Myths. Council on foreign relations. http://www.cfr.org/china/us-china-economic-relationship-separating-facts-myths/p20757. Accessed 1 May 2013.

Francois, J. (2010). Deconstructing Sino-US codependence: Revaluation, tariffs, exports, and jobs. VOX-EU. http://www.voxeu.org/index.php?q=node/4968. Accessed 3 June 2012.

Gao, G. P, & Cai, X. C. (2008). *Theory and practice of international trade* (p. 17, 20, 23). Wuhan: University of Technology Press.

Garg, H. (2004). *WTO and regionalism in world trade*. Spring: New Century Publications.

Hsu, A. (2012) Taiwan ranked 18th in world trade report. Taiwan Today. http://www.taiwantoday.tw/ct.asp?xitem=189276&CtNode=415. Accessed 22 May 2012.

Hufbauer, G. C. (2008). Answering the critics: Why large American gains from globalization are plausible. On-line commentary copyright by The Peterson Institute for International Economics. http://www.iie.com/publications/papers/print.cfm?Researchid=929&doc=pub. Accessed 3 June 2012.

Jones, J. (2012). Americans still view China as world's leading economic power. http://www.gallup.com/poll/152600/americans-view-china-world-leading-economic-power.aspx. Accessed 1 May 2013.

Mikic, M. (1998). *International trade*. New York: St. Martin's Press.

Ministry of Commerce (MOFCOM). (2009, 2011). *The China commerce yearbook*. Beijing: China Business Press.

Ministry of Commerce. (2010). China became the world's largest goods exporter. http://www.mofcom.gov.cn/aarticle/i/jyjl/l/201002/20100206771253.html. Accessed 30 May 2012.

Miroudot, S., Sauvage, J., & Shepherd, B. (2010). Measuring the cost of international trade in services. Working Paper, Grouped' Economie Mondiale (GEM).

Nanto, N., Illias, S., & Donnelly, J. M. (2009). US international trade: Trends and forecasts. Congressional research service 7-5700, RL 33577. http://assets.opencrs.com/rpts/RL33577_20090306.pdf. Accessed 19 March 2014.

Pew Research Center for People and the Press. (2009). The United States seen as less important, China as more powerful: Isolationist sentiment surges to a four-decade high. Report 569, December 3. http://people-press.org/report/569/americas-place-in-the-world. Accessed 3 June 2010.

Trade Stats Express. (2010). State export data (interactive mapping program). http://tse.export.gov/MapFrameset.aspx?MapPage=SEDMapStateDisplay.aspx&UniqueURL=asalc455obr20q45zb4zvo45-2010-6-6-14-10-56. Accessed 10 May 2012.

United Nations Commodity Trade Statistics Database. http://comtrade.un.org/. Accessed 30 May 2012.

United Nations Service Trade Statistics Database. http://unstats.un.org/unsd/ServiceTrade/. Accessed 30 May 2012.

United States Bureau of Economic Analysis (BEA). http://www.bea.gov/. Accessed 30 May 2012.

U.S. Census Bureau (2012) The 2012 statistical abstract. http://www.census.gov/compendia/statab/cats/foreign_commerce_aid/exports_and_imports.html. Accessed 22 May 2012.

United States Department of Commerce Website. http://www.commerce.gov/. Accessed 30 May 2012.

United Nations. (2012). World economic situation and prospects 2012, Chapter 2: International trade. http://www.un.org/en/development/desa/policy/wesp/wesp_current/2012wesp.pdf, Accessed 19 March 2012.

Wikipedia. (2010). http://en.wikipedia.org/wiki/List_of_the_largest_trading_partners_of_the_United_States. Accessed 3 June 2010.

Workman, D. (2010). Top U.S. Imports and exports, 2009. Suite 101.com, http://import-export.suite101.com/article.cfm/top-us-imports-exports-2009. Accessed 6 June 2010.

World Trade Organization. (2012). International trade statistics. http://www.wto.org/english/res_e/statis_e/statis_e.htm. Accessed 6 June 2012.

Yao, S., & Luo, D. (February 2010). Chinese economy 2009: Leading the world economy out of crisis. China Policy Institute, The University of Nottingham. http://www.nottingham.ac.uk/cpi/documents/briefings/briefing-59-sy-econ-review-2009.pdf. Accessed 30 May 2012.

Chapter 7
Megaregions of China and the U.S.

Russell M. Smith, Yuejing Ge, Rudi Hartmann, Xiaoping Dong and Yang Cheng

7.1 Introduction

A new geography is transforming how a large quantity of the world's population lives and works. The Megaregion, a vast expanse of urban, suburban and sometimes even rural territory connected through economic, social, cultural and environmental linkages is a global reality. From the Amsterdam-Brussels-Antwerp Megaregion of Europe to the Greater Tokyo Megaregion of Asia, these massive urban agglomerations are paving a new pathway to economic supremacy. Replacing the era of nations as the most important economic entities on the planet, Megaregions are the future of urban and economic life.

Megaregions consists of more than 10 million inhabitants and can stretch across national and international political boundaries. These 'endless cities' can be found on North America, South America, Europe, Africa and Asia. These polycentric urban agglomerations account for a large portion of the world's economic

R. M. Smith (✉)
Department of Social Sciences, Winston-Salem State University, Winston-Salem, USA
e-mail: smithrm@wssu.edu

Y. Ge
School of Geography, Beijing Normal University, Beijing, China
e-mail: geyj@bnu.edu.cn

R. Hartmann
Department of Geography and Environmental Sciences,
University of Colorado Denver, Denver, Colorado, USA
e-mail: Rudi.Hartmann@ucdenver.edu

X. Dong
School of Chinese Language and Literature,
Beijing Normal University, Beijing, China
e-mail: dongxpzhh@hotmail.com

Y. Cheng
School of Geography, Beijing Normal University, Beijing, China
e-mail: chengyang@bnu.edu.cn

R. Hartmann, J. Wang (eds.), *A Comparative Geography of China and the U.S.*,
GeoJournal Library 109, DOI 10.1007/978-94-017-8792-5_7,

activity and contain an increasingly large percentage of the world's population. According to one report, the top 40 Megaregions account for almost 1/4th of the world's poulation, more than 66 % of the world's economic output and 85 % of global innovation (Florida et al. 2007). This accumulation of population, wealth and knowledge is creating a new reality in which countries are not competing with each other, but rather vast colossal regions are challenging each other for global domination.

In this chapter we will explore several Megaregions located within China and the United States of America. These two countries offer interesting case studies from which to begin to understand the importance of Megaregions and the impact they have on the planet. The United States is home to what many would identify as the first Megaregion, Gottmann's 'Megalopolis'. Gottmann's research on the 500 mile stretch of the eastern seaboard from Boston to Washington, DC ushered in a new era of urban geography and is home to more than 50 million residents and an economic output of $2 trillion. Meanwhile, according to a recent United Nations report China is home to the largest Megaregion based on population. The Pan Hong Kong-Shenhzen-Guangzhou Megaregion's population is estimated to be in excess of 120 million residents (UN-Habitat 2010). Additionally, China's Megaregions account for approximately 43 % of China's economic output and deserve further examination (Florida 2008).

The discussion that follows will attempt to define the unique characteristics and provide a better understand of the evolving urban geography of Megraregions in each country. It is interesting to analyze this emerging geographic phenomenon in light of the drastically different political and economic realities of the United States and China. As you will discover, both countries have witnessed the development of similar Megaregional structures, but through vastly different circumstances and processes. While not explaining why Megaregions have become so important for each country, the discussion that follows does provide an excellent overview of this global experience in selecting Megaregions of China and the United States.

In the context of discussion the capitals of the two countries (Beijing and Washington D.C.) and major financial/ commercial centers (Shanghai and New York) are compared. Obviously, even if they cannot represent China and the U.S. in the largest expansion and the broadest definition, they are still regarded as the most typical cities in the countries. They were planned and built based on different purposes. The biggest difference between these two pairs is they don't have complementary organs of authorities. Though Beijing and Washington D.C. continue to represent their countries' political cultures in distinguish ways as capitals, Shanghai and New York have become more international as financial/ commercial centers. Shanghai and New York show the epoch-making trend in the progress of urbanization and high-tech development.

7.2 Defining "Urban", Urban Development Patterns and Trends in the United States

7.2.1 Defining "Urban"

Defining the term urban can be difficult. However, most accepted definitions include a population size and population density component. The next three subsections provide the reader with a basic understanding of urban terminology at the Federal, State and Academic levels as it relates to the United States. While not all territory included in a megaregion is urban, urbanity provides the beginning point for any discussion on the topic.

7.2.1.1 Federal Designations: U.S. Bureau of the Census

The United States Bureau of the Census provides several designations of "urban" that are comparable across the entire country and which are very helpful for national analysis. According to the U.S. Census Bureau, an Urbanized Area (UA) must have a population of at least 50,000 people. To qualify as an Urbanized Area, the area does not have to be a legally incorporated entity. An Urban Cluster (UC) has a lower population requirement of between 2,500 people and 50,000 people. A density requirement of at least 1,000 people per square mile at the core is required for both definitions (U.S. Bureau of the Census 2010).

The building blocks of these Urbanized Areas and Urban Clusters are Census Tracts, Block Groups and Census Blocks. Census Tracts are geographic areas that remain relatively constant over time to allow for comparison purposes. They typically contain between 1,200 and 8,000 residents, but can vary greatly in geographic size due to density patterns. Block Groups are a smaller Census geography than Census Tracts and generally contain between 600 and 3,000 people. Census Blocks are the smallest statistical unit for which Census data is available and usually correspond to a city block in urban environs and less regular patterns in more suburban and rural locales (U.S. Bureau of the Census 2010).

These definitions served the United States well for numerous decades. However, as America's urban form changed, the U.S. Census Bureau responded with new terms to better understanding America's shifting urban geography. The Standard Metropolitan Area (SMA) concept was introduced during the 1950 U.S. Census in response to the new urban reality of the U.S. This new concept, which was later renamed to Standard Metropolitan Statistical Area (SMSA) for the 1960 Census and subsequently to just Metropolitan Statistical Area (MSA) in 1983, incorporates population, employment and commuting thresholds into its definition.

Currently, the U.S. Census Bureau relies on four major statistical designations to define metropolitan America: Core Based Statistical Areas (CBSAs), Combined

Statistical Areas (CSAs), Metropolitan Statistical Areas (MSAs) and Micropolitan Statistical Areas (μSAs). CBSAs are the basic building blocks of metropolitan designation and consist of a county or county equivalent with at least one UA or UC of 10,000 residents and surrounding counties with significant social and economic integration. MSAs contain at least one urbanized area with a minimum of 50,000 residents and adjacent counties with high levels of employment and commuting overlap. μSAs have an urban cluster of between 10,000 and 50,000 residents and surrounding counties with a high degree of integration. Finally, CSAs consist of two or more adjacent CBSAs with significant employment interchange (U.S. Bureau of the Census 2012a; b). CSAs may be formed from a combination of MSAs and/or μSAs.

7.2.1.2 Sub-Federal Urban Designations: Cities, Towns and Villages

In the United States each State (e.g. Alabama, New York, Wyoming) has the legal authority to define 'urban' differently. In general, most states rely upon a few common urban designations to accomplish this task. The terms city, town, and village are the most common manifestations developed to help define and categorize 'urban' at the State-level. In sum, these terms can be grouped together and called municipal. Municipalities are geographic entities that are legally incorporated under the laws of their respective State.

In some states, these terms are used interchangeably and do not signify a difference in population size (e.g. North Carolina). However, in other States the difference between a city, town and village could include population requirements and legislative powers. For example, Kentucky, Minnesota, and New Jersey all use a population classification system to define differences among cities based on population size. The city class system then can be used to determine which laws are applicable to a particular size city.

Another peculiarity of the U.S. system is the use of township in much of the Midwestern United States. The usually large geographic areas can provide municipal like services including land use planning, road maintenance, and even police and fire protection. Unlike cities, towns and/or villages—townships tend to have lower population densities due to the large land area that they occupy. In some cases, townships eventually disappear due to encroaching annexation activity from nearby municipalities. In other instances, a township may go through a process of incorporation to become a municipality (U.S. Bureau of the Census 2007).

7.2.1.3 Academic Classifications

Over the last fifty years the academic literature has provided three major concepts that help define and understand urban America as it has taken on a more polycentric urban morphology. These terms include Jean Gottmann's—Megalopolis (1961),

Lewis's-Galactic Metropolis (1983) and Carbonell and Yaro's (2005)/Lang and Dhavale's (2005)—Megaregions/Megapolitan Areas.

Megalopolis is a concept that was developed by the French geography Jean Gottman (1961) in his classic book Megalopolis: The Urbanized Northeastern Seaboard of the United States. Megalopolis in its original usage was the name of a city in the Peloponnese that existed between 371–368 BC. Later both Patrick Geddes (1854–1932) and Lewis Mumford (1895–1990) used the word with a negative connotation in describing an urban agglomeration that was at the later stages of development and headed into decline.

Gottmann's Megalopolis described the urban corridor of the Northeastern United States seaboard between Boston and Washington. In Gottmann's view the entire urban corridor had high levels of interaction among people and businesses. As a result, seven separate metropolitan areas with more than 1 million residents and 30 smaller metropolitan areas all coexisted and congealed into a single functioning urban entity unlike anything that had been previously seen. Since 1961 the term has been expanded and used to describe polycentric, multi-jurisdictional economic and social regions around the globe.

Lewis's Galactic Metropolis was another attempt to describe a new and evolving urban form that was appearing on the landscape. Building upon Gottmann's concept, Lewis highlighted the role of the globalization of capitalism in fragmenting the urban landscape. Lewis also discussed the importance of edge cities and suburban commercial and office centers on the metropolitan landscape. The declining importance of the traditional downtown or core of the city is also an additional characteristic of galactic metropolis's (Knox et al. 2012).

Finally, Megaregions and Megapolitan Areas are two concepts that build upon the megalopolis foundation of Gottmann. Carbonell and Yaro's (2005) Megaregion and Lang and Dhavale's (2005) Megapolitan Areas both describe a new metropolitan reality taking shape. America 2050, a project of the Regional Plan Association, has advanced Carbonell and Yaro's Megaregion concept that seeks to describe the coalescing of multiple metropolitan areas which is similar to the Megapolitan Areas concept, although with less specific boundary delineation. Megaregions are defined through the layers and relationships that connect urban areas. These connections can include: environmental systems and topography, infrastructure systems, economic linkages, settlement patterns and land use, and shared culture and history (Carbonell and Yaro 2005).

Lang and Dhavle's examination and creation of Megapolitan Areas has moved the discussion of polycentric metropolitan areas further ahead. Their research focused on developing a new overlay category that could be incorporated into the existing U.S. Census Bureau's terminology and work much like the existing CSAs, MSAs and μSAs work. According to Lang and Dhavale's research, the United States contains 10 megapolitan areas each with a population that will exceed 10 million residents by 2040. Additionally, these 10 megapolitan areas can be found in each region of the United States and extend into 35 of the 50 states. To qualify as a megapolitan region, a geographic area must have the following characteristics:(Lang and Dhavale 2005)

1) combine at least two metropolitan areas,
2) have a population that is currently or projected to be in excess of 10 million residents by 2040,
3) have a shared cultural identity across the region,
4) share a similar physical environ,
5) be united by major transportation corridors,
6) work as an economic entity.

Clearly, this new form of urban development will have profound impacts on the world.

7.2.2 Urban Development Patterns and Trends

Despite the romanticized view of America as a country of wide open spaces, the majority of Americans live, work, and play in urban settings. Since 1920, the United States of America has been an urban country with more residents living in urban environs than rural places. According to the latest figures from the U.S. Census Bureau, 80.7% of Americans reside in an urban area (U.S. Bureau of the Census 2012a). While the United States has been considered an urban country for almost a century, the pattern of urbanization has gone through several cycles. In the next section we will explore the metamorphosis of America's urban tradition from early urban developments to metropolitan supremacy (Hartshorn 1992).

Early urban America (1760–1860) developed primarily around coastal and river locations as a necessity for extracting natural resources and as a means to facilitate trade by water. Coastal locations with good port potential made for excellent urban settlements (e.g. New York, NY; Philadelphia, PA; Boston, MA; Charleston, SC). As trade increased and the growing population of America sought more opportunities for development, river cities emerged on the interior of the continent (e.g. New Orleans, LA; Cincinnati, OH; Albany; NY; Richmond; VA). During these early days of urban development and city building, urban populations were relatively small with the largest city (New York City) only containing approximately 33,000 residents at the time of the first Census in 1790 (Gibson 1998). However, in subsequent decades and with the beginning of the Industrial Revolution, American cities were on their way to larger populations and greater importance.

The period from roughly 1860–1950 can be viewed as the heyday for American cities or an era of Central City Domination. At the start of the Industrial Revolution, large influxes of immigrants to urban areas and the improvement in transportation technology ushered in an era in which American cities reach unprecedented heights. During this era New York became the first city to have a population of 1 million residents in 1880 (Gibson 1998). Within ten years, Chicago and Philadelphia both surpassed the 1 million person population plateau. This era of urban development saw large swells in urban populations, large increases in urban economic importance and the development of the present day national urban system. The development of a national-city system featured some of the oldest municipalities in the country (e.g. New York, Boston, and Philadelphia) as well as some of the newer cities in the United States: Los Angeles, Denver, and Salt Lake City.

Since the 1950s, urban America has seen a dramatic shift in urban form. It has shifted from a period of concentration of economic resources and population in major cities that marked the previous era to a period of decentralization and fragmentation. The period of Metropolitan Supremacy was ushered in following World War II. The return of large numbers of service men and women, the development of the federal interstate system and the advent of a federal mortgage system all aided in the de-concentration of urban America to the periphery (Jackson 1985). This redistribution of population and economic opportunities resulted in a new behavioral model that for the first time separated residence and employment. The notion of working and living in the same city began to diminish as residents followed the American dream into the suburbs while still commuting into the city for employment. To better account for this new urban form and in response to the suburbanization of America, the Office of Management and Budget developed a new classification system to understand America's urban form: Metropolitan Statistical Areas.

This new metropolitan reality was further underscored after the 1961 release of Jean Gottmann's classic analysis of the Northeastern seaboard of the United States and the subsequent coining of the term 'Megalopolis' (Gottmann 1961). Gottmann, who brought a European lens to his analysis, rightly characterized the corridor between Boston, MA and Washington, D.C. as one large urban agglomeration. Unlike the European model of distinct urban and rural boundaries, Gottmann viewed the Northeast United States as one unified entity, from which one could not clearly identify where one city began and another ended.

Since Gottmann's pivotal work, the United States has witnessed acceleration in metropolitan agglomeration. Megaregions can now be found in all regions of the country (Regional Plan Association 2005). In many instances, these Megaregions stretch across multiple political jurisdictions (e.g. municipalities, counties, states). Three of these megaregional entities, the Northeastern (Boston-Washington), Southern California (Los Angeles—Las Vegas) and the Piedmont Atlantic (Raleigh—Birmingham), will be discussed in greater detail in a later section.

7.3 Mega-regions in the United States

In the next forty years the Mega-regions (see Fig. 7.1) in the United States are expected to capture two thirds of the United States economic growth and add millions of residents (Georgia Institute of Technology 2006). As a result, the following section explores three Mega-regions located within the United States in an attempt to provide the reader with a better understanding of the current patterns and trends of urban development in the United States. Each region was chosen for its

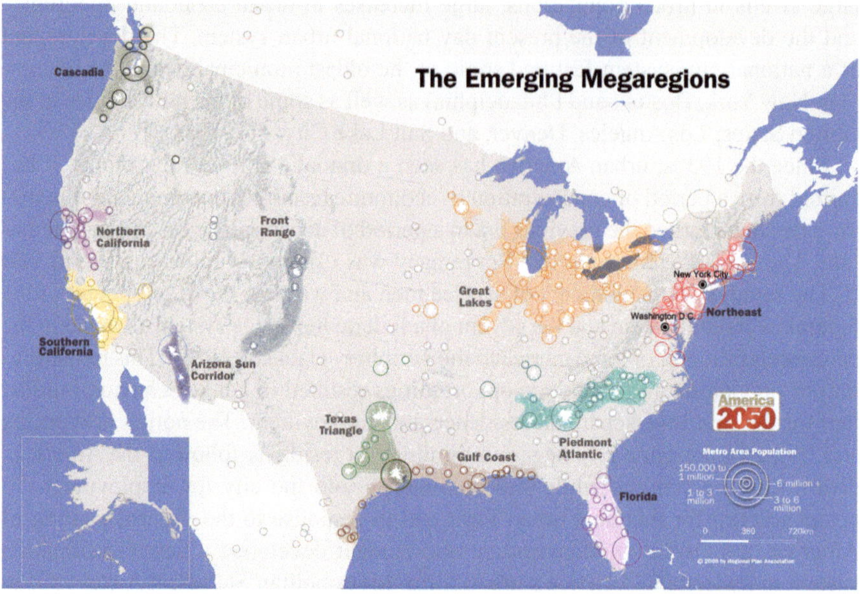

Fig. 7.1 Mega-regions of the United States. (Regional Plan Association 2012)

unique stage of development and to show some of the differences and similarities that exist between the Mega-regions. An overview of the geography and population, transportation system, economy, and future prospects and challenges is provided for each Mega-region.

The descriptions and characteristics of each of these Mega-regions come from a combination of research conducted on Mega-regions, Megapolitan Areas and original research. It should be noted that the Mega-region concept espoused by America 2050 is utilized more frequently due to its pliable boundaries. However, Megapolitan Areas as a unit of analysis are quite helpful and much more exact due to their reliance on U.S. Census Geographies. For more information on the differences between the two concepts please see Mega-regions: Planning for Global Competitiveness edited by Catherine L. Ross (Ross 2009).

7.3.1 *Northeast Mega-region*

7.3.1.1 Geography & Population

The Northeast Mega-region (see Fig. 7.2) consists of more than 52 million people and stretches along the Northeastern seaboard of the United States for more than 500 miles. Beginning in southern Maine and ending in northern Virginia, the Mega-region extends into 12 states and is bounded on the east by the Atlantic Ocean and

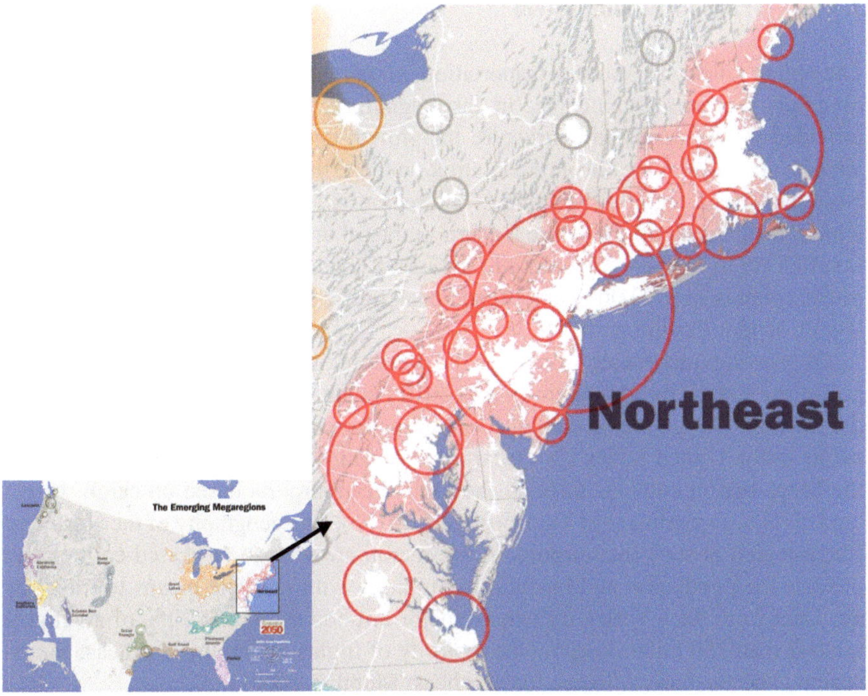

Fig. 7.2 Northeast Mega-region as Conceptualized by America 2050. (Regional Plan Association 2012a)

the west by the Appalachian Mountains (Regional Plan Association 2012a). The Northeast Mega-region is the oldest and most mature Mega-region in the country having been first identified in 1961 by Jean Gottmann.

The region has experienced steady growth over the last fifty years, growing from roughly 32 million people in 1950 to almost 49 million in 2000 and over 52 million people today. While the overall population of the region grew 62.5 % between 1950 and 2010, the share of the United States population residing in the Northeast Mega-region has declined from 21 % to approximately 17 %. While fewer US residents call the Northeast Mega-region home compared to sixty years ago, one in six Americans still lives in this region. Meanwhile, population density has increased from 610 person per square mile in 1950 to 931 person per square mile in 2000 (Vicino et al. 2007). By 2050, the region is expected to have a population over 70 million.

The region includes the major population centers of Baltimore, Boston, New York, Philadelphia and Washington D.C..While many of these cities have seen a decrease in the share of the region's population, they still serve as important economic and cultural hubs within the region. These cities began as independent entities that have coalesced over the decades into an interconnected urban agglomeration. While still maintaining their political autonomy, these cities have created strong economic, cultural and social linkages. The Northeast Mega-region is the most mature urban region in the United States and is the financial and governmental center for the country.

7.3.1.2 Transportation System

The region relies on a strong transportation infrastructure to facilitate the movement of goods, services and people. In the beginning, the sea and navigable rivers provided the necessary infrastructure for growth and development. Today, roads, airports and rail service are of crucial importance to the growth and prosperity of the Northeast Mega-region. The spine of the region is I-95, a limited access highway which stretches along the entire eastern seaboard and provides a vital transportation link for the region. According to the I-95 Corridor Coalition, the average daily traffic count is over 72,000 vehicles and peak traffic counts can reach over 300,000 (I-95 Corridor Coalition 2012).

Of growing importance globally is the presence of quality air facilities to handle passenger and air freight. According to 2010 Federal Aviation Administration data, the Northeast Mega-region contained six of the top twenty five airports for enplanements in the United States US Department of Transportation 2010a. Additionally, the Mega-region contains three of the top sixteen airports based on cargo weight landed (U.S. Department of Transportation 2010b). The geography of the Northeast Mega-region is such that airports in the region are viewed as shared entities and provide the population and business community with multiple means to travel by air. Data from New York City's three major airports show that 20% of all flights are less than 350 miles, keeping the majority of flights within the Northeast Mega-region (The Business Alliance for Northeast Mobility 2009).

Finally, the rail infrastructure located in the Northeast Mega-region is the oldest and most heavily traveled of America's passenger rail system. The population density and relatively short distances between major population centers allows the system to serve more than 750,000 passengers every day. These passengers depend upon the rail service to provide reliable and efficient service to support the economic vitality of the Mega-region. However, the rail infrastructure is aging and is in need of major investment. Without new funding the region could see a decline in performance and a decline in economic output.

Gottmann's original exploration of the region sixty years ago, the Northeast Mega-region has continued to be the "hinge of the US economy". According to estimates, the Northeast Mega-region is responsible for one-fifth of the nation's gross domestic product while occupying only two percent of the country's land area (Regional Plan Association 2012). Part of this economic wealth can be attributed to New York City, the financial capital of the world with almost 70 corporate headquarters and first level subsidiaries (Godfrey and Zhou 1999).

While the economic muscle associated with the region has been stable, there has been a restructuring around the type of employment generated in the region. The region which once dominated manufacturing employment for the country has seen that position erode from 50% of all jobs in 1900 to only 12% in 1997 (Vicino et al. 2007). The decrease in manufacturing employment was offset by a rise in service sector opportunities recently. The growth of finance and insurance, securities intermediation, information and data processing, and professional/scientific and technical jobs has been pronounced. As Vicino et al (2007) noted in their examination of the Megalopolis 50 Years on, "Megalopolis has become more of an information processing center than a metal bashing economy" (349).

7.3.1.3 Prospects and Challenges

The future outlook for the Northeast Mega-region is generally optimistic. The global and national importance of New York and Washington D.C. cannot be overstated. These two growth poles will continue to be financial and government magnets for economic development. Additionally, Boston serves as a pseudo "Silicon Valley" for the East Coast. With excellent access to national and international markets, a growing population and the financial capital of the world all located within its borders, the Northeast Mega-region has a comparative advantage over most other Mega-regions. As a result, the Northeast Mega-region should continue to be an economic engine for the United States into the foreseeable future.

While the overall prospects for the future are encouraging, challenges do exist. The region's population is expected to reach 70 million by 2050 and while that growth brings opportunities, it also has many challenges associated with adding an additional 20 million people. Two of the largest obstacles facing the Mega-region are transportation system deficiencies and a degrading environmental landscape.

Maintaining and improving the transportation infrastructure will be critically important to the health of the region. One of the major advantages of the region is the transportation connections that exist. As previously discussed, the highway, air and rail linkages reinforce the existing connections between the urban areas of the Northeast Megaregion. This transportation infrastructure is some of the oldest and most heavily traveled in the United States. The highways, airports and rail lines all need massive investments in order to maintain existing levels of service. Additionally, improvements are needed to stay competitive globally. A recently released report titled "The Future of the Northeast Corridor" highlights some of the needs related to rail service in the region including $5 billion in basic infrastructure, improvements to enhance safety, expand capacity and reduce trip times and upgrading equipment and facilities (The Business Alliance for Northeast Mobility 2009).

A host of environmentally related problems pose a major challenge to the future of the Northeast Mega-region. A growing population will result in an overall loss of habitat for plant and wildlife, as well as reduce the amount of available open space. Additionally, a growing population will place further strains on the existing water supply and water quality. In the end, a regional approach to conservation must be developed to allow for the management of shared resources across political boundaries. This idea is currently being espoused by the Regional Plan Association in a recently released study titled "Landscapes: Improving Conservation Practice in the Northeast Mega-region".

Fig. 7.3 Southern California Mega-region as Conceptualized by American 2050. (Regional Plan Association 2012b)

7.3.2 Southern California Mega-Region

7.3.2.1 Geography & Population

The Southern California Mega-region is a diverse mosaic of physical landscapes that include coastline, mountain and desert environs. The Mega-region stretches from the Pacific Ocean in the west to the desert borders of Nevada and Arizona in the east. The Sierra Nevada Mountains serve as the northern boundary of the region while San Diego and the Mexican border are the southern terminus of the region (see Fig. 7.3). In sum, the region encompasses more than 50,000 square miles of territory but only includes two states (California and Nevada) and has only eight county governments within its boundaries. Interestingly, approximately two-thirds of the region's land is owned by the government (Kern County Council of Governments, San Diego Association of Governments, and Southern California Association of Governments 2005).

According to recent estimates the Southern California Mega-region has a population rapidly approaching 25 million (8 % of US Population) and is expected to add an additional 15 million residents by 2050. This represents a growth rate of approximately 62 % and will place additional pressures on development within the region (The Business Alliance for Northeast Mobility 2009). Already, the limited availability of water, government ownership of land, and environmentally sensitive areas have resulted in population densities as high as 15,000–20,000 people per square mile in some coastal communities (Kern County Council of Governments et al. 2005).

The population that comprises the Southern California Mega-region is the most diverse in the nation. According to research conducted by Brenner and Pastor (2008), the Southern California Mega-region is the only Mega-region where the majority of the population is composed of a non-white population. Additionally, 29 % of the population in the Southern California Mega-region is foreign born. However, the percentage of foreign born residents is expected to decrease in the coming decades as the next wave of immigrants locates to other areas of the country (Benner et al. 2008). The major population centers for the region include Los Angeles, CA, San Diego San Diego, CA and Las Vegas, NV which have all experienced significant population growth over the last century. These communities serve as important centers of commerce, employment, and culture from which urban growth has dispersed.

7.3.2.2 Transportation System

The transportation system of the Southern California Mega-region is of vital importance to the region and the nation. The roads, rail, air and sea facilities serve to move people and goods within the region and around the globe. Of particular importance is the port system which acts as a global gateway between the nation and the rest of the world. The ports of Los Angeles and Long Beach combined are the fifth busiest in the world and the two busiest container facilities in the United States (Kern County Council of Governments et al. 2005). San Diego's seaport is also a top thirty port facility in the United States for containers.

The road infrastructure located in the region is also of major significance. The region has more than 10,500 lane-miles of highway and was a national leader in highway construction following World War II. Specifically, Interstate 5 and Interstate 15 serve as important highway corridors within the Mega-region and facilitate trade. However, due to traffic congestion it is estimated that Southern California loses almost $14 billion annually. Traffic congestion is not forecast to decrease in the coming decades. Population growth, increases in vehicle miles traveled (VMT) and truck traffic will all add to the traffic congestion problem. However, it should be noted that carpool rates are the highest in the country and intelligent transportation systems are being implemented in the Mega-region.

The construction of additional highways and the widening of existing freeways have become exceedingly difficult in recent decades. As a result, high speed rail service has become an important issue for the Mega-region. As early as 1996 the State of California created a rail authority whose mission is to develop a high speed rail system for the State. While the Authority's jurisdiction is state-wide, the project could have major implications for the Southern California Mega-region linking the population centers and airports with high speed rail connections.

Finally, the Mega-region is served by more than 90 public use airports and almost 15 commercial service airports. This aviation cluster makes the Mega-region the busiest in the country for air traffic. According to FAA data in 2010 the Los Angeles International Airport served almost 29 million passengers (#3), McCarran International Airport in Las Vegas, NV served close to 19 million passengers (#9) and San Diego International Airport served 8.5 million passengers (#28). In addition to the high passenger volumes of many of the Megaregions airports, two of the airports are also located in the top fifteen nationally for air cargo: Los Angeles International Airport (#7) with close to 4 million pounds of landed cargo and Ontario International Airport (#14) with approximately 2.3 million pounds.

7.3.2.3 Economy

The Southern California Mega-region has an estimated Gross Domestic Product (GDP) of between $900 billion and $1 trillion, making it the 10th largest economy in the world and representing approximately 7 % of the United States GDP (Regional Plan Association 2012). The general public might perceive that the Mega-regions' economic might is the result of the entertainment industries of Los Angeles and Las Vegas. However, logistics and the movement of goods are vital components to the overall economy of the Southern California Mega-region.

Due in part to its relative geography of being located on the west coast of North America and fronting on the Pacific Ocean and its absolute geography which includes excellent harbors for port facilities, the Southern California Mega-region serves as a global gateway to the Pacific Rim and the growing markets of Asia. According to the Los Angeles Economic Development Corporation, international trade is responsible for one out of every fifteen jobs. The heavy reliance on international trade for employment has not always been so pronounced in the region. For example, in 1972 approximately 25 % of jobs were classified as manufacturing. Today, roughly 10 % of the Mega-regions' employment is generated by manufacturing.

The economic future of the Mega-region will be heavily influenced by global patterns and trends. Maintaining and enhancing the transportation related infrastructure that makes the Mega-region a global gateway for trade is imperative. The Mega-region must maintain its global competitiveness or risk losing international trade to other facilities on the west coast. Additionally, the role of the entertainment industry in the overall economic output can be fickle. The latest recession has resulted in a decrease in disposable income for most Americans and a decrease in the dollars spent on entertainment related purchases.

7.3.2.4 Prospects and Challenges

The future of the Southern California Mega-region is complicated. Having the 10th largest economy in the world that is based on entertainment, defense and agriculture provides for a diverse stable foundation for future growth and prosperity. Serving as the gateway to the Pacific Rim and as a major international trade center are envious attributes that any Mega-region around the globe would welcome. Finally, the diversity of landscapes and populations make the Southern California Mega-region truly unique. However, challenges exist in the form of transportation congestion, environmental concerns and sustainability.

First, transportation related congestion is a major problem. The Mega-region loses billions of dollars annual due to road congestion and the situation is not much

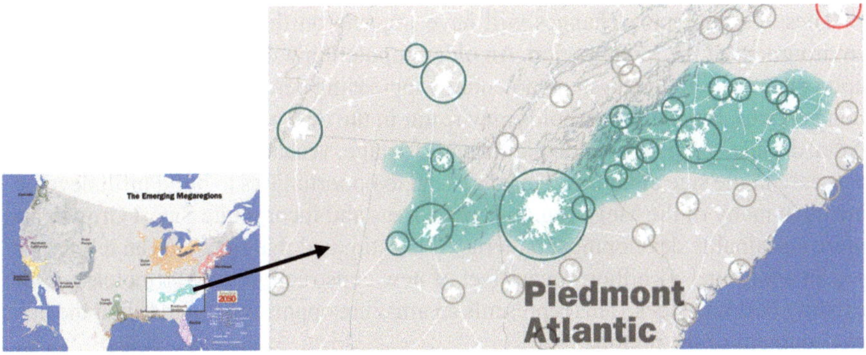

Fig. 7.4 Piedmont Atlantic Mega-region as Conceptualized by America 2050. (Regional Plan Association 2012c)

better for the Mega-regions airport and seaport facilities. Many of these facilities are reaching capacity. Secondly, environmental concerns are a challenge. Poor air quality has plagued the Mega-region for decades. While steps have been taken in the right direction future growth and development will continue to pose many problems for air quality. The general sustainability of the Mega-region is questionable. The Mega-region suffers from shortages of energy and water. Recent conservation efforts have resulted in changing consumer behavior and increasing reserves. However, future population growth and development pressures will place a large burden on the existing energy and water infrastructure of the Mega-region.

7.3.3 Piedmont Atlantic Mega-region

7.3.3.1 Geography and Population

The Piedmont Atlantic Mega-region (PAM) is located in the southeast United States and encompasses an area of more than 240,000 square miles. The Mega-region spans from the Atlantic Ocean in the east to the Appalachian Mountains in the West. The Gulf of Mexico serves as the southern boundary of the Mega-region and the northern boundary is approximately the North Carolina/Virginia border (see Fig. 7.4). In total, PAM includes territory in six states including: Alabama, Florida, Georgia, North Carolina, South Carolina, and Tennessee. PAM is a relatively young Mega-region—in fact Gottmann himself commented on the burgeoning Mega-region developing around Atlanta, GA in 1987 (Gottman 1987).

According to research conducted at the Georgia Institute of Technology's Center for Quality Growth and Regional Development, more than 34 million people reside in the Piedmont Atlantic Mega-region, which is almost 11 % of the total U.S. population. By 2050 the Mega-region is expected to be home to over 57 million people, a 68 % growth rate. The population growth in the region will lead to the Mega-region becoming older, more diverse and is expected to result in smaller household sizes.

These demographic changes will have impacts on the urban growth and built environment of the Mega-region. An older population will place different demands upon the transportation system, potentially increasing demand for public transportation services. Smaller households may result in the desire for smaller living spaces. After decades of rising residential square footage, recent trends seem to predict a decrease in the size of homes. This may lead to opportunities to build infill developments in many of the Mega-regions urban areas and incorporate Smart Growth and more sustainable development patterns. It is estimated that by 2030 the region will see an additional 84 billion square feet of new construction, a 54 % increase in the existing building stock. This represents an amazing opportunity to redesign the built environment of the region.

7.3.3.2 Transportation System

Interstates are of critical importance to the Piedmont Atlantic Mega-region and it could be argued that interstates bind PAM together more than any other Mega-region previously discussed. The Mega-region is crossed with a multitude of highways that provide links between the multiple cores of the Mega-region and between the periphery and core. In particular, I-85 and I-95 provide important north and south corridors for the movement of goods and services. Meanwhile, I-40 and I-20 serve as east and west conduits of automotive activity.

The seaport facilities that are located on the periphery of the Mega-region provide an important link to the core urban areas. Jacksonville, FL, Savannah, GA, Charleston, SC and Wilmington, NC all provide an opportunity for the import and export of goods. According to a 2012 Special Report on the top 30 U.S. Ports, Savannah, GA ranks 4th in the country for container shipments, Charleston SC ranks 9th, and Jacksonville, FL and Wilmington, NC rank 17th and 18th respectively (Burnson 2012). The key to maintaining and improving these rankings may lie in the Mega-regions ability to handle mega-vessels that are anticipated to increase in importance after the expansion of the Panama Canal is 2014.

Rail provides an essential component to the overall transportation system in the Piedmont Atlantic Mega-region. Railways serve an especially important role linking the inland ports of Atlanta, GA and Charlotte, NC with the seaport facilities located on the periphery of the Mega-region. Additionally, planning for the future extension of high speed rail is underway. The Southeastern High Speed Rail Corridor would connect the major cities of the Mega-region with Washington, DC and subsequently the Northeast Mega-region (Southeast High Speed Rail Corridor 2012). With speeds expected to approach 110 mph, this enhanced rail service would offer an alternative mode of transportation to combat traffic and air congestion within the Mega-region.

Finally, Atlanta and to a lesser degree Charlotte provide important air transportation connections for the Mega-region. Atlanta's Hartsfield Airport is the nation's largest airport measured by enplanements, with more than 43 million occurring in 2010 (22). Charlotte, the 11th busiest airport in the nation, served more than

18 million passengers in 2010. Both of these airports serve as important hubs for nationally based air service providers. The Mega-region is also served with numerous smaller airports including: Birmingham-Shuttlesworth International Airport (AL), Columbia Metropolitan Airport (SC), Greenville-Spartanburg International Airport (SC), Piedmont Triad International (Greensboro, NC) and Raleigh- Durham International (NC).

The transportation infrastructure of the region is a critical issue for PAM. Continued growth and prosperity requires an efficient and reliable transportation network. According to a participant of a forum hosted by Georgia's Institute of Technology, "Transportation is an intergovernmental activity…It involves the cities, it involves the states, it involves the regions, and we need to develop policies and funding systems that reflect everyone's needs" (America 2050 Forum 2009).

7.3.3.3 Economy

The Piedmont Atlantic Mega-region has undergone a dramatic economic transformation over the last century and as of 2010 represented 4 % of the total U.S. GDP (Regional Plan Association 2012c). An economy that was once based in low skilled manufacturing (i.e. textile, furniture) and agriculture (tobacco) has been forced to reinvent itself to meet the challenges of a global economy. Traditionally, the Piedmont Atlantic Mega-region was able to depend on low cost workforce, low cost of living and few regulations to attract and retain business. Industries that once benefited from those characteristics have taken advantage of opportunities overseas in recent decades.

The economic spine of the Piedmont Atlantic Mega-region is the I-20/I-85 corridor that runs through the major population and employment centers of Atlanta, GA, Birmingham, AL, Charlotte, NC, and Raleigh, NC. These urban centers serve as important growth poles for the Mega-region. While these centers of economic advancement share some similarities, they all cater to different economic sectors. Atlanta, GA has developed into the pseudo-capital of the Mega-region and serves as an important inland port and center of logistics and trade. Birmingham, AL, once dependent on the iron and steel industries has more recently developed a growing healthcare sector. Charlotte, NC is the second largest center of finance and banking in the country. Finally, Raleigh, NC and the Triangle (Raleigh-Durham-Cary, NC) have utilized the presence of several institutions of higher learning to become a hub of research and development for the technology and pharmaceutical industries.

These centers of economic success are juxtaposed to the realities faced by many of the medium and smaller metropolitan areas of the Mega-region. Many metropolitan areas still rely on manufacturing to provide jobs for their residents which results in these areas having higher levels of unemployment and poverty. For example, poverty rates can range from 41 to 80 % in the periphery of the region. It is clear that the economic success of the Mega-region lies with the core communities along the I-20/I-85 corridor, but attention must be given to the growing disparities of the outlying areas.

7.3.3.4 Prospects and Challenges

Over the last fifty years the Piedmont Atlantic Mega-region has witnessed explosive growth fueled by the migration of people and businesses out of the Northeast and Midwest and into the Sunbelt. This migration has transformed a region once perceived as "backward" or a "backwater" into a major force nationally. The strengths of the Mega-region lie in the strong record of economic development that has seen Charlotte, NC become a national banking hub; Atlanta, GA develop into a logistics and trade center for the southeast; and Raleigh, NC become a national center of research and development. Additionally, other smaller urban centers in the region have been successful in attracting automobile manufacturing plants and other industries to replace lost textile, furniture and tobacco employment generators. The Mega-region must build from this strength to ensure a successful future. However, the Piedmont Atlantic Mega-region is still developing and faces many challenges.

Some of the challenges confronting the Mega-region include political fragmentation, land consumption and an economic divide between the larger urban centers and surrounding communities. PAM covers 6 states, more than 500 counties and 4,000 municipalities that range in size from a few hundred residents to more than half a million (Hennie 2008.). These political organizations make it difficult to complete any regional planning that would benefit the entire Piedmont Atlantic Mega-region and this excludes the plethora of Council of Governments, Metropolitan Planning Organizations, and Chambers of Commerce (Godschalk 2007).

The rate at which land is consumed is another potential obstacle. The Mega-region is known for its suburban pattern of development and has experienced high levels of land consumption. On average, 650,000 acres of land are consumed by development each year. Atlanta witnessed an 81 % increase in land that was used for urban purposes while its population only increased 61 %. Likewise, Charlotte had a 71 % increase in urbanized land to support only a 39 % increase in population. These consumption patterns are not sustainable and need to be addressed.

Finally, there is a growing economic divide between the major growth centers of the region and the surrounding communities. The Atlanta, Birmingham, Charlotte and Raleigh Metropolitan Areas all witnessed healthy annual growth rates over the last twenty years. However, many of the medium to smaller sized communities that have had a more difficult time transitioning away from the manufacturing economy of old have not been as fortunate. These communities can act as a burden on the overall Mega-region and undermine the potential of the Piedmont Atlantic Mega-region.

7.4 Core Cities in Mega Regions of the U.S.

7.4.1 Washington D.C.

Washington D.C., formally the District of Columbia and commonly referred to as Washington, "the District", or simply D.C., is the capital of the United States of

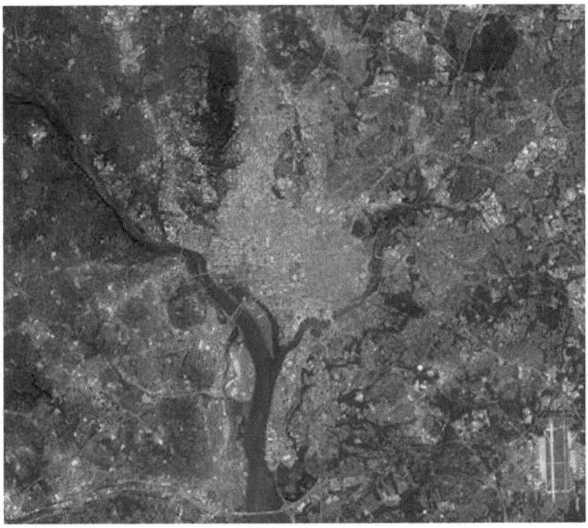

Fig. 7.5 Remote-sensing image of Washington D.C.

America. The signing of the Residence Act of 1790 approved the creation of the capital district. As permitted by the U.S. Constitution, the District is under the exclusive jurisdiction of the United States Congress and is therefore not a part of any U.S. state.

Washington D.C., located at 39°N, 77°W., is in the humid subtropical climate zone and exhibits four distinct seasons. Its climate is typical of Mid-Atlantic U.S. areas removed from water bodies. The average temperature in June and August is 25/26 °C, thunderstorms are frequent; during a typical year, the city averages about 37 days at or above 32 °C. Spring and fall are warm and long, while winter is cool with average temperature of 1.4 °C in January (Fig. 7.5).

The capital area was selected near the Potomac River and comprised parts of Maryland and Virginia. In 1791, the name of the capital city was chosen as Washington- in the honor of the first U.S. President George Washington. It was placed in a new "District of Columbia" which remained under federal authority (Bowling 1991). The original plan for the new capital city and federal district was designed by Pierre (Peter) Charles L'Enfant in accordance with contemporary European ideals of planning a capital city (Boquet 2010). The plan divided the city into a grid of North-South and East-West running streets as well as broader diagonal thorough ways. L'Enfant included in his layout the "President's house" (the later White House), the "Congress house" (U.S. Capitol) and the wide East West running garden area which eventually would become the National Mall. The plan was slightly revised and completed by Andrew Elliott after L'Enfant's dismissal in 1792. It was not until 1800/01 that the Congress began to convene in the new capital city. During the time period 1789 to 1800 the deliberations of the First and Second U.S. Congress were held in New York. The first U.S. President who moved into the "White House" and conducted his official functions there was John Adams.

Over the past two hundred years Washington, D.C. has changed considerably not only politically, economically and socially but also in terms of its physical appearance and the quality of the space. A multitude of new government offices, cultural and educational institutions and numerous memorials of national importance have been added on over the years. The Washington population continues to grow and to change. In 2010, the District of Columbia is home to 632,000 people. The Washington D.C. Metropolitan Area counts 5.58 million people who live in D.C. and surrounding counties of the States of Maryland, Virginia and West Virginia. In the past three decades changing transportation needs have contributed to a new subway system, two airports and the Capital Beltway. The latter is now widely considered the de facto boundary of the capital city displaying a specific political culture "inside the beltway". With an area size of only 68 square miles (177 km^2), the District of Columbia makes up a relatively small territory for a major capital city. Though, the core area of Washington was well planned and organized; it was able to contain all the functions of a capital throughout the 1800s. In later years, in particular after WW II, federal offices, such as the Department of Defense (Pentagon Building in Arlington, VA completed in 1945), were relocated beyond D. C. to nearby communities.

Washington is the unique center of political power with a distinct political landscape. It was the McMillan Plan of 1901 which formed an important extension of the symbolic lay-out of the capital city. The plan, an expression of the City Beautiful Movement, would considerably add to the aesthetic qualities of Washington and helped to establish among others the National Mall in the current form. It places the U.S. Capitol and White House—connected by a diagonally running grand avenue (Pennsylvania Avenue)—into the wider context of the D.C. power landscape (Boquet 2010). The National Mall and Memorial Parks includes such prominent structures as the Washington Memorial, the Lincoln Memorial and the Jefferson Memorial. The Washington and Lincoln memorials are visually bound together by a reflecting pool while the specific location of the Jefferson Memorial on the southern end of the Tidal Basin provides another important Washington vista. Meanwhile several other monuments have been erected in this garden and park portion of D.C., most notably the Vietnam Memorial (1982), the President Franklin Delano Roosevelt Memorial (1997) and a memorial for Civil Rights leader Martin Luther King Jr. (2011). Another addition to the power landscape of Washington is the Federal Triangle, with 10 key buildings built mostly in the 1930s. They fill the triangle shaped space between 15th Street, Constitution Avenue (which marks the northern boundary of the National Mall) and Pennsylvania Avenue. The largest federal office building, the Ronald Reagan Building, was dedicated in 1998.

On the northern and southern boundary of the National Mall several outstanding museums, such as the National Museum of American History, the National Gallery of Art and the Smithsonian, are located. Other buildings near the Mall honor the role of ethnic minority groups in American society such as the Jewish Community (in the U.S. Holocaust Memorial Museum established in 1993) and the Native Americans (in the National Museum of American Indian opened in 2005). All the above mentioned museums and memorials are part of a heavily visited national heritage

Fig. 7.6 The White House

in D.C. The power landscape of Washington, D.C. takes center stage during the presidential inauguration ceremonies which happens every four years (the last time January 20, 2013). The president is sworn in front of the U.S. Capitol and after a parade along Pennsylvania Avenue reaches his office, the White House. The public is invited to attend, and the whole core area of D.C., from the western edge of the National Mall to the Supreme Court Building and the Library of Congress East of the Capitol, is populated by large crowds of spectators. This carefully crafted D.C. event signifies a celebration of American Democracy (Figs. 7.6, 7.7).

7.4.2 New York City

New York City, the most populous city in the United States, is the center of the large New York Metropolitan Region, which is one of the biggest mega regions in the world.

New York City is located at 40°N, 74°W in a prominent coastal situation of the Mid-Atlantic Region. The region's climate is strongly affected by the nearby Atlantic Ocean, and it is classified—despite cold winters—as humid subtropical. Summer temperatures can go up to close to 40 degrees Celsius (above 100 degrees Fahrenheit).The City was first settled at the southern end of Manhattan commonly referred to as Lower Manhattan (with the 'Financial District'). During the 1790s and throughout the 1800s, New York's population moved North past Canal Street to Greenwich Village and the East Village direction 'Mid-Town' Manhattan and

Fig. 7.7 Remote-sensing image of New York city

eventually to the Upper East and Upper West Side separated by Central Park, the first large planned urban park established in the U.S. (1857–1873). Midtown Manhattan serves as a second financial hub of the city, with the bulk of the skyscrapers including the Empire State Building erected there in the 1920s/1930s. The local geology, with hard schist rock formations, allowed the safe construction of Manhattan's unique skyline. While the foundations of the two towers of the World Trade Center in Lower Manhattan were built, the excavated soil was used for land reclamation—to add on Battery Park City to the most southern tip of Lower Manhattan.

New York was founded as New Amsterdam, a Dutch trading post on Manhattan Island, in 1626, more than 150 years earlier than Washington, DC. When the British took control of the settlement in 1664 it was renamed New York. The city has an excellent natural harbor which favored trade from Great Britain and the transfer of goods along the Hudson River into the interior. By 1790 New York surpassed Philadelphia by population volume. New York served as capital for the young nation from 1785 to 1800 and continued to grow by leaps and bounds. In 1898, the five boroughs Manhattan, Brooklyn, Queens, the Bronx and Staten Island were consolidated into New York City (Burrows and Wallace 1999) which evolved as the major commercial center in North America. In the 1920s, New York attained the status of a world city even surpassing London as the world's leading financial center. The current population of New York City is estimated at 8.3 million (2012); the New York Combined Statistical Area (CSA) comprising adjoining counties in New York State, New Jersey, Connecticut and Pennsylvania has 22.1 million residents (2010).

New York City may well be the most culturally diverse city of the world. It is estimated that over 800 languages are spoken in the five boroughs of the city. Queens and Brooklyn, with a multitude of ethnic neighborhoods and enclaves, show racially and culturally the greatest diversity within the city, while Staten Island has remained the only borough with a white non-Hispanic majority. The two largest ethnic/racial groups in the Bronx are Americans with Hispanic ancestry, mostly from Puerto Rico, and Afro-Americans. Early on, many distinct national or ethnic groups, first from Europe then from other parts of the world, made their home in New Amsterdam/New York depending on immigration laws enacted during the different time periods. Manhattan serves as an example of longtime persisting as well as ever changing ethnic neighborhoods in New York. Manhattan which reached with 2.3 million its population peak in 1910 (compared to 1.6 million in 2010) has had a large number of distinct ethnic neighborhoods during the 1800s and 1900s. Little Italy, Chinatown, the Lower East Side (with a large community of Eastern European Jews), Korea town in Midtown East, Hell's Kitchen (a predominantly Irish neighborhood), Yorkville on the Upper East Side (with a large community of Germans, Czechs and Hungarians), East Harlem (with Puerto Rican/Latino neighborhoods also called 'Spanish Harlem') and Harlem, the longtime center of the Black community in the city, are past and present names of such ethnically/racially defined places in Manhattan.

New York is the leading financial center nationally and worldwide. The New York Metropolitan Area economy currently ranks highest by GDP in the United States and is second only to Tokyo worldwide. A spectacular rise of the city economy occurred in the 1920s while the mid/late 1970s saw New York in a crisis and in economic decline, in particular its manufacturing sector. New York's harbor location and far reaching distribution system helped to build a well-diversified industrial complex in the region which eventually attracted a large number of corporate headquarters. The beginnings of a stock exchange market go back to the 1790s. Currently, the city houses, in particular in the Financial District of Lower Manhattan, the New York Stock Exchange (NYSE) and NASDAQ, the two world largest stock ex-changes by market capitalization. Many leading companies in the financial sec-tor have chosen to have their headquarters either in the Financial District or in

Fig. 7.8 Skyline of Manhattan

Midtown Manhattan. Financial services account for about 35 % of New York City's employment income. Manhattan is home to altogether six major stock, commodities and futures exchanges. The NYSE building along Wall Street in the Financial District has become a tourist attraction. In a wider sense, Wall Street is considered a symbol of the financial power New York exerts nationwide and worldwide. Wall Street has generated many positive and negative associations, from the often quoted contrast of 'Wall Street' versus 'Main Street' in the public debate in the U.S. to the recent 'Occupy Wall Street' movement.

Another location in New York City with high place recognition is Broadway, with Times Square in its center. Manhattan is home to several dozen theater companies and stages for musicals and shows 'on' and 'off Broadway'. Besides its prominence in the entertainment business New York is also widely recognized for its outstanding role in the media and advertising, with the "New York Times" and the "Wall Street Journal" leading in the national print media, many influential advertising agencies and publishing houses, and the major national TV broadcasting corporations such as NBC housed in Rockefeller Center. The visual and classical performing arts have great museums and stages in New York, with among others, the Metropolitan Museum of Art, the Lincoln Center for the Performing Arts and Carnegie Hall. New York has also become an important location for film ventures and productions. Last but not least it should be mentioned that New York continues to be one of the world's leading fashion centers with many implications for the local economy (Fig. 7.8).

7.5 Defining "Urban", Urban Development Patterns and Trends in China

7.5.1 Defining "Urban" China

Cities and regions in developed countries have experienced a process that has transformed individual cities and metropolitan areas into massive urban agglomerations and Mega-regions. Most research focuses on the state of individual city and urban agglomeration and ignores the intermediate process of urban development, which leads to the confusions of concepts, such as urban agglomeration, metropolitan area, and Mega-regions.

In the literature on urban development in China, there are various opinions on the development process of urban area and urban structure. For example, Cui divided the structure of urban areas into three types: urban region, urban agglomeration, and Mega-region (Cui 1991). Zhu divided the evolution model into four stages: dispersed individual core city, urban clusters, the expansion stage of urban agglomeration, and the formation stage of Mega-region (Zhu et al. 2002). Wang and Wu proposed a five-stage process: individual city, urban clusters, urban groups, urban agglomeration and Mega-regions (Wang and Wu 2008). Zhang argued that a central city and its surrounding urban areas experienced four stages of development: urban area, urban clusters, metropolitan circle and megalopolis. A megalopolis is composed of several urban agglomerations and an urban agglomeration includes several metropolitan areas. Metropolitan areas are the sub-units of a megalopolis (Zhang 2009).

The concept of urban belt refers to a huge urban region that is constituted by a number of socially, economically, and culturally interconnected metropolitan areas. Many researchers in China treat the concepts of urban belt, urban cluster and megalopolis the same. Zhou (1995) comprehensively compared and analyzed these concepts and proposed that the concept of megalopolis is the same as the concept of Metropolitan Interlocking Region (MIR) that he defined. An MIR is composed of various cities as its core area, with one or more transportation corridors. The major cities and the surrounding areas located within an MIR develop into a giant urban-rural integrated region with strong interaction and close socio-economic connections. Liu (2006) argued that an urban belt is organized and coordinated by one or a few core cities in space during urbanization process. Core cities play a significant role in coordinating a number of cities at various scales to constitute an urban network by highly developed transportation and information networks and spatial interactions.

Currently, the concept of urban agglomeration in the book under the same title published in 1992 is relatively agreeable by researchers in China (Yao et al. 2006). In this book, urban agglomeration is defined as a considerable number of cities with different nature, types and scales in a specific geographical region. Based on the natural environment, one or two megacities form the growth poles for the regional economy and constitute an integrated urban "agglomeration" with access to the transportation network and internal links between cities. Three conditions are necessary for an urban agglomeration: First, there are a considerable number of cities of different types; second, there is more than one megacity as a regional center; third, there are internal links between cities.

An urban agglomeration has various characteristics: First, an urban agglomeration is not only a number of cities that are densely distributed in space, but also an organic integration that links with modern means of transportation and highly developed transportation and information networks. Second, an urban agglomeration is composed of a number of cities with one or more core cities. It is a giant polycentric urban system with continuity and strong internal interactions, and each core city has its own complete urban system. Third, there is a series of interactions between urban and rural areas as the cities grow. The non-agricultural industries grow rapidly in suburban and rural areas and result in a rising level of urbanization. Fourth, the polarization and spillover effects of core cities are obvious in urban agglomerations. The dual effects strengthen the regional core structure and constantly improve the functions and scale of urban agglomerations (Yao et al. 2006).

The combination of internal and international networks in urban belt promotes the convergence of population, ideas, capital, material and information flows, which influences the policy making of the national socio-economic and cultural development strategies. With the interaction of various flows, urban belt constantly produces new ideas, methods, technologies and products which promote progresses and innovations in society.

The six megalopolises in the world identified by Gottmann (1961) are all directly connected to the open sea except the urban belt of the Great Lakes in the United States. It is, however, connected to the Atlantic Ocean through the St. Lawrence River. Portal location of urban belts enables the flows of resources, technologies, and ideas to other regions and countries, which affects the development of the world economy (Yuan et al. 2007).

7.5.2 Urban Development Patterns and Trends

The history of Chinese cities date back to the Western Zhou Dynasty (1046–771BC). The spatial pattern of urban agglomerations in China, however, was dramatically changed after the foundation of the People's Republic of China in 1949. From 1949 to 1957, both China's urbanization rate and number of cities were significantly increased due to the national economic recovery from the Anti-Japanese War and the War of Liberation. The number of cities under 500,000 population increased from 98 in 1949 to 140 in 1957 (Lu 2007). Industrial cities, such as Baotou, Lanzhou, Xi'an, and Chengdu, emerged in central and western China, while some existing cities in the Yangtze River Delta, Central Liaoning Province, and Beijing-Tianjin area were further developed. The "Great Leap Forward" and the "Cultural Revolution" in China effectively slowed down the process of urban development in China after 1958. Since the end of the "Cultural Revolution" and the opening of China to the West, the urbanization rate has significantly increased. The central government proposed a series of national development strategies, such as the "T"-shape development strategy

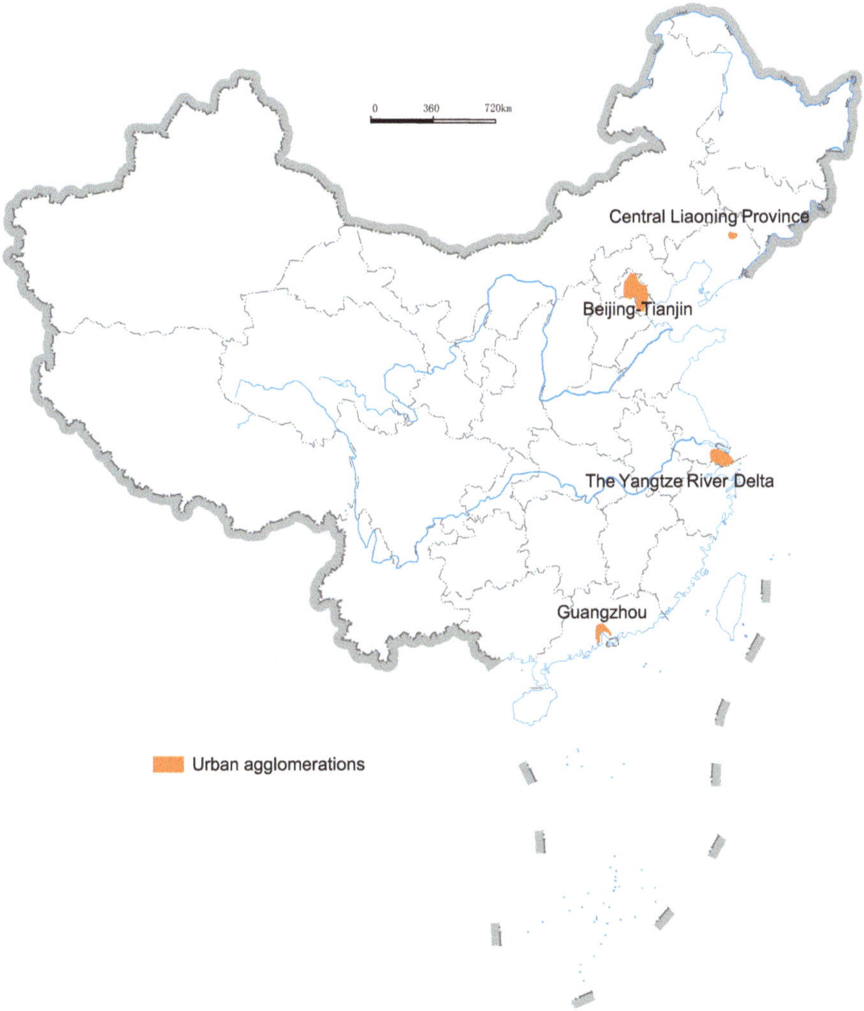

Fig. 7.9 The urban agglomerations in China in 1949

in Yangtze River Delta, the development of West China, and the revitalization of the traditional industrial base in the northeast (Miao and Wang 2005).

The development process and spatial distribution pattern of urban agglomerations in China from 1949 to 2003 can be divided into five stages (Guand Pang 2007).

7.5.2.1 Germination

In 1949, Beijing-Tianjin, Central Liaoning Province and the Yangtze River Delta started to develop urban agglomeration, even though the influence each area had on each region was small at this stage (see Fig. 7.9).

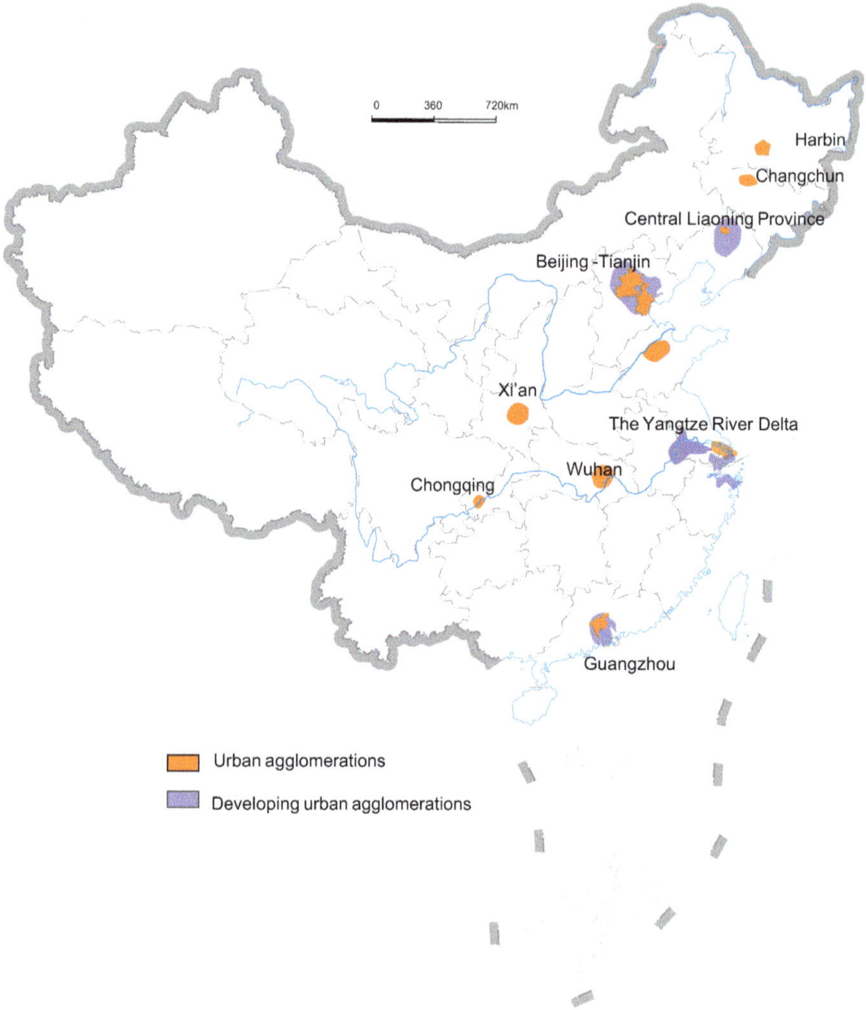

Fig. 7.10 The urban agglomerations in China in 1975

7.5.2.2 The Second Stage

By 1975, urban areas in Beijing-Tianjin, Central Liaoning Province, and the Yangtze River Delta further developed and expanded. Central Liaoning Province and the Yangtze River Delta formed urban agglomerations at a certain scale. Population began to centralize in Harbin, Changchun, Wuhan, Chongqing, Guangzhou and Xi'an (see Fig. 7.10). By the end of this stage, there were two urban agglomerations and seven developing urban agglomerations.

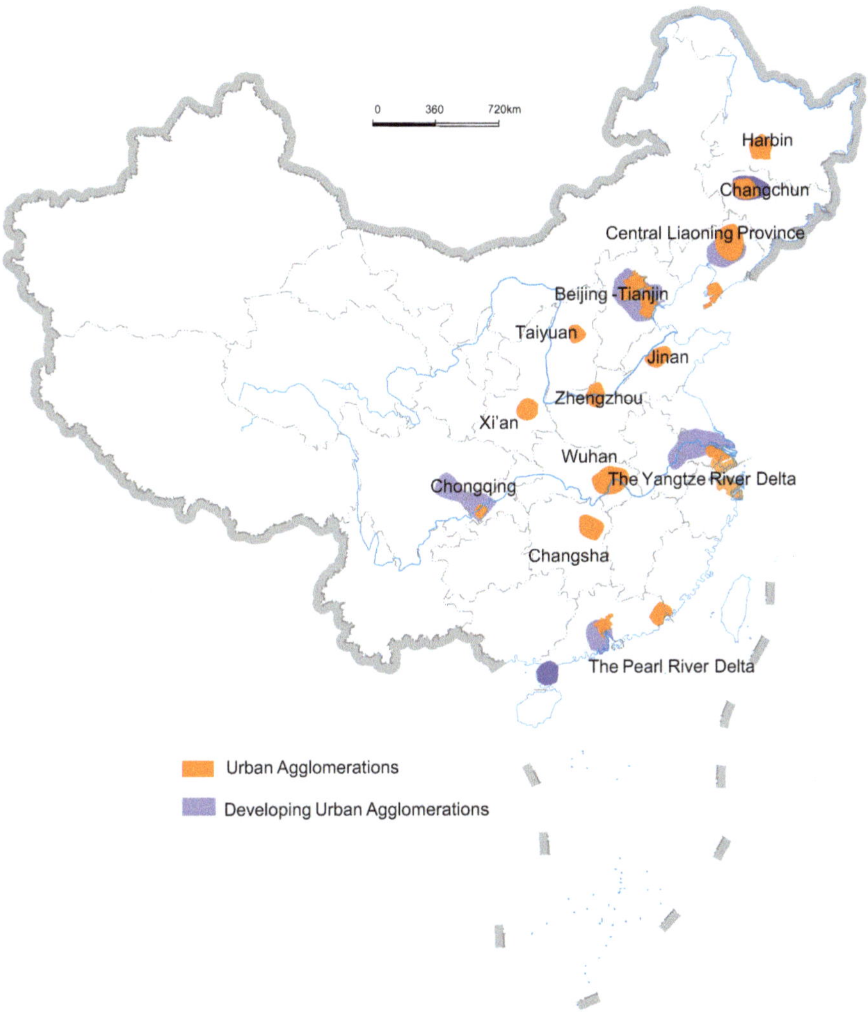

Fig. 7.11 The urban agglomerations in China in 1985

7.5.2.3 The Third Stage

By 1985, the Beijing-Tianjin area and the Pearl River Delta formed two new urban agglomerations. The urban agglomerations in the Central Liaoning Province and the Yangtze River Delta continued to expand, and population continued to centralize in Harbin, Changchun, Wuhan, Chongqing, Guangzhou, and Xi'an. Jinan, Taiyuan, Zhengzhou and Changsha were four emerging cities that attracted population concentration (see Fig. 7.11). At the end of this stage, there were four urban agglomerations and nine developing urban agglomerations.

7.5.2.4 The Fourth Stage

By 1995, multiple urban agglomerations were developed at various scales in China. The larger ones were Central and South Liaoning Province, Beijing-Tianjin-Tangshan area, the Central Plains, Shandong Peninsula, the Yangtze River Delta, Wuhan, Chengdu-Chongqing, and the Pearl River Delta urban agglomerations. Among the eight areas, the Central Plains-Central Shanxi- Central and South Hebei, Shandong Peninsula, the Yangtze River Delta and the Pearl River Delta were relatively more developed and had a potential to form Mega-regions. The Central Plains agglomeration stretched towards north to Shijiazhuang and Taiyuan with Zhengzhou as its core area. Shandong Peninsula agglomeration stretched southwest towards the border of Jiangsu, Shandong, and Anhui provinces with Jinan and Qingdao as its core area. The Yangtze River Delta agglomeration expanded towards the west to Central Anhui Province with Shanghai-Nanjing-Hangzhou as its core area. The Pearl River Delta agglomeration stretched towards the two wings. In addition, Changsha-Zhuzhou-Xiangtan, Lanzhou, Central and South Hebei Province, North Jiangxi Province, the border of Jiangsu, Shandong, and Anhui provinces, Central Yunnan Province, Central Guizhou Province, and the North Slope of Tianshan Mountains started to develop urban agglomerations (see Fig. 7.12). By the end of the stage, there were eight urban agglomerations and eleven developing urban agglomerations.

7.5.2.5 The Fifth Stage

By 2003, Southeast Fujian Province and Baotou in Inner Mongolia were two new developing urban agglomerations. The existing urban agglomerations and their core areas continue to expand. The "Yangtze River Delta, Shandong Peninsula, the border of Jiangsu, Shandong, and Anhui provinces", "Beijing-Tianjin-Hebei, Central Plains, Central Shanxi Province" and the Pearl River Delta formed larger scale Mega-regions (see Fig. 7.13). Currently, there are eleven urban agglomerations and nine developing urban agglomerations.

7.6 Mega-Regions in China

Currently, the urban agglomerations can be divided into three levels based on GDP, because the economic scale of an urban agglomeration directly impacts its radiation area (Table 7.1). The highest level is the developed urban agglomerations of national significance, such as the three major urban agglomerations on the east coast of China. The GDP of the three urban agglomerations on the top level contribute for around 40 % of the national GDP. Urban agglomerations on the second level

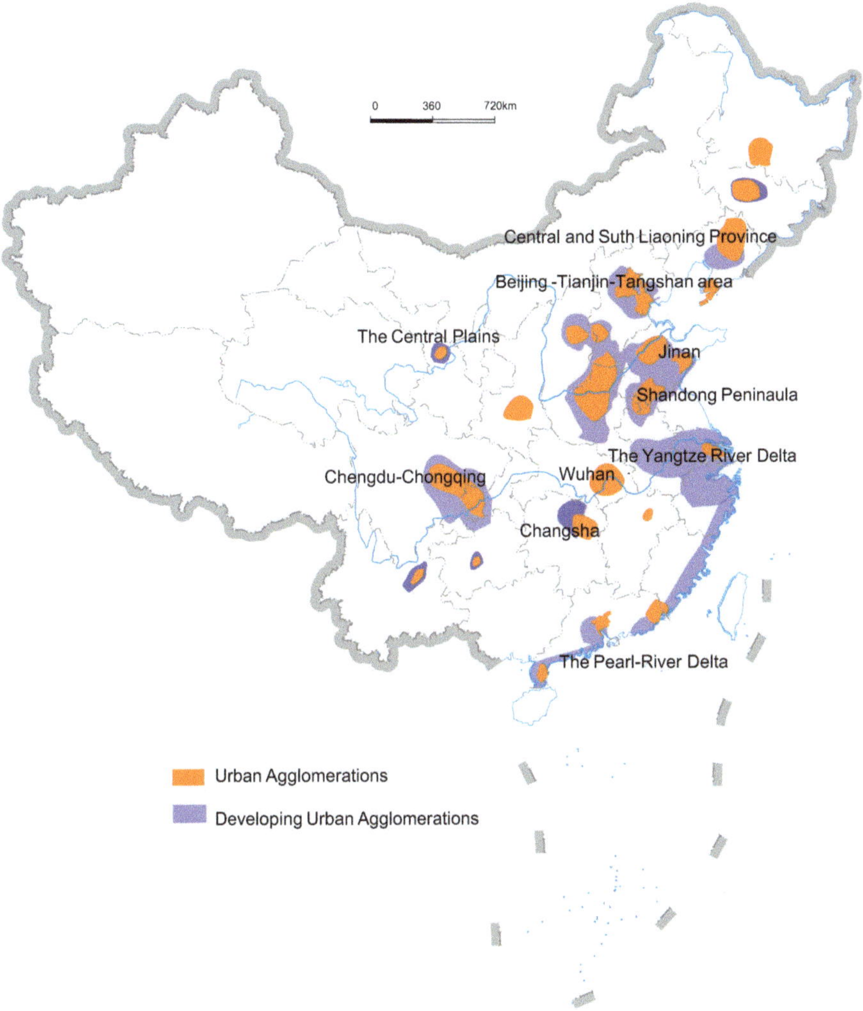

Fig. 7.12 The urban agglomerations in China in 1995

play important roles in regional economy. The urbanization level of the urban agglomerations on the third level is relatively low. The contribution rate to the national GDP is lower than one percent in the third level urban agglomerations, but they are significant at the provincial level. The following section focuses on the three Mega-regions that are the highest level urban agglomerations in the country. An overview of the geography and location advantages, economy, regional industrial specialization and cooperation, and spatial structure is provided for each Mega-region.

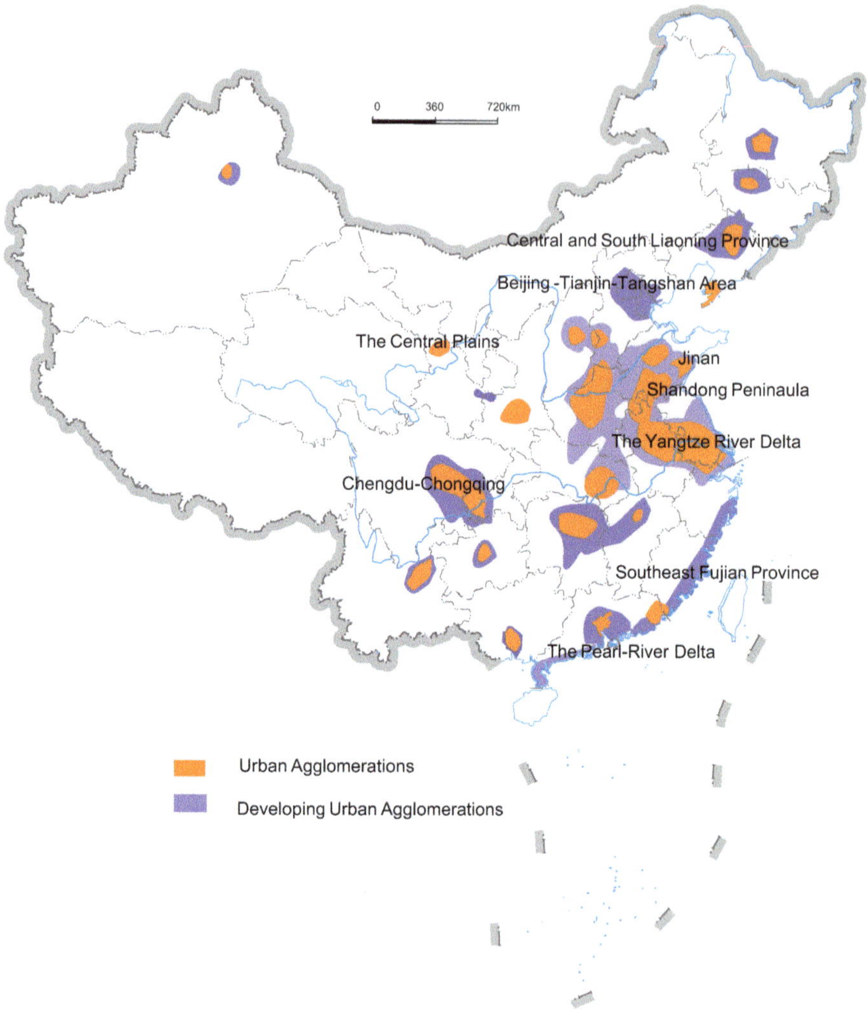

Fig. 7.13 The urban agglomerations in China in 2003

Table 7.1 Population, Area and GDP of the three Mega-regions. (National Bureau of Statistics of China 2010)

	Population (10,000)	Proportion (%)	Area (km^2)	Proportion (%)	GDP (10,000 Yuan)	Proportion (%)
Beijing-Tianjin-Hebei	7344.49	5.50	182601	1.90	335325554	9.85
Yangtze River Delta	11527.87	8.64	167521	1.75	685380121	20.13
Pearl River Delta	2967.02	2.22	55036	0.57	321470045	9.44
Total	21839.38	16.36	405158	4.22	1342175720	39.42

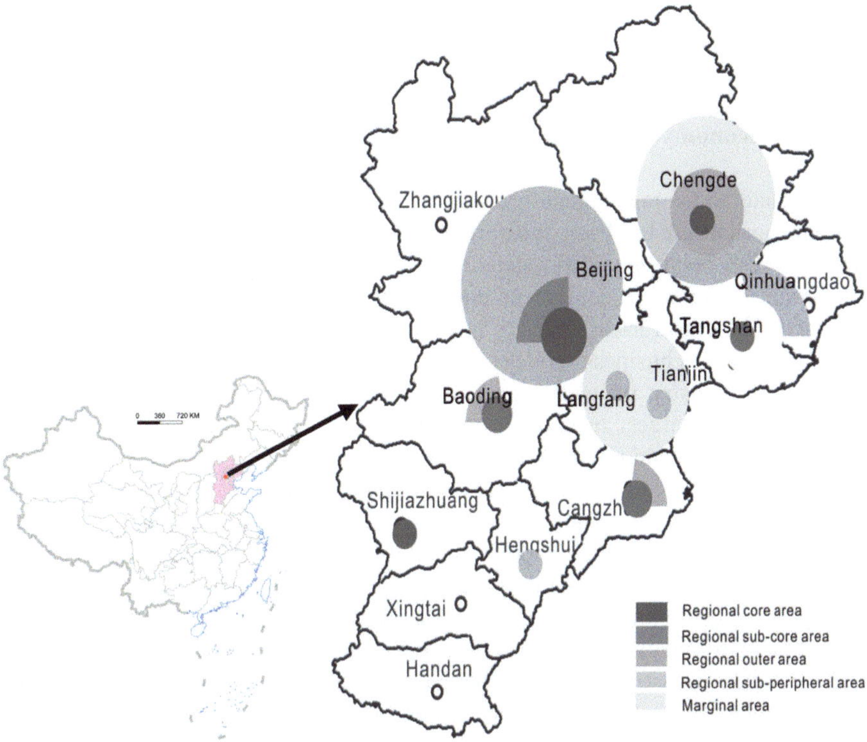

Fig. 7.14 The location and spatial structure of Beijing-Tianjin-Hebei Megaregion

7.6.1 Beijing-Tianjin-Hebei Mega-region

7.6.1.1 Geography & Location Advantages

The Beijing-Tianjin-Hebei Mega-region was formed in the middle 1980s. Its administrative area includes 10 cities, which are Beijing, Tianjin and the other eight cities in Hebei Province (Shijiazhuang, Tangshan, Qinhuangdao, Baoding, Zhangjiakou, Chengde, Cangzhou, and Langfang). The Mega-region is located at the junction of the North, Northeast and East China with an area of 185,000 km^2(1.9 % of the total area in China) and 73.4 million population (5.5 % of the total population in China in 2010). It is the political and cultural center of China with a high density of population and economic activities (see Fig. 7.14).

Since the "Eleventh Five-year Plan" (2006–2010), the national development strategy has planned to develop Tianjin costal area, Pudong in Shanghai, and Shenzhen as the three poles for regional growth (Zhu 2009). Beijing is positioned as "the National Capital, an international city, a cultural city, and an age friendly city", while Tianjin is entitled as "an international port city, the economic center of North China, and an eco-city". Beijing and Tianjin share complementary advantages.

7.6.1.2 Economy

Beijing had an average annual growth rate of nearly 11 % during the 1990s. Traditional agriculture has been transformed to modern agriculture, and the industrial structure has been upgraded. Metallurgy, petrochemical industries, automotive, electronics, machinery, equipment, and high-tech industries have been rapidly developed. The construction industry has a leading position in the country. Service industries such as business services, financial sector, tourism, transportation, communication industry, real estate, and information services are flourishing. Beijing is also China's largest technology and intellectual-intensive area supported by many universities and research institutions. The vast majority of the national bank headquarters are located in Beijing and foreign banks also set up offices in Beijing to carry out extensive domestic and international business, which enable Beijing to take advantage of capital markets. In addition, Beijing is a megacity with more than 17 million people which provides a large consumer market. The main challenge for the economic development in Beijing is that the industrial structure and layout have not met all the requirements of a capital city. The major industries in the secondary industry grow at a slow pace and a large proportion of the tertiary industry provides traditional services. Emerging industries are lagging behind economic development due to severe resource constraints, especially water shortages, limited land resources, and environment pollution.

The economic scale of Tianjin is much smaller than Beijing. Secondary industry has played an important role in Tianjin's economy and its average growth rate has been the highest in North China since 2000. Tianjin is an important port that serves the north and northwest provinces in China. Tianjin Port has established long-term shipping services and trade with many countries and regions. It is also the starting point of the Eurasian Continental Bridge by railway transportation. Tianjin has various types of natural resources including oil, crude salt, coal, natural gas, geothermal, mineral and marine resources. The abundant land resources in Binhai New Area of Tianjin provide favourable conditions for its economic development. In recent years, the four pillar industries in Tianjin have been automobiles, electronics, chemicals and metallurgy. The challenges for the development of Tianjin are that the tertiary industry lags behind and some state-owned enterprises have financial difficulties for operation. The mode of economic growth is relatively extensive and people's income level is relatively low compared to Beijing.

Hebei province actively cooperates with Beijing and Tianjin to boost the regional economic development of this Megaregion over the past 20 years. Tangshan, Shijiazhuang and Langfang are cities that have achieved rapid economic development. Some central cities, such as Zhangjiakou, Qinhuangdao, and Baoding also significantly enhance the economic strength in the region. The development of these central cities balances the economic development in the Beijing-Tianjin-Hebei Megaregion and narrows the gap with other developed coastal provinces (Liang et al. 2009).

7.6.1.3 Regional Industrial Specialization and Cooperation

Regional industrial specialization is based on the conditions and advantages of the leading industries. Hebei Province's primary industries are agriculture and animal husbandry, which provide needed agricultural products for Beijing and Tianjin. Tianjin focuses on the development of secondary industries to the region. Meanwhile, Beijing has advantages in the development of the tertiary industry, such as transportation, post and telecommunications industry, finance and insurance, real estate, wholesale and retail, and catering industries (Zhang 2008).

Currently, the Beijing-Tianjin-Hebei Megaregion is at a stage of integration. The industrial cooperation within the region is based on new industrial specialization and gradually forms an industry chain, which will be a benefit for joint development, dislocation competition, and the win-win industrial pattern (see Fig. 7.11). In the future, Beijing will focus on the development of the tertiary industry, and secondary industry will be shifted to the high-end development. The most dynamic industries in Beijing will be transnational headquarters, high-tech industries, financial management, intermediary services, cultural and creative industries, and logistics. Tianjin will focus on the development of heavy chemicals, high-tech industries and logistics. The most dynamic industries in Tianjin will be the automotive industry, electronics industry, the petrochemical industry, the aerospace industry, the shipbuilding industry, the pharmaceutical industry, and warehousing logistics industry. Hebei Province will develop ten pillar industries. The most dynamic industries will be the iron and steel industry, the pharmaceutical industry, petrochemical, equipment manufacturing, building materials industry, and textile industrial. Beijing-Tianjin-Hebei Megaregion has all types of industries and a considerable advantage for industrial development.

7.6.1.4 Spatial Structure

The spatial structure and economic structure of the Beijing-Tianjin-Hebei Megaregion fit into Burgess's Concentric Zone model (see Fig. 7.15) (Burgess 1924). The industries of the core area, the sub-core area, peripheral area and marginal area from the center outwards are producer services, general services, industry and agriculture, forestry, animal husbandry and fishery, or the general service sector, industry, agriculture, forestry, animal husbandry and fishery. Cities such as Baoding, Zhangjiakou, Chengde, Tangshan, Cangzhou, and Qinhuangdao are gradually connected to the Beijing-Tianjin core area, while some other cities such as Hengshui, Xingtai, Shijiazhuang are dispersed located (Yu and Wu 2006).

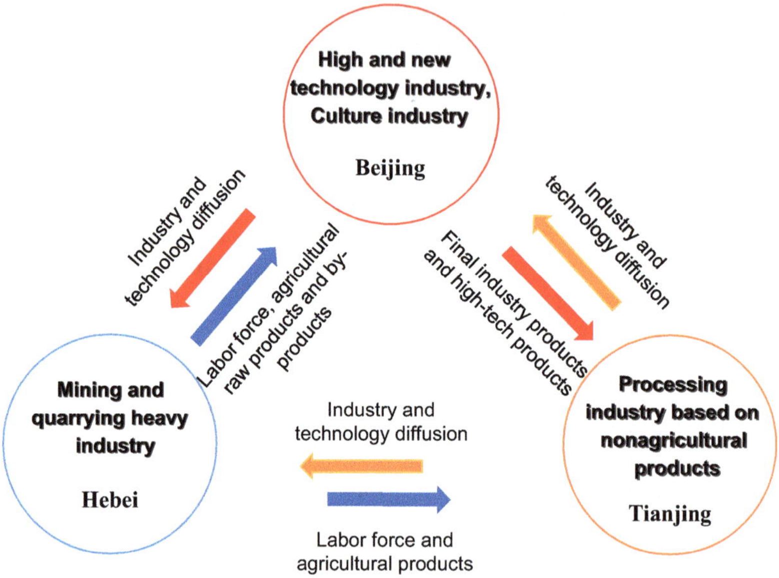

Fig. 7.15 Regional cooperation model of Beijing-Tianjin-Hebei Megaregion

7.6.2 Yangtze River Delta Mega-region

7.6.2.1 Geography & Location Advantages

The Yangtze River Delta Mega-region is the most developed and influential Mega-region in China, which is also known as the sixth largest Mega-region in the world. The administrative area of this Mega-region includes Shanghai, nine cities in Jiangsu Province (Nanjing, Suzhou, Wuxi, Changzhou, Zhenjiang, Yangzhou, Xuzhou, Nantong, and Taizhou) and six cities in Zhejiang Province (Hangzhou, Ningbo, Huzhou, Jiaxing, Shaoxing, and Zhoushan) with Shanghai as the core city. In 2010, another four cities in Jiangsu and Zhejiang province (Yancheng, Huai'an, Jinhua, and Quzhou) and two cities in Anhui province (Hefei and Ma'anshan) joined the Mega-region (see Fig. 7.16). The entire area covers 168,000 km^2 (1.75 % of the total area in China) with 115.2 million population (8.64 % of the total population in China in 2010). In recent years, with the improvement in transportation and economic links in this region, the Yangtze River Delta Megaregion refers to a much greater region than the original region, which included Shanghai, Jiangsu Province and Zhejiang Province.

7.6.2.2 Economy

The Yangtze River Delta is the most important economic and trade region in the east coast of China. In 2004, Shanghai proposed to establish an international economic,

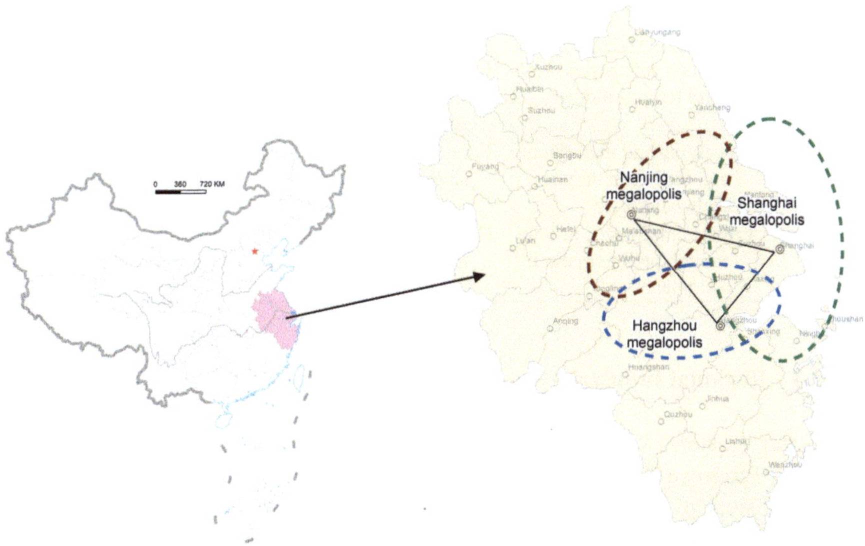

Fig. 7.16 The location and spatial structure of Yangtze River Delta Megaregion

financial, trade and shipping center. A significant number of large and medium-sized cities in the Yangtze River Delta region and the hinterland of non-agricultural industries joined in the Shanghai metropolitan area (Lu and Zhu 2004). In 2010, the State Council officially approved "Regional Planning of the Yangtze River Delta" and made strategic planning for the development of the Yangtze River Delta region. The Yangtze River Delta region is positioned as an important international gateway to the Asia-Pacific region, the center for service and advanced manufacturing industries, and a competitive Megaregion in the world.

Currently, the Megaregion is in a transitional phase from developing manufacturing industries to industrial specialization. Beginning in 2000, Shanghai gradually relocated labor-intensive industries and some capital-intensive industries to Jiangsu and Zhejiang provinces. Shanghai put forth an effort to develop port and urban industries, such as petroleum processing and coking industry and the printing industry, and reduced the share of manufacturing industries in the region. Zhejiang Province steadily absorbed the labor-intensive industries that were relocated from Shanghai and Jiangsu provinces; thereby it increased the share of manufacturing industries in the region. Jiangsu Province was balanced by absorbing and relocating the manufacturing sector and the share remained unchanged. After years of adjustment in the Yangtze River Delta Megaregion, industrial relocation and replacement contributes to the optimal configuration of the industry chain and the formation of industrial specialization.

Table 7.2 The industrial specialization and advantages of Jiangsu, Zhejiang and Shanghai

Region	Economic characteristics	Priorities	Advantages
Shanghai	High-tech enterprises, high-level tertiary industry	Economy (services), financing, trade, shipping	Technology, information, research and development, service personnel
Jiangsu	Processing and manufacturing, high-tech, large-scale enterprise	Economy (manufacturing), shipping	Industry, technology, management talents, skilled workers
Zhejiang	Private enterprises, private capital, private market, and talents	Economy (manufacturing), trade, shipping	Structures, entrepreneur human resources

7.6.2.3 Regional Industrial Specialization and Cooperation

Industrial clusters were developed in Shanghai, South Jiangsu Province and Zhejiang Province though their industrial structures are quite different. The industry clusters of services and financial services are located in Shanghai, and the industry clusters of manufacturing are located in South Jiangsu Province and Zhejiang Province. The specialization, collaboration and interaction among industrial clusters greatly contribute to the international competitiveness of the region. The industries in Shanghai and the surrounding areas are complementary. Shanghai gathered a large number of high-skilled talents and has strong technological strength, which enable Shanghai to develop high level service industry and build high-tech research and development base. Jiangsu and Zhejiang provinces have a strong basis for the processing industry and have formed a unique and sizable industry cluster. South Jiangsu Province has attracted a large amount of foreign investment and rapidly developed as the manufacturing center of IT products, especially computer equipments and integrated circuits. Zhejiang Province is also a big manufacturing province and has advantages in secondary industries, such as communications, software, traditional textile, clothing, and machinery. The Yangtze River Delta Mega-region has developed "three centers" with different functions (Table 7.2).

7.6.2.4 Spatial Structure

The Yangtze River Delta Mega-region has formed around a polycentric urban spatial pattern with Shanghai, Nanjing and Hangzhou as the core cities (see Fig. 7.16). The inner ring is the urban agglomeration with Shanghai, Suzhou and Wuxi as the core cities and has formed Shanghai-Suzhou-Wuxi Growth Golden Triangle. The influential areas extend to Changzhou and Nantong in Jiangsu Province and Jiaxing and Huzhou in Zhejiang Province. The second circle covers the area of Nanjing and Hangzhou and has formed Shanghai-Nanjing-Huangzhou Growth Triangle. As the three corner cities of the triangle and the provincial capitals, Nanjing, Hangzhou and Ningbo have powerful economic strength and have developed two relatively independent metropolitan areas. Nanjing metropolitan area includes Zhenjiang and Yangzhou in Zhejiang Province and Ma'anshan, Chuzhou and Wuhu in Anhui Province. The influential areas extend to Changzhou, Taizhou, and Huai'an in Jiangsu province. The Hangyong metropolitan area includes four cities- Hangzhou, Shaoxing, Ningbo and Zhoushan. The influential areas extend to Jiaxing, Huzhou and Taizhou City. The peripheral sphere is the radiation area of the Shanghai-Nanjing-Hangzhou Golden Triangle, including Yancheng, Huai'an in North Jiangsu Province, Jinhua and Quzhou in South Zhejiang Province, and other cities along the Yangtze River in Anhui Province (Li et al. 2006).

The links between each city and Shanghai decreases from the central area to the peripheral area of the Yangtze River Delta in a circle structure. The population density, level of economic development, and industrial structure, however, do not show the same distribution pattern that decreases from the center outwards. A series of secondary cities with relatively good development basis, large population size and rapid economic development are located in the peripheral area of Shanghai Metropolitan Area, such as Nanjing, Hangzhou, Suzhou, Wuxi, and Ningbo. These cities become dynamic economic growth poles in the Megaregion.

7.6.3 Pearl River Delta Mega-region

7.6.3.1 Geography and Location Advantages

In 1994, Guangdong Province established the Pearl River Delta economic zone, which included Guangzhou, Shenzhen, Zhuhai, Dongguan, Zhongshan, Foshan, Zhaoqing, Jiangmen and Huizhou. In 2005, the government planned "the Small Pearl River Delta—the Greater Pearl River Delta—the Pan-Pearl River Delta" triple strategy as the goal for strategic integration of the Pearl River Delta Mega-region (see Fig. 7.17). The "Small Pearl River Delta" includes seven cities—Guangzhou, Shenzhen, Foshan, Zhuhai, Dongguan, Zhongshan, and Huizhou with an area of 55,000 km^2 (0.57 % of the total area in China) and 29.7 million population (2.22 % of the total population in China in 2010). The "Greater Pearl River Delta" includes the "Small Pearl River Delta", Hong Kong and Macau. The "Pan-Pearl River Delta" includes the Pearl River basin and the surrounding nine provinces (or autonomous regions), which are Guangdong, Guangxi, Hunan, Fujian, Jiangxi, Hainan, Sichuan, Yunnan, and Guizhou, and the two special administrative regions (Hong Kong and Macau). The "Pan-Pearl River Delta" is also referred as the "9+2" region. The development of the Pan-Pearl River Delta created a bottom-up model in accordance with the demand for regional cooperation. It was a milestone in China's regional economic development and an innovation in economic cooperation.

The formation of the Pan-Pearl River Delta economic cooperation zone was affected by natural, economic, social and political factors (Li et al. 2005). The region is located in the tropical and subtropical regions. The cities in this region have

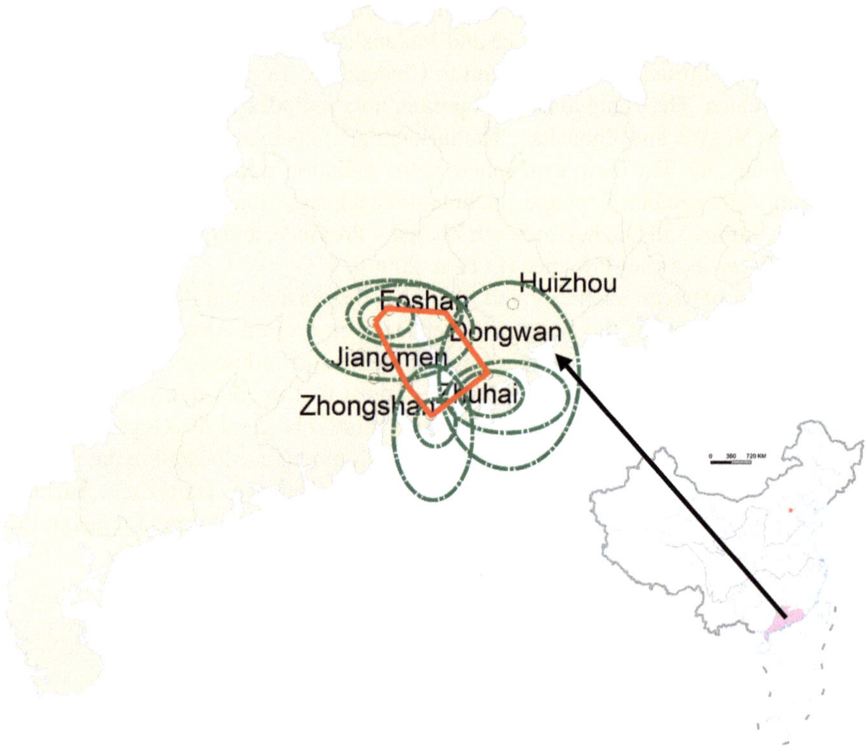

Fig. 7.17 The location and spatial structure of Pearl River Delta Mega-region

advantages in spatial proximity and rich natural resources. The spatial structure of the dominant industries is complementary although the levels of economic development in these cities are different. The economic development in this region promotes population migration and logistics flows in recent years, which also constitutes a basis for economic cooperation. In addition, the "Closer Economic Partnership Arrangement" (CEPA) between mainland China, Hong Kong and Macao, and the Association of Southeast Asian Nations (ASEAN) Free Trade Area provide great opportunities for the regional development.

7.6.3.2 Economy

The regional cooperation in Pan-Pearl River Delta Mega-region is based on the regional growth in Hong Kong and Guangdong Province. The optimized combination of production, capital, technology, human resources, information and natural resources enable the region to achieve a higher level of regional economic development. The Pearl River Delta has become the production base for the IT industry and more of a focus has been placed on quality economic growth and not just quantity. The goals of the "Tenth Five-year" plan (2001–2005) for the Pearl River Delta Mega-region were: to enhance the competitiveness of the Pearl River Delta Mega-region and become the major cities of the Asia-Pacific Region; to strengthen the function of the central cities, such as Hong Kong, Shenzhen, and Guangzhou, and improve the international influence of these modern cities; to develop a number of large cities and promote the development of smaller size cities around them; and to actively develop small and medium-sized cities and attract population migration to these cities.

7.6.3.3 Regional Industrial Specialization and Cooperation

Regional cooperation in the Pearl River Delta began in the 1980s. Hong Kong, Macao, and the Pearl River Delta experienced disorderly and excessive competition in investment and in the development of export-oriented economies. The service sector in Hong Kong failed to enter the Pearl River Delta as the manufacturing sector did due to structure limitation. The industrial structure of each city was identical. For example, electronic communications manufacturing, electrical machinery manufacturing, chemicals, plastics, pharmaceuticals, textile and garment, and food industry accounted for more than 60 % of the total industrial output value in each city. The duplicate construction in industrial development resulted in a waste of resources, low level of industrial technology, and a weak complementary of industrial structure between cities. All these factors reduced regional economic competitiveness.

The "Pan-Pearl River Delta" Megaregion breaks the traditional constraints of the economic zone and the local protection barriers. It has changed the economic link that relied solely on capital and resource inputs in the past to regional integration. The latter has promoted regional economic cooperation and the formation of a unified market in Pan-Pearl River Delta. It has increased consumption, export, investment, the efficiency of resource allocation, and the capability of gathering production elements.

7.6.3.4 Spatial Structure

Major cities in the Pearl River Delta Mega-region are highly concentrated in the estuary area of the Pearl River. It forms two urban circles with different levels of development and characteristics (see Fig. 7.17). The inner circle includes six prefecture-level cities: Shenzhen, Dongguan, Guangzhou, Foshan, Zhongshan and Zhuhai. Most of the ports, airports, highways, railways and other infrastructures in the region are located in this inner circle. Shenzhen-Hong Kong, Macao-Zhuhai, and Guangzhou-Foshan constitute three cores of the Golden Triangle in the Pearl River Delta Mega-region. The outer circle includes three prefecture-level cities: Zhaoqing, Huizhou and Jiangmen.

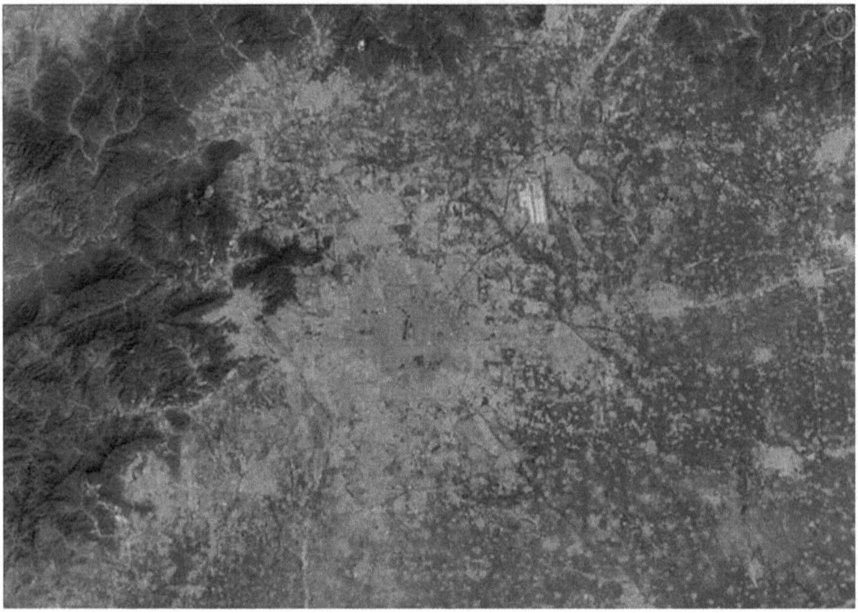

Fig. 7.18 Remote-sensing image of Beijing

7.7 Core Cities in Mega-regions of China

7.7.1 Beijing

Beijing ("Jing" for short, sometimes Romanized as Peking), is the capital of the People's Republic of China, the center of political, cultural, transportation and international communication. It also plays a role as the center of national economical/financial decision-making and management. Beijing is one of the "Four Ancient Capitals" with six UNESCO World Heritage Sites (Fig. 7.18).

Beijing is located at the northwestern edge of the North China Plain, at latitude 39° to 41° N and longitude 115° to 117° E. West of Beijing is the Xi Shan of the Taihangshan Mountains, while the Yanshan Mountains protect the city from the north and northeast. The Bohai Sea is about 150 km southeast of Beijing. The area has a semi-dry monsoon influenced humid continental climate. The annual precipitation is 571.8 mm. Beijing has four distinct seasons. Spring can bear witness to sandstorms. Summer is hot with pretty much rainfall. Fall is clear and dry. Winter is cold and windy. The spring and fall are rather short for two months at most. The summer and winter are quite long for three months and more than five months. Beijing has a typical monsoon climate with 60 % of precipitation in July and August. The annual temperature is 12.3 °C. The month daily average temperature in January (the coldest) is −3.7 °C, while in July (the hottest) is 26.2 °C.

Bei Jing, meaning Northern Capital, has been the capital city of the country for more than 700 years. It was first chosen by Kublai Khan in 1264 as capital for the Mongolian led Yuan Dynasty (1279–1368). Originally, it was called Da Du ("Grand Capital"). With the exception of two interludes (1368–1421, 1928–1949) Beijing served as seat of the government for two imperial dynasties (Ming Dynasty 1421–1644, Qing Dynasty 1644–1911), for the Republic of China and The People's Republic of China since 1949.

Beijing (and Da Du) was built according to the layout of the Chinese capital in traditional China as expressed in the Confucian classic text Zhou Li (Sit 1995). Some of the major principals for building the capital were the centrality of the capital, its square or rectangular shape and a central axis with a North South orientation. The capital city was started during the reign of the Yongle Emperor (1402–1424). The Ming Dynasty capital city of Beijing consisted of four parts: the Palace City (or 'Purple' Forbidden City which now houses the Palace Museum) where the emperor and his family resided; the Imperial City which comprised the two main ceremonial buildings (Ancestral Temple and Altar of Grains and Soils) to the south, the Coal Hill to the north, and a group of royal mansions and gardens placed in the three lake area of Shishahai to the northwest; the Inner City and the Outer City. The latter area to the south of the Inner City was added in the late Ming Dynasty. The layout and architecture of the capital city reflected the feudal society during the last two dynasties in Chinese history which ended in 1911. In a wider sense, the royal gardens in the outskirts of the city, the Ming and Qing Tombs outside the city and the nearby sections of the Great Wall in the Yanshan Mountains were part of an elaborate system of structures.

In modern Beijing, the built up areas have expanded more and more to the suburbs which are well connected to the downtown areas of the Inner and Outer City. The Municipality of Beijing, with 14 urban districts and 2 rural districts, has a most modern transportation infrastructure including an efficient subway system and a new national airport (PEK Capital International Airport). Automobile traffic has in-creased tremendously, and several concentric beltways surround the city, one immediately outside the old city walls and gates of the Inner City. The beltways have gone beyond third and fourth ring roads to a fifth ring road (with a sixth ring road partially completed) where the new Olympic Stadium, the Bird's Nest, is located. Not surprisingly, urban planners and architects deliberately placed the main sites of the 2008 Summer Olympic Games at the northern end of the Central Axis to integrate these pre-eminent contemporary additions in the city within the timeless frame of the capital city.

Beijing is the political, cultural, scientific, technological and educational center of modern China. The capital houses the headquarters of the State Council, the National People's Congress (NPC), the Chinese People's Political Consultative Conference (CPPCC), the central ministries and other main national institutions. The offices of these state offices, state agencies as well as other representative buildings of the state are mostly found on or near Tiananmen Square, the power center of the city. It was here that Chairman Mao Zedong proclaimed the People's Republic of China in 1949 and it is in the center of the 1958/59 largely expanded square where

Fig. 7.19 Tian'anmen Square and the axises

his mausoleum was established after his death in 1976. On the west side of the square towers the Great Hall of the People where the National People's Congress and the CPPCC National Committee convene as well as other important state meetings are hosted. The east side of the square is dominated by two museums of highest national importance (Museum of the Revolution and the Museum of National History). Tiananmen Square is the continuation of the symbolic North South running Central Axis whereas the wide Chang'An Jie Street on the northern end of the square marks the East West Axis of modern-day Beijing (Fig. 7.19).

Beijing hosts many prestigious cultural institutions including the China Peking Opera Company, the State Conservatory of Music, the Central Academy of Fine Arts and the National Centre for the Performing Arts. The Centre, an ellipsoid dome of titanium and glass surrounded by an artificial lake dubbed The Giant Egg was opened in a prominent location just one block West of Tiananmen Square in 2007.

It is also important to note that a large majority of China's universities and leading national research institutions are found in Beijing. The greatest cluster of universities, colleges and institutions of higher learning is in Haidian District in Northwest Beijing which has become a neighborhood for students, artists and intellectuals. Here were the beginnings of Beijing's "Electronic Street" (see more on the Beijing Zhongguancun Science Park in Chapter 9). This high tech core area continues to thrive in the immediate neighborhood of the two highest ranked universities in China, Peking University and Tsinghua University.

Beijing is an ancient city occupied by many ethnic groups and practitioners with a diversity of religious beliefs. The main groups which make Beijing their home are the majority ethnic group of the Han and the minority groups of the Manchu (once the leading group occupying the Inner City during the Qing Dynasty), Mongolian, Hui, and Korean. Among the smaller ethnic minority communities in town are the Uyghur, Moslems from Northwest China, and Asian groups such as the Vietnamese which, among others, showcase their distinct culture and cuisines in a multitude of local restaurants. Major religions represented in the capital city are Buddhism, Taoism and Islam, but there are also sizable religious communities which practice Christian beliefs (Catholics, Protestants, and Orthodox Christians). There are many temples in the city. A leading and nationally recognized Buddhist temple is the Lama Temple. There are also several historic sites in Beijing which honor the philosopher Confucius and his disciples. The Niujie Mosque, the oldest mosque founded in the 10th century and reconstructed during the Qing Dynasty, is the spiritual center for 10,000 Muslims living in or near Beijing. Among the established Christian churches in Beijing is the Wangfujing Catholic Church. With the opening up of China in the 1980s and the 1990s the goal of a harmonious development of religious beliefs within the society of the People's Republic of China has been more widely supported; it is a trend that can be observed currently in the practices of many residents of Beijing.

Among the unique features of the Beijing urban landscape are several thousand hutongs, narrow and winding alleys which crisscross the old sections of the Inner and Outer Cities of the Ming and Qing Dynasties. Some of the hutongs even predate these time periods. In fact, the term hutong was coined during the Mongolian led Yuan Dynasty. In general, the alleys run in an East West direction, but some also North-South. Sanmiaojie (Three Temple Alley) is considered the oldest one (900 years old); the longest alley is the East West running Dongjiaominxiang, with a length of 3,250 m. The shortest alley is about 10 m; the narrowest alley is Qianshi (Money Market Alley) in the Dashilan area south of Qianmen Gate, which is only 0.7m wide. The hutong alleys, with quadrangle house complexes called siheyuan, usually have their entrance gates to the South. Historically, the more respectable larger siheyuans were in the areas West and East of the Forbidden City; they were mostly occupied by high ranking officials and merchants. In recent years, hundreds of hutong alleys have fallen victim to waves of modernization, and new emerging high rise areas have frequently re-placed them. The pressure is greatest in the poorer sections in town. In the large courtyard complexes, residents plant trees and flowers and even raise chickens and ducks. Some yards have Chinese flowering crabapple and lilac, showing the owner's cultural accomplishments. In the Shichahai historic area (northwest of the Forbidden City), in South Luoguxiang and North Luoguxiang, from Xisi to Xinjiekou Avenue, and even from Dongdan to the Lama Temple, the siheyuan courtyard complexes have remained relatively intact. Many of them are already listed as Beijing Architectural Conservation Areas. As the hutongs display traditional Beijing ways of life they have been discovered more and more by tourists, and tours on foot and by bicycle are offered. Preservationists and historically minded residents keep a record of these endangered environments to secure the survival of the hutongs into the future (Fig. 7.20).

Fig. 7.20 Hutongs in Beijing

7.7.2 Shanghai

Shanghai ("Hu" for short, "Shen" for another name) is the largest city in China. It is one of the three national integrated gateway cities which are given the title by the State Council. It is also one of the largest cities in the world (Fig. 7.21).

The city is located at 31 to 32°N and 121 to 122 °E, near the mouth of the Yangtze River, China's longest, navigable river. Its original port location is near Suzhou Creek and the Huangpu River, a tributary to the Yangtze twelve miles from its outlet, which made for a well-protected harbor. Shanghai has a humid subtropical climate and experience four distinct seasons with abundant sunshine and precipitation. The annual temperature is 15.7 °C. The flood season here is from May to September with 60 % of the whole year's precipitation for spring rain season, plum rain season and fall rain season. The summer drought comes every July and the weather becomes hotter and wetter with average temperature of 27.8 °C.

Fig. 7.21 Remote-sensing image of Shanghai

The port town of Shanghai became a major city much later than Beijing. Although it is now, with over 23 million people (2011), the most populous city in China, its urban history is relatively short. It was not until the mid-Qing Dynasty, that it became a big city in East China thanks to the advantageous location of the port. It offered access to internal routes along the Yangtze River into the interior of China as well as international routes to Japan, Korea and Southeast Asian countries. The rise of Shanghai as a leading port city and commercial center in Asia from the 1840s to the 1930s is intricately connected to external influences. The Treaty of Nanking, the first of several "unequal treaties" China had to sign after the Opium War (1839–1842), allowed European powers to establish sovereign "concessions" in Shanghai eventually resulting in the Shanghai International Settlement. By the 1920s/30s, Shanghai's economy flourished, and the Chinese and foreign population surpassed 3 million thus making the city one of the largest urban centers of the world. "Old Shanghai", dubbed the Paris of the East (though with a rampant opium trade, gambling and prostitution), came to an end during the Japanese occupation of the city 1937–1945 and with the foundation of the People's Republic of China in 1949. In the following decades, the international trade advantages were lost to Hong Kong. It was in the 1990s that Shanghai's economy rebounded. Later, the administrators helped to reduce the tax burden of the city and to greatly encourage foreign and domestic investment into the Pudong Area, a special economic zone established in 1990. Shanghai became a major economic hub in East Asia with a stock market and hundreds of international, national and regional corporate headquarters.

Shanghai is the economic center of China. Its GDP ranks number one among the cities in Greater China and number two in Asia. In 2009, Shanghai's GDP surpassed Hong Kong's, and its container port was leading worldwide in volume. By the 1950s and 1960s, Shanghai had become a leading industrial city in China. Its heavy industry comprised the Shanghai Baosteel Group, the largest iron and steel manufacturer of the nation, shipbuilding plate companies and also eventually automobile manufacturers (Shanghai Automobile Industrial Company) in the mid-1990s. Because of its heavy industry and other industry sectors Shanghai became a major contributor of tax revenues to the central government. Shanghai's manufacturing base was soon extended to the petroleum and chemical industries and eventually to household electronic appliance manufacturing, equipment assembly and the pharmaceutical/biomedicine industry in the 1990s and early/mid 2000s. Shanghai's economy went through major structural changes from 1978 to 1990 and 2006, from agriculture and a dominating industry sector in 1978/1990 to a leading service sector in 2006. While in 1978 the percentage of GDP (out-put) was 4% (agriculture), 77.4% (industry) and 18.6% (services), the proportion of the sector outputs was 4.3%/63.8%/31.9% in 1990. By 2006 the relative importance of the three sectors switched to 0.9% (agriculture), 48.5% (industry) and 50.6% (services). The changes were even more dramatic in terms of employment: from 34.3% in agriculture, 44.0% in industry and 21.6% in services in 1978 to 6.2% (agriculture), 37.0% (industry) and 56.8% (services) in 2006 (Chen 2009).The structural changes resulted in a decrease in agricultural activities and in the output of goods in traditional industries like the textile industry which were moved more and more to the periphery of the City (Chongming Island) or outside Shanghai. The remarkable increase in service industry employment as well as out-put/contribution to the Shanghai economy was closely tied to the success of a new special economic zone in the Pudong Area. By the mid-2000s the "rising Shanghai" (Balfour 2007) had the attention of the national and international business community. For 15 years in a row, from 1992 to 2007, Shanghai's GDP rose by 12% plus annually which was even above the impressive national trend for China. A visual expression of the paramount changes in Shanghai was the new skyline towering over the Pudong side of Shanghai (east bank of the Huang-pu River), frequently showcased in the international press as the "new face of China".

Till the mid-1980s, the Pudong District of Shanghai was largely an agricultural area in town which could be reached by ferry service only. This would change in the late 1980s and 1990s when the New Pudong Area Special Economic Zone including the Lujiazui Financial and Trade Zone was introduced in 1990. It allowed foreign direct investment and encouraged massive economic development in this part of town. A corner stone of the changes was the re-opening of the Shanghai Stock Exchange in November/December 1990 (after the longtime prominent role the Shanghai stock market had played from 1920/21 to 1941). The SSE is now the leading stock market on the mainland and the 6th largest stock exchange world-wide in terms of market capitalization. The largest stocks traded at the SSE include Petro China, Sinopec, China Life and several leading banks in China. Dozens of regional headquarters of international corporations relocated to the Lujiazui Financial and Trade Zone area. With the establishment of the New Pudong Area the growth of hi-tech and service industries in Shanghai accelerated. The Pudong District is the location of many new ventures in the information industry, most notably in the 17 km^2 large Zhangjiang Hi-Tech Park area, in the Jingqiao Export Processing Zone and the Waigapoquiao Free Trade Zone. The financial and insurance industry, real estate and tourism service sectors gained enormously both in the Pudong District and city wide. It was also in the Pudong area that the 2010 World Expo was held. The event attracted seventy million of visitors and showcased new developments in China and worldwide. The World Expo was in the tradition of Shanghai's leading role in international trade, technology, the media, film production and fashion in China. Another major Pudong project in the entertainment field is Shanghai Disneyland to be opened in 2015. It is a joint venture between The Walt Disney Company and the Shanghai Shendi Group, a state owned company working in collaboration with the Shangha Lujiazui Group, Shanghai Radio, Film and Television Development Company and Jinjiang International Group Holding Company. The Pudong District's population has risen to over 5 million, with a growth of nearly 2 million new residents during the past decade 2000 to 2010. Most of the new residents come from other provinces in China (Fig. 7.22).

Shanghai is the railway hub and the airline hub in China. It owns the largest container port in the world. Since its inception as a special economic zone the New Pudong Area has been more and more effectively connected with Shanghai's Downtown area (on the west side of the Huangpu River) with four bridges and several tunnels. The citywide highway system has seen substantial improvements, to currently close to 400 miles in total length, as car ownership in the city has dramatically increased. At the same time the Municipality of Shanghai has made major strides in expanding and modernizing the subway system. The Shanghai Metro's number of lines and current length is comparable to that of the subway systems in London and New York. Several new lines are in the planning and/or implementation phase. Since 2000 Shanghai has two international airports, the older Shanghai Hongqiao International Airport (west of the Downtown area) and the Pudong International Airport. The latter has seen considerable increases in passenger volume, to more than 41 million in 2011. Both Shanghai airports have a joint volume of about 75 million—comparable to that of the new Peking Capital Airport and significantly more than that of the Hong Kong airport. The Pudong airport is connected to the financial area with a 30 km (18.6 miles) long magnetic levitation train ("Shanghai Maglev Train") that reaches a maximum speed of 430 km/hour (268 miles/hour). This

Fig. 7.22 Skyline of Pudong

innovative link to the airport and the subway system underlines the willingness of the city officials and their planning departments to break new ground in transportation technology.

Needless to say Shanghai is also well connected within the national high-speed railway system. The Beijing-Shanghai line was opened in 2011 reducing the travel time for the 1,300 km (800 miles) distance to less than 5 hours.

Shanghai is in the process of fully developing its Yangshan Deep Sea Water Port, south of the Municipality in the Hangzhou Bay. The port is connected to the city/mainland with the 32.5 km (20 miles) long Donghai Bridge which opened in 2005. The new Yangshan Port has the potential of making the Shanghai port the leading shipping center in Asia and worldwide.

Lilong Shikumen is a traditional housing structure in Shanghai which was very common in mid/late 19th century and in the first half of the Twenth century. Shikumen along an alleyway are two or three story buildings with a stone gate ("shikumen") and a courtyard allowing sunlight and rain to reach the ground. This type of building represents a blend of American type townhouses and homes built in the tradition of the Lower Yangtze architecture. By 1949 there were more than 9,000 shikumen style buildings in the Municipality of Shanghai comprising about 60% of the housing stock. In the 1950s and 1960s, when Shanghai's population grew by leaps and bounds, many shikumen were subdivided and sublet which resulted in very crowded living conditions in town.

In the reform period after 1978 Shanghai families began to move to larger apartments and eventually in the 1990s and the first decade of the 2000s to the ubiquitous modern high rise residential towers in town. Nevertheless, shikumen continue to stand for an important Shanghai tradition and way of life. In recent years, locally produced films and TV series have portrayed this side of Shanghai from the

1920s/30s; the production of "Shanghai Shanghai" and other series have found a wide distribution within China and in Asia responding favorably to a romanticized picture of 'Old Shanghai' including a recreation of the main shopping street Nanjing Road and Shikumen of that era.

As the local economy restructured and became more service and consumption oriented in the past years the shikumen environments were socially and structurally affected as well. Many lilong shikumen areas lacking modern amenities were bulldozed and replaced by high rise structures. The most prominent case of a redesigned and rebuilt shikumen lane is Xintiandi ("New Heaven and Earth") once hailed as the site of the first congress of the Communist Party of China. The neighborhood was redeveloped by Shui On Land (from Hong Kong) which hired American architect Benjamin Wood and Nikken Sekkei International (from Japan) to redesign Xintiandi in the mid-2000s. It is a widely recognized rejuvenation project that introduced "the concept of adaptive reuse" (Rutcosky 2007) thus helping to maintain a preexisting vernacular architecture to survive in a rapidly changing city environment. As adaptive reuse of Shikumen, Xintiandi contributes towards a more sustainable architecture otherwise endangered of being eradicated. Visitors and locals alike are drawn to this new type of a mixed-use urban setting combining traditional housing elements with new boutique style stores and cafes/bistros/restaurants (Balfour 2007). It has become the home of local elites and expatriates as most Shanghainese were priced out. The southern block alleyway includes an Open House Shikumen Museum. The economically successful urban renewal project led to the displacement of 3,500 Shanghai families—an implication which has been one of the many consequences of Shanghai's rapid transformation to a more cosmopolitan city.

7.8 Conclusions

In this chapter we have explored six Mega-regions in the United States and China. An examination of the geographic, economic, transportation and spatial structures of these Mega-regions has resulted in a broader understanding of the similarities and differences that exists among Mega-regions in the two countries. Interestingly, the path to the development of Mega-regions in each country are vastly different. The formation of America's Mega-regions was the result of a market-oriented economy with little interference from the national government. Cities in the United States tend to be incorporated around economic concerns and urban competitiveness rather than administrative exercises. The development of urban agglomeration is based on individual and cooperative development of the cities, while the central cities play the key role in coordination and management. The industrial structure and spatial layout of these cities are the results of market competition. The type and scale of the cities are diverse and together form an organic entirety.

There are, however, some limitations in the American model. Lack of effective coordination between cities, states and the federal government has resulted in relatively weak regulatory capacity related to urban planning. For example, cities develop policies to improve their own urban competitiveness which inevitably causes redundant construction, waste of resources and excessive competition. The internal stakeholders in American cities have a strong awareness of their rights. As a result of strong property rights, there is an intense conflict of interests and tensions in the American model, which makes planning difficult. For example, during the reconstruction of an old industrial area in San Francisco, poor coordination and a lack of communication have led to delays and poor use of resources.

In China, cities are developed in accordance with State led administrative initiatives. The dual social structure of urban and rural areas restricts the process of urbanization. For example, the identity system, education system, employment system, public service, and public finance systems have huge gaps between urban and rural areas. In fact, urban development policy in China "strictly controls the size of large cities and rational development of medium-sized and small cities". This policy was developed based on the household registration system management rather than from the perspective of economic and social development.

The current dual structure of the rural-urban division in China is gradually changing through modernization processes. The large rural migration into urban areas is an irresistible trend that will continue to occur in China into the foreseeable future. The development of urban agglomerations in China needs to break through the restrictions of administrative divisions and begin to adjust through market competition. This in the end will help to optimize the allocation of internal resources in urban agglomerations. However, the government should also still have a role in determining the various functions of cities and the division of labor according to regional advantages and resources of the cities.

In the end, the Mega-regions of China, the United States and the world face many of the same challenges from traffic congestion, escalating land rents, and environmental degradation. These challenges are not easily dealt with and due to the multi-jurisdictional nature of these issues cooperation among varying political entities at the local and national level will be of critical importance. How countries manage these urban agglomerations will determine if we are able to minimize the negative side effects of Mega-regions and take advantage of the economic opportunities that Mega-regions afford.

References

America 2050 Forum. (2009). The case for a national infrastructure policy: The Piedmont Atlantic Megaregion in the global economy. http://www.america2050.org/PAM%20Forum%20Summary%20of%20Proceedings.pdf. Accessed 18 June 2012.

Balfour, F. (2007). Shanghai Rising (February). *Business Week, 19*(2007), 50–55.

Benner, C., & Pastor, M. (2008). Fractures and fault lines: Growth and equity in California's megaregions. http://www.rpa.org/america2050/pdf/2008ResearchSeminar_1_Benner_Pastor.pdf. Accessed 18 June 2012.

Boquet, Y. (18 April 2010). *Paris, Washington, Brasilia, Beijing: Power landscapes of capital cities*. paper presented at the AAG Meeting 2010. Washington, D.C.

Bowling, K. (1991). Creating the federal city. 1774–1800: Potomac fever. Washington, D.C.: American Institute of Architects Press

Burnson, P. (2012). Top 30 U.S. ports: Finding the right balance. *Logistics Management, 5,* 50A–55A.

Burgess, E. W. (1924). The growth of the city: An introduction to a research project. *American Sociological Society Publications, 18,* 85–97.

Burrows, E., & Wallace, M. (1999). *Gotham—A history of New York city to 1898*. New York: Oxford University Press.

Carbonell, A., & Yaro R. D. (2005). American spatial development and the new megalopolis. *Land Lines, 17*(2).

Chen, X. (ed.). (2009). *Shanghai rising—state power and local transformations in a global megacity*. Minneapolis: University of Minnesota Press.

Cui, G. (1991). *Urban geography*. Nanjing: Jiangsu Education Press.

Florida, R., Gulden, T., & Mellander, C. (2007). The rise of the mega-region. Cambridge. *Journal of Regions, Economy and Society, 1*(3), 459–476.

Florida, R. (2008). The rise of the mega-region. *Cambridge Journal of Regions, Economy and Society, 1*(3), 459–476.

Georgia Institute of Technology. (2006). *Center for quality growth and regional development*. Emerging megaregions: Studying the southeastern United States. http://hdl.handle.net/1853/13128. Accessed 17 Mar 2014

Gibson, C. (1998). *Population of the 100 largest cities and other Urban places in the United States: 1790 to 1990. U.S.* Washington, D.C.: Bureau of the Census.

Godfrey, B. J., & Zhou, Y. (1999). Ranking world cities: Multinational corporations and the global urban hiearchy. *Urban Geography, 20*(3), 268–281.

Godschalk, D. R. (2007). Southeast growth and planning: Trends and issues. http://ncsu.edu/iei/programs/growth/resources/Southeast%20Growth%20and%20Planning%2021%20March.pdf. Accessed 18 June 2012.

Gottmann, J. (1961). *Megalopolis: The urbanized northeastern seaboard of the United States*. New York: The Twentieth Century Fund.

Gottman, J. (1987). *Megalopolis revisited: 25 years later*. Baltimore: University of Maryland Institute for Urban Studies.

Hennie, M. (24 Nov 2008). *Global implications of our growing megaregion*. Atlanta Business Chronicle.

Hartshorn, T. A. (1992). *Interpreting the city: An Urban geography*. New York: Wiley.

I-95 Corridor Coalition. (2012). I-95 facts and stats. http://www.i95coalition.org /i95/Home/I95CorridorFacts/tabid/173/Default.aspx. Accessed 18 June 2012.

Jackson, K. (1985). *Crabgrass frontier*. New York: Oxford University Press.

Kern County Council of Governments, San Diego Association of Governments, and Southern California Association of Governments. (2005). The Southern California mega-region—a case study of global gateway regions: America's third century strategy. http://www.cqgrd.gatech.edu/program_areas/megaregions/pdfs/southern_california_2005.pdf. Accessed 18 June 2012.

Knox, P. L., & McCarthy, L. (2012). *Urbanization: An introduction to Urban geography*. Boston: Pearson.

Lang, R. E., & Dagavale, D. (2005). *Beyond megalopolis: Exploring America's new "megapolitan" geography*. Metropolitan Institute Census Report Series, Census Report 05:01, May. Virginia Polytechnic Institute and State University: Alexandria.

Li, J., Ning, Y., & Shi, S. (2006). Urbanization development and megalopolis reconstruction in the Yangtze Delta Region. *Urban Planning Forum, 3,* 16–21.

Li, K., Li, J., & Cao, X. (2005). Adjustment of regional economic co-operation in northwest China based on basis and characteristics of extensive Zhujiang River Delta economic zone. *Economic Geography, 25*(2), 165–168.

Liang, X., & Xie, J. (2009). Integration of Beijing, Tianjin and Hebei: Evolution, current status, and development. *Journal of Hebei University of Economics and Trade, 30*(6), 60–69, 74.

Liu, D. (2006). The integration and spatial structure optimization of the Pan-Pearl river delta megaregion. *China Opening Herald, 2,* 63–65.
Lu, D. (2007). *2006 China regional development report: The process of urbanization and spatial expansion* (p. 16). Beijing: Commercial Press.
Lu, L., & Zhu, H. (2004). The research of the four center of the Changjiang Delta and the construction of metropolitan Shanghai surrounding area. *Economic Geography, 24*(6), 753–757.
Miao, C., & Wang, H. (2005). The analysis on developmental situations of China's urban agglomerations. *Urban Studies, 4,* 11–14.
National Bureau of Statistics of China. (2010). *2010 China city statistical yearbook.* Beijing: China Statistics Press.
Regional Plan Association. (2005). *America 2050: A national strategy for global competitiveness.* Concept paper. New York: Regional Plan Association.
Regional Plan Association. (2012a). America 2050: Northeast. http://www.america2050.org/northeast.html. Accessed 18 June 2012.
Regional Plan Association. (2012b). America 2050: Southern California. http://www.america2050.org/southern_california.html. Accessed 18 June 2012.
Regional Plan Association. (2012c). America 2050: Piedmont atlantic. http://www.america2050.org/piedmont_atlantic.html. Accessed 18 June 2012.
Ross, C. L. (2009). *Megaregions: Planning for global competitiveness.* Washington, D.C: Island Press.
Rutcosky, K. (2007). *Adaptive reuse as sustainable architecture in contemporary shanghai.* Lund: Lund University.
Sit, V. (1995). *Beijing—The nature and planning of a Chinese capital city.* NewYork: Wiley.
Southeast High Speed Rail Corridor. (2012). http://www.sehsr.org/. Accessed June 18 2012.
The Business Alliance for Northeast Mobility. (2009). *The future of the northeast corridor.* New York.
UN-Habitat. (2010). *The state of the world's cities report 2010/11: Bridging the Urban divide.* London: Earthscan.
U.S. Bureau of Census. (2007). Census of governments—individual state descriptions. http://www2.census.gov/govs/cog/all_ind_st_descr.pdf. Accessed 18 June 2012.
U.S. Bureau of the Census. (2010). Census geographic terms and concepts. http://www.census.gov/geo/www/2010census/gtc_10.html. Accessed 18 June 2012.
U.S. Bureau of the Census. (2012a). Growth in Urban population outpaces rest of nation, census Bureau reports. http://2010.census.gov/news/releases/operations/cb12-50.html. Accessed 18 June 2012.
U.S. Bureau of the Census. (2012b). Core based statistical areas and related statistical areas. http://www.census.gov/geo/www/2010census/gtc/gtc_cbsa.html#mesa. Accessed 18 June 2012.
U.S. Department of Transportation. (2010a). Enplanements at primary airports. http://www.faa.gov/airports/planning_capacity/passenger_allcargo_stats/passenger/media/cy10_primary_enplanements.pdf. Accessed 18 June 2012.
U.S. Department of Transportation. (2010b). All cargo landed weight. http://www.faa.gov/airports/planning_capacity/passenger_allcargo_stats/passenger/media/cy10_cargo.pdf. Accessed 18 June 2012.
Vicino, T. J., Hanlon, B., & Short, J. R. (2007). Megalopolis 50 years on: The transformation of a city region. *International Journal of Urban and Regional Research, 31*(2), 344–367.
Yao, S., Chen, Z., & Zhu, Y. (2006). *The Urban agglomerations of China* (3rd ed). Hefei: China University of Science and Technology Press.
Yu, T., & Wu, Z. (2006). Construction and reconstruction of Beijing-Tianjin-Hebei region. *Urban Planning Review, 9*(30), 36–41.
Yuan, J., Tan, C., & Chang, X. (2007). Research on megalopolis and cultivation of coastal megalopolis in China. *Economic Management, 15,* 85–90.
Zhang, G. (2008). *Research on actuality, problems and countermeasures of the Beijing-Tianjin-Hebei regional economic integration.* Master Thesis. Chongqing University.

Zhang, P. (2009). The definitions of metropolitan and related concepts. *Journal of Inner Mongolia Agricultural University* (Social Science Edition), *3,* 106–108.
Zhou, Y. (1995). *Urban geography* (pp. 41–43). Beijing: Commercial Press.
Zhu, E. (2009). Production upgrading and integration in the process of the integration of Beijing-Tianjin-Hebei. *Economic Geography, 6,* 881–886.
Zhu, M., & Yao, S., & Li, Y. (2002). Study on theory of regional structure of urban agglomerations in China. *Urban Research, 6,* 50–52.
Wang, S., & Wu, C. (2008) Definition and analysis of cities group and relevant concepts. *Modern Urban Research, 3,* 6–13.

Chapter 8
Regional Urban Economic Clusters

Susan Walcott and Huasheng Zhu

8.1 Introduction

Successful industrial clusters are socially constructed by human activities, building on a base of physical location attributes. Their economies thrive on an "innovation habitat", an interlinked network of relationships among people, companies and regulatory institutional practices that promote job and personal mobility, educational attainment, community infrastructure and related attractive arrangements such as cooperation and environmental preservation. A key "lighthouse" industry attracts other firms and needed levels of skilled labor to a regional sustaining culture in a closely located, distance diminishing synergy. It should be noted that all of the metropolitan areas featured benefit from an economic base utilizing innovative strategies, whether product or process related, and survive due to leadership that binds political and economic factors maximizing spatial interaction and social networks (Fig. 8.1).

This chapter is organized into three pairs of case studies comparing the economic trajectory of six major Chinese and US urban areas in different parts of the country and in different stages of development (Sies and Silver 1996; CSSB 2002; CRSUD 2011). The format for each city is similar. Fixed factors such as location and economic base are first identified, followed by a look at the development strategies, problems, and policies involved in their human geography. The initial set looks at the cities of Harbin-Dalian and Pittsburgh, Pennsylvania that grew on the basis of traditional heavy industries, in some cases leading to a rustbelt revival. They are located on the east coast corridor of early industrial development, and their lessons have to do with preserving old strengths while dealing with new opportunities in changed economic circumstances. The second set features the mid-continental

S. Walcott (✉)
Department of Geography, University of North Carolina-Greensboro,
233 Graham Building, PO Box 26170, 27402-6170 Greensboro, NC, USA
e-mail: smwalcot@uncg.edu

H. Zhu
School of Geography, Beijing Normal University, Beijing, China
e-mail: zhuhs@bnu.edu.cn

R. Hartmann, J. Wang (eds.), *A Comparative Geography of China and the U.S.*,
GeoJournal Library 109, DOI 10.1007/978-94-017-8792-5_8,

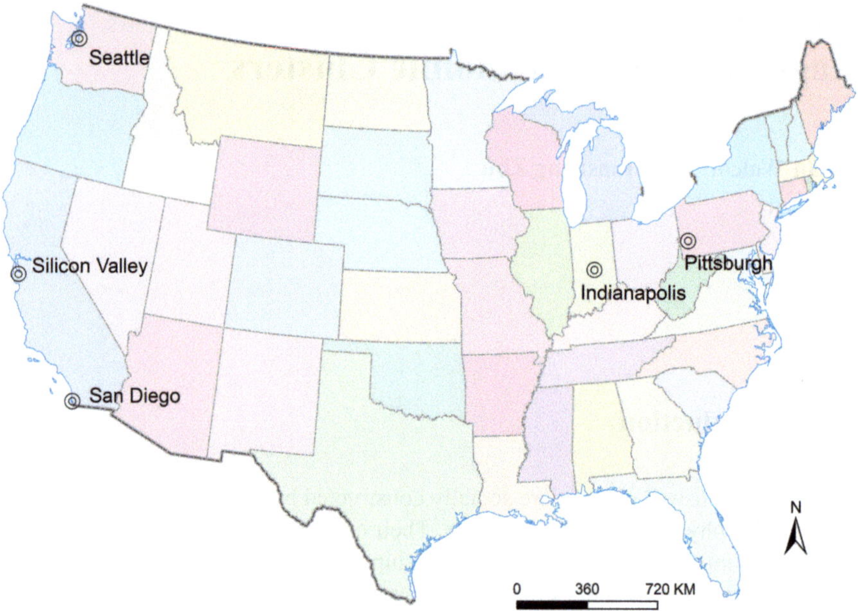

Fig. 8.1 Location of case study cities in the United States of America

inland transportation hubs of Wuhan and Indianapolis that leveraged their location advantage for inter-modal growth. Economic opportunity lies with providing a service moving goods between origin and destination points, as well as innovative improvements and diversifying economically with services, as in the two case study clusters. A contrasting pair of high tech habitats wraps up the comparison of models that seem similar but differ in important respects, such as Beijing's Zhongguancun and the U.S. west coast's San Diego, Silicon Valley, and Seattle. The concluding section summarizes major lessons from metropolitan experiences (Fig. 8.2).

8.2 Harbin-Dalian: Traditional Industrial Cluster —NE China Steel Industry and Transportation Infrastructure

8.2.1 Location and Industrial History

The Harbin-Dalian industrial cluster lies in the central area of Northeast China. In total, this region includes 21 cities along the Harbin-Dalian railway line from the most northern province of Heilongjiang through Jilin Province in the middle to Liaoning province on the southern end. This region covers 354.5 thousand km^2, occupying 28.6 % of Northeast China and 45 % of the three provinces listed above.

Fig. 8.2 Location of case study cities in China

Large parts of Northeast China were subjected to colonization from the late nineteenth to the mid-twentieth century. To take advantage of the abundant natural resources, colonizers constructed the Harbin-Dalian Railway and several harbors in order to transform raw materials and to ship products overseas. After 1931, when Japan used the Northeast Region as an ordnance and ammunition manufacturing base, several cities and heavy industries began to develop such as steel and iron making, machinery, and the chemical industry.

With the establishment of the People's Republic of China in 1949, regional clusters developed quickly by implementation of development plans. During the period

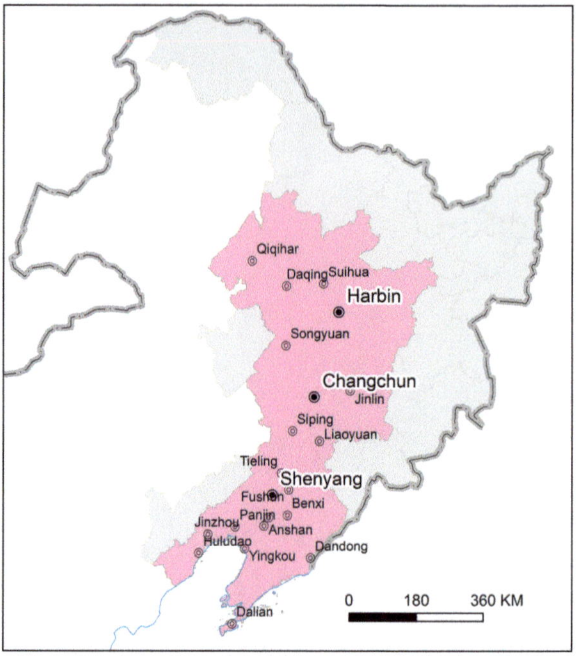

Fig. 8.3 Geographical location of the corridor

of the first Five-year Plan, 54 of 156 national projects were laid out in northern China. These included building up Anshan Steel Corporation, Benxi Steel Corporation, Fushun Coal Corporations, Fulaerji Heavy Machinery Plant, China First Automobile Corporation in Changchun, Jilin Chemical Industry corporation, and three power plants in Harbin. During the period of the Second Five-year Plan, Daqing Oil Field, Jilin Oil Field, Qinghe Power Station, and Liaoyang Chemical Fiber Corporation were built up. The Northeast is known as China's "industrial giant" since this region contains more than 1,700 state-owned large and medium enterprises,accounting for 1/7 of China's total number. Meanwhile, this area also integrated an extractive base of planting, farming, forest and fishing.

Since the late 1980s the northeast region experienced a period of slow growth. In 1990, the industrial output value of the northeast region increased by only 0.6 % over the previous year, far below the national average of 7 %. The profits of enterprises fell by 25–45 %, a significantly higher decline than the national average of 18.5 %. The serious recession experienced by manufacturing is called in China the "northeast phenomenon". This region's clusters played a leading role and made great contributions to the renaissance of Northeast China. At present, the region contains the most important agglomeration of population and industries in China, and its general regional product accounts for more than 70 % of that of Northeast China (Zhao and Chen 1999) (Fig. 8.3).

8.2.2 State of the Local Economy

By 2001, the GDP of the Harbin-Dalian cluster reached $ 95.04 billion, accounting for 69.15 % of the total for Northeast China. Per capita GDP was $ 1,782.05, nearly double the national average of $ 911.32. The main industries are machinery manufacturing and energy, which mostly focus on the early stage of processing of raw materials. Different city clusters feature different types of leading industries. Rich mineral resources, agricultural and forestry development, the construction of infrastructures such as seaports, railway and other kinds of transportation lines and the national development strategy all promoted the prospects of this area for becoming a new industrial center. Northeast China rests on five bases in steel, energy, machinery manufacturing, forestry, and commodity grains to lead economic development in this region. The northeast region's cultural facilities, education popularization and graduation rates are among the best in China. The popularity of higher education in Liaoning province ranks first in China. The northeast region is China's fourth economic growth pole (Wang 2008).

8.2.3 Urban-Industrial Transformation Strategy

Contemporary administrative segmentation restrains integrated development of the region. Blending the industrial structure is a big problem in the Harbin-Dalian cluster. Each province puts forward its own development plan emphasizing industrial competition between cities. Liaoning Province focuses on two industrial bases and three sectors. One of the two industrial bases consists of an equipment manufacturing industrial base including transportation equipment manufacturing, fundamental equipment and complete equipment manufacturing, and military equipment manufacturing. The raw material industrial base includes the petrochemical industry, steel and iron, and construction material. Three other clusters consist of high-tech and new technology industries, agricultural product processing and modern services. Shenyang continues to be the core of the urban agglomeration in the middle of Liaoning. Greater Dalian is designed to become the international shipping center of Northeast Asia.

Jilin Province aims to construct five industrial bases: a national level automobile industry, petrochemicals, agricultural product processing, modern Chinese medicine and biomedicine, and a high-tech and new industry base. Another goal is to improve the metallurgy industry and the promotion of tourism and other advantageous industries. The city of Changchun is the core in the urban agglomeration of central Jilin. Heilongjiang province intends to set up six industrial bases: equipment manufacturing, petrochemicals, energy manufacturing making use of coal, agricultural product processing, green food production, a medicine industry base in the north of China, and forest product processing (Wang and Zhao 2004).

Since 2003, China redeveloped traditional industries in the Northeast. Industrial transformation and upgrading, especially the sustainable development of resource-based cities, constitutes one of the most crucial issues. The fundamental way for the development of Northeast China is economic system reform and transforming the mode of economic growth. The key measures of reform include the following:

- Cultivating superior industry clusters and actively extending the industrial chain. The northeast region should focus on raw materials and subsequent processing industry, equipment manufacturing industry, intensive processing of agricultural products and high-tech industry clusters.
- Promoting opening up a wide range of multi-directional and multi-level developments.
- Accelerating the restructuring and reorganization of state-owned enterprises by improving the investment environment, removing restrictions of administrative divisions to promote steel, automotive power and other key industries across restructured regions.
- Strengthening infrastructure facilities by construction projects such as traffic equipment, water conservancy and energy in order to attract more domestic and foreign investments.
- Developing education and cultivating the quality of human capital to promote the combination of "production, teaching and research" and accelerate the transformation of high-tech achievements and industrialization.
- Strengthening ecological construction to improve the capacity for sustainable development by making good use of the comparative advantage of ecological environment and resources to develop an eco-economy industry.

8.3 Pittsburgh: Traditional Mid-Atlantic Heavy Industrial Cluster

8.3.1 Location and Industrial History

Pittsburgh lies at the confluence of the Allegheny and Monongahela rivers, which form the mighty Ohio River system and gave rise to its nickname of "City of Bridges". It was also known as the "Steel City" since, like Detroit, access to water transportation fed the city's strength as a historic steel manufacturing center. Industry titans such as Carnegie, Heinz and Mellon bequeathed amenities from libraries to museums that generated a strong civic pride. To regenerate their town, twenty-first century city leaders raised taxes and built on funds from a newly profitable natural gas industry in order to restore sullied facades and sustain landmarks, prioritizing pride of place for retaining urban labor talent as advocated by Richard Florida, local Carnegie Mellon University geographer-author of the bestseller Creative Cities.

European colonizers set up a fur trading station in the late seventeenth century at this confluence of major rivers such as the mighty Monongahela and Ohio. The desirable physical location made it a contentious area for Indian tribes, British and French colonizers, and later during the American War for Independence. Following

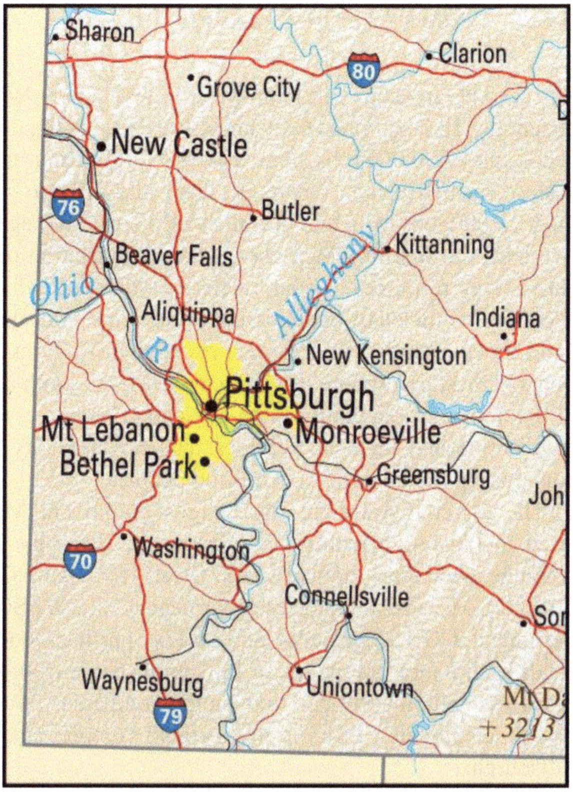

Fig. 8.4 Greater Pittsburgh region (www.nationalatlas.gov)

the expulsion of the French from their fort at the site, British General Forbes constructed a new fort which he named Fort Pitt after the then-Secretary of State William Pitt; the name stuck to this settlement at a strategic site. In the mid-nineteenth century completion of the Pennsylvania canal and the Pennsylvania Railroad confirmed its importance as a transport hub. In the early twentieth century, Pittsburgh's industry and commerce developed rapidly and the city served as a center for the industrial revolution. Steel king Andrew Carnegie started Carnegie Steel Company, which built on the innovative high-temperature Bessemer furnace to monopolize the North American steel industry. Andrew Mellon's financial firms provided funding to further develop key Pittsburgh based industries such as aluminum, coke ovens, and industrial abrasives. Pittsburgh native Henry Heinz organized what became the global food industry giant Heinz Company in the late 1800s. The Heinz brand is best known for condiments such as ketchup ("57 Varieties") and side dishes such as baked beans. A century later Pittsburgh sought a new economic base to supplant its faded industries and became a model of reform. By the dawn of the twenty-first century Pittsburgh became a US urban economic success reform model, drawing in part on the foundations set up by its wealthy citizens of an earlier era (Fig. 8.4).

8.3.2 State of Local Economy

The second largest city in Pennsylvania with a population of 334,563 in 2000, and occupying 58.3 sq. miles (151 km^2), Pittsburgh's population was estimated to have declined by 100,000 jobs since the 1980s, reflecting its faded prospects as a northeastern rustbelt center. However, from mid-2011 to mid-2012 the city added almost 15,000 jobs to signal its comeback status. The regeneration strategy is built on two legs, the first of which uses existing transportation infrastructure strengths to ship energy resources of coal, nuclear, solar, wind, oil and gas from shale. A "new economy" is fueled by healthcare and medicine, education (University of Pittsburgh and Carnegie Mellon among other research institutions) in a metro area with the nation's highest percentage of young professionals with a graduate degree and a third of the above-25 population with a bachelor's degree, robotics technology, nuclear engineering and financial services. The seven-county metropolitan area currently contains a population of 2.35 million people.

The universities are the city's largest employers. Major corporations headquartered in Pittsburgh include PNC Financial Services, PPG Industries (diversified coatings, chemicals, optical), U.S. Steel, H.J. Heinz Company (ketchup and condiments), Mylan Laboratories (pharmaceutical), WESCO International (engineers, designers), CONSOL Energy, and Dick's Sporting Goods. Other companies in these fields with major operations in Pittsburgh range from Alcoa (aluminum) to Bayer (a German pharmaceutical) and energy companies. Western Pennsylvania's largest private employer, UPMC is the medical complex anchored by the University of Pittsburgh.

8.3.3 Urban-Industrial Transformation Strategy

According to the president of Pittsburgh-based Future Strategies LLC, Pittsburgh's experience holds two lessons: "… you can come back successfully from major economic decline with the right kind of leadership and support" and "… don't try to hang on too long or wait for what you lost to come back". Several city leaders also ascribe Pittsburgh's resurgence to a fortunate match between the research strengths of its universities and the "hot" areas of the new economy, such as robotics, that began to emerge in the late 1980s through the 1990s. By the end of the next decade, in both 2009 and 2010 Pittsburgh was named "Most Livable City in the U.S." by several major magazines and had earned a new nickname: "The Paris of Appalachia". In one of the poorest, mountainous regions of the country, Pittsburgh used its leadership skills to find an economic way back as an urban star, successfully transitioning from brawn to brain based job skills and building a broad-based economy to sustain future growth.

Key areas of reform include:

- Strengthening infrastructural facilities with more construction projects such as traffic equipment and industrial park land to attract more domestic and foreign investments. Tax abatements downtown fueled a post-recession building boom of office skyscrapers and residence apartments.
- Developing a high technology economy by shifting to new management mechanisms, providing funds and other support to cultivate the new technology industry and manufacturing industry promotion for a new enterprise foundation in energy resources and supporting service providers.
- Developing education to cultivate human capital based on the large number of local institutions of higher education, particularly focusing on the most employable science, technology, engineering, and mathematical (STEM) fields. Additionally, at the K-12 level the "Pittsburgh Promise" extends a $ 40,000 college scholarship to graduates with a good attendance record and B- grade point average.
- Improving entertainment such as high quality sports and performing arts facilities to increase amenity attractions. A state film tax credit attracted investment in movie industry facilities, transforming a steel mill and vacant spaces to studios.

8.4 Wuhan: Inner China Transportation Hub

8.4.1 Location and Industrial History

Wuhan is situated in central Hubei province, east of the Jianghan plain, intersecting the middle reaches of the Yangtze and Han rivers which divide Wuhan into three parts: Hankou, Hanyang and Wuchang, which are generally known as Wuhan's "Three Towns". Wuhan is an important strategic supporting point of Central China. It connects the east with the west, channels the north to the south, and links rivers with seas by means of its developed water, land and air traffic. Because of Wuhan's excellent transportation links the city is relatively close to foreign countries such as Japan and the Republic of Korea. China's major metropolises such as Beijing, Shanghai, Guangzhou, Chengdu, and Xi'an are all within a circle around the center of Wuhan with a radius of 1,000 km (FAOHPPG 2011).

Wuhan's economy was at the forefront of Asia in the late Qing dynasty and the Republican period. As the largest treaty port in inland China, Hankou had the reputation of the "Chicago of the East". After the founding of the People's Republic of China in 1949, a large number of enterprises were built up such as Wuhan Steel, Wuhan Boiler Factory (WBC), and the Wuhan heavy machinery plant, which tremendously promoted Wuhan's economic status and the city's comprehensive strength. From 1959 to the beginning of the Reform and Opening Up period in the late 1970s, Wuhan's industrial output was in the top fourth in China. In the 1980s, Wuhan failed to keep up with the pace of reform. However, in the 1990s Wuhan established Wuhan Economic and Technological Development Zone in Hanyang, the Wuhan East Lake Hi-tech Development Zone in Wuchang, the Wuhan Wujiashan Taiwan-investment Area in DongXiHu, and the Yangluo Development Zone in Xinzhou.

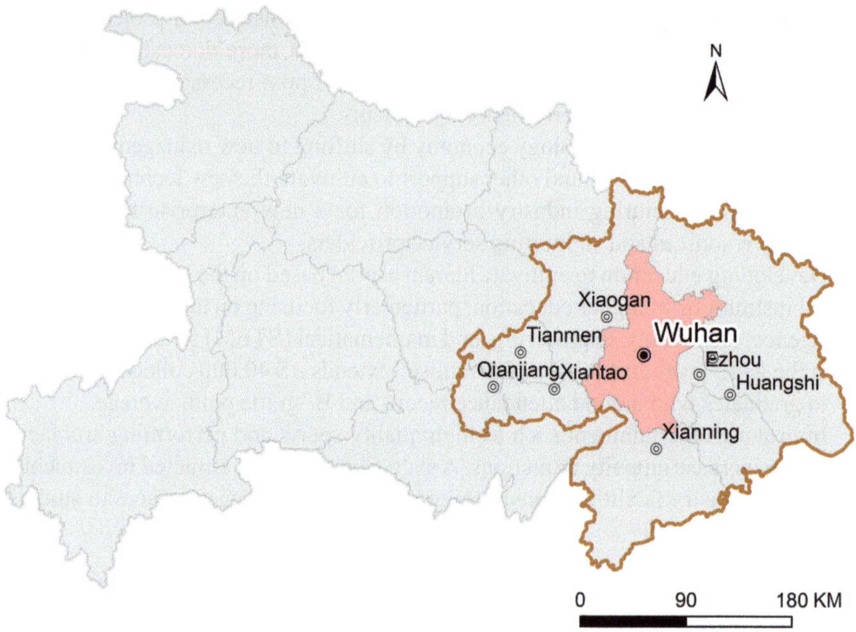

Fig. 8.5 Geographical location of Wuhan cluster

As one of the origins of China's modern industry, Wuhan is China's traditional base of manufacturing and one of the biggest automobile industry bases in China; domestically, nine out of a hundred cars are produced in Wuhan Citroen-Dongfeng. Wuhan is also one of the biggest iron and steel industry bases in China; Wuhan Iron and Steel Group is the third largest iron and steel consortium. In 2007, the State Council approved Wuhan City Circle as a comprehensive experimental zone for resource efficient and environmentally friendly reforms. The Wuhan 8 + 1 City Circle includes Wuhan city, Xiaogan city, Tianmen city, Qianjiang city, Xiantao city, Xianning city, Ezhou city, Huangshi city and Huanggang city. The main projects of Wuhan's comprehensive experimental reforms include strengthening industries, finance, transportation and reducing the differences between rural and urban areas. In recent years Wuhan focused on the development of high-tech industries and became the biggest optoelectronic information industry base in China; Wuhan's China Optics Valley has become a leading new high-tech industrial agglomeration (Fig. 8.5)

The Wuhan city circle is situated on the Yangtze River economic belt, at the Jing-Guang Railroad and Beijing highway intersection point. This constitutes the north-south development axis, located at the Midwestern intersection with the Yangtze valley's middle reaches and central five provinces, which historically is always referred to as the "nine province thoroughfares". At present Wuhan boasts one of the four largest railway stations in the nation, serves as a first-class highway hub, a mid-Yangtze River transportation center, the third biggest communication service command and adjustment center in the nation, a telecommunication fiber optics ring net connection node, and holds a superior transportation and communication position with which other central China regions cannot compete.

8.4.2 State of the Local Economy

Wuhan is one of the 15 subsidiary cities in China. The largest city in central China at 9.1 million in 2010, on 8,494.41 km^2, Wuhan's urban population has been in the fourth position for a long time. Wuhan is known as one of China's important industrial bases, a science and education base and a comprehensive transport hub. Wuhan also has the reputation of the "biggest port in central China", the "Aviation center in central China", and "One of China's four railway hubs". As the first choice of foreign investment in central China, Wuhan is also the top location for French investment in China.

Wuhan possesses strong economic and regional advantages. The headquarters of Wuhan Iron and Steel Group company (one of the top three Iron and Steel Group companies) and Dongfeng Motor Corporation (one of the top three Automobile Works) are located in Wuhan. The city attracted a Microsoft innovation technology center, an IBM global service center, an EDS global service center, a France telecom software R&D center and a well-known domestic enterprises software service outsourcing and financial service center. Wuhan is abundant in technological resources, with 76 universities or colleges, one state laboratory, 12 state key laboratories, 56 state-level and province-level research institutes, 10 state engineering and technology research centers, three national technology centers in enterprise, 1,050,000 university students and over 350,000 graduates every year.

8.4.3 Urban-Industrial Transformation Strategy

Wuhan is the primate city of Inner China, thus limiting the development potential of other cities. It deters the flow of information, energy, personnel, materials and the development of alternate clusters. Market barriers still exist. The territory's transportation and corresponding network system lags in its economic development speed, and significant damage to the environment from air and water pollution persist in the region. Key strategies for improvements are as follows:

- Establish a cross regional benefit coordination mechanism. All levels of urban government should agree on the cross region interest distribution in order to break specialization divisions and best utilize regional resources.
- Implement unified management.
- Actively construct a unified legal system platform. Every city must locally implement the legislation that is exercised by the National People's Congress or the provincial government.
- Establish an urban circle coordinated advancement mechanism through an expert consultant committee for the Wuhan city circle.

8.5 Indianapolis: Midwestern "Silver Buckle on the Rustbelt"

8.5.1 *Location and Industrial History*

Formerly known as "Indiana-no-place" and "Naptown", this typical American "Heartland" capital of Indiana is the only city at the confluence of three of Joel Garreau's Nine Nations: the rustbelt (upper Midwest), the bread basket (Plains), and the bible belt (South). Its location on the southern edge of the flat glacial plains connecting the American industrial heartland, from Chicago and Detroit around the southern edge of the Great Lakes, made it an important railroad city and the site of the first "Union Station" conjunction of major intersecting lines. The automobile was invented near Indianapolis, so it was fitting that a former airport became the famous "Indianapolis Speedway" for showcasing high tech car races. The city's economy really took off with the birth of pharmaceutical company Eli Lilly & Company, only the second major "Big Pharma" firm (along with Minnesota's Merck) outside of the east coast. Lilly's decision in the 1970s to stay in Indianapolis, as a "big fish in a little pond", saved its future and fueled its rebirth as the Rustbelt's "silver buckle" (Fig. 8.6).

8.5.2 *State of Local Economy*

Indianapolis' city population is 798,382 (2010), with a metro population of 1,715,459. The city remains the bright spot of the state's struggling economy. A Midwest transportation hub, Indianapolis straddles six major interstate highway systems and is the hub of the state's spoke configuration of roadways. A major deficit is the lack of a strong local research university; for historic reasons the two major universities (Indiana University and Purdue) are located in different parts of the state.

The driver of Indianapolis' economy is health care, accounting for the largest proportion (13.6 %) of the city's workers. This sector utilizes a synergy among the area's hospitals and Lilly's research facilities, as well as medical device manufacturers and other medical services in the life science cluster. Biotechnology research tends to be located either in-house at Lilly or within its nationally and globally dispersed biotech affiliates. Construction and manufacturing continue to decline, as the city searches for a way to leverage its strengths and escape from its blue collar past.

8.5.3 *Urban-Industrial Transformation Strategy*

Stone's famous "Growth Machine" book about the power of congenial, socially networked business leaders to transform their city was built on the model of Atlanta

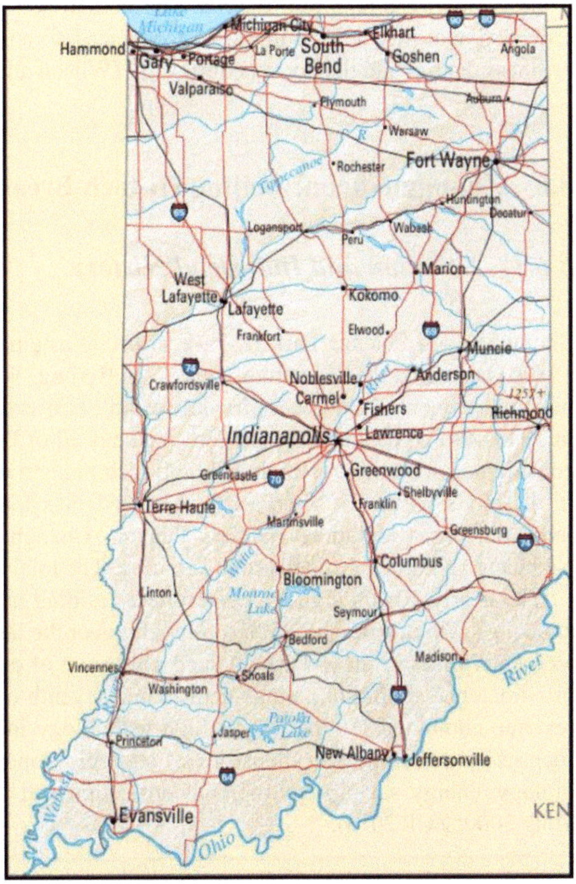

Fig. 8.6 Indianapolis, capital of the Midwestern "Heartland", transportation nexus of an interstate highway and railroad network (www.nationalatlas.gov)

and Indianapolis. A growth coalition formed in the early 1970s, launched by the innovative "Unigov" that stretched the urban boundary out to encompass the more prosperous inner suburbs. A group of businessmen who had attended similar colleges formed a civic planning group outside of the city bureaucracy (Greater Indianapolis Progress Committee) to reimage their town as a healthy place—building on the presence of Lilly and the Raceway to attract major sports events and convention traffic. New sports and arts facilities and museums were constructed, financed with public-private partnerships to make Indianapolis the kind of place where footloose high tech employees (principally Lilly's) would want to live. This in turn attracted similar companies.

Indianapolis leverages its transportation position in several ways. As a major interstate hub it has developed numerous warehouse and distribution centers around the bypass highway circling the city and linked to its airport. Business generated by the expanding Fed Ex hub fueled the construction of a recently completed new airport. The former Union Station was turned into a downtown shopping center, anchoring a new strip that sports the façade of the old buildings that a major new enclosed mall has replaced. Baseball, basketball and football stadiums also bring crowds downtown, as do cultural performances, to enliven the heart of the central business district and the pockets of the city (Wilson and Lindsey 2005).

8.6 Zhongguancun: Beijing Hi-tech Breakthrough

8.6.1 Location and Industrial History

Zhongguancun Science Park (Z-Park in brief) originated from what in the early 1980s was called the "Electronic Street" in Beijing. With the approval of the Chinese government, Z-Park was officially established as the first national science park in 1988. Zhongguancun Science Park includes all of Beijing's highly concentrated educational, scientific research and high-tech industrial zones. With its centerpiece in the city's northwest, the Park features a landscape of multiple sub-parks and industrial bases including Haidian, Fengtai, Changping, "the Electronics City", Yizhuang and others located around Beijing. The total area of the Park is 232 km^2.

The nearly 20,000 high-tech enterprises residing in Z-Park have maintained an annual growth rate of over 25 % continually over the last decade. In 2006, the high-tech enterprises in the Z-Park attained a revenue of over 600 billion Yuan (or approximately $ 80 billion), equivalent to one seventh of the nation's science parks' revenue combined. As the largest high technology industry base, Zhongguancun focuses on fostering new industries of strategic importance such as information, biology, energy saving, environment protection and new energy (Ma, Jiang and Tong 2000; ZCP 2013).

8.6.2 State of Local Economy

As the high-tech center of Beijing, Zhongguancun is known as "China's Silicon Valley". In 1988, foreign exchange earned through export was 100 million $, increasing to 19.7 billion $ in 2007. Per capita foreign exchange earnings through export increased from one thousand dollars in 1988 to 22 thousand dollars in 2007. In 2000, there were 6,186 firms in Z-park, producing 3,941 types of products and 775 patents. In 2007, 20,125 firms produced 14,880 types of products, and 3,214 patents. Among 14,880 kinds of products, those made by firms' own technology reached 85.6 %. That demonstrates the rapid increase of innovation capability. Zhongguancun features abundant human resources and is the most intellectual-intensive region in Beijing. Population related to science and technology activities increased from 71 thousand in 2000 to 340 thousand in 2007 (Fig. 8.7).

R&D was 48 thousand in 2000, and increased to 179 thousand in 2007. Until 2007, Z-park attracted 9,527 overseas people, and 32 thousand graduate students. More than one million professionals work on innovative activities, with skills in

Fig. 8.7 Geographical location of Beijing high-tech cluster

venture capital, business management and commercialization of high and new technology. China's most prestigious and selective colleges and universities are located in the capital city, providing a ready source of high skill labor.

8.6.3 Urban-Industrial Transformation Strategy

The central government's leading role in the early construction of the Park was very important. However, as the park continues to expand, an intricate administrative network restricts the development of Zhongguancun. As the driving force of the Zhongguancun Science Park, universities and research institutions naturally become the main body of technological innovation in the region. However, the relationship between industry and research lacks sufficient interaction. Financing of safeguards has become a bottleneck to the improvement of the technological innovation capacity of Zhongguancun. The management system should adopt the principle of "minimum institution, maximum service" which can advance the government's management and improve the interaction of industry and research. The government should establish a venture investment mechanism to improve the overall investment environment (Wang and Wang 1998).

Zhongguancun Science Park now must make full use of high-tech industry to support the development of multi-risk investment funds and guarantee funds for risky small and medium enterprises (the current size of the total capital of 10 billion), focus on the commercialization of Zhongguancun high-tech achievements. In line with international norms, the venture capital market operation mechanism should be established with the following steps:

- Set up the relevant investment legal system;
- Establish a finance service guarantee for high-tech corporations;
- Adopt related policies and measures to encourage high-tech enterprises to actively utilize foreign capital;
- Focus venture capital on cultivating venture capitalists and venture capital firms to meet the needs of high-tech enterprise development.

8.7 Three West Coast Axes of Excellence: San Diego, Silicon Valley, and Seattle

8.7.1 Location and Industrial History

These three cities became high technology urban growth poles centered on different industries—biotechnology in San Diego, IT in Silicon Valley, and aerospace/computer software development in Seattle—but driven by a similar cultural wind of innovation. The predominance of the west coast cities of the United States as models of profitable creativity, birthing new industries from research bench to competitive market, was no coincidence. As a seaport city, San Diego, California is a major federal navy and marine site supported by the state-funded University of California at San Diego. Silicon Valley is a product of private Stanford University and other San Francisco Bay area colleges. Seattle, Washington is a major aviation, transportation and internet services center fuelled by the public state-funded University of Washington. All three profit from numerous grants and attract a vibrant, interactive scientific and venture capital community supported by financial institutions, consultancies and major accounting and legal firms that undergird regional strengths.

Fig. 8.8 Southern California borderland and biotech center city of San Diego

8.7.2 *State of Local Economies*

San Diego is the second largest city in California and the second largest technopole in the state, after the Silicon Valley region. Similar to the situation of Pittsburgh, this seems to inspire a positive "try harder" attitude. Its population of 1.3 million (2010 est.) in 324 square miles (963.6 km^2) lies 15 miles north of the border with Mexico, and slightly south of its research rich suburbs such as La Jolla. Drawing on an excellent harbor location and the post-WW II presence of the navy, San Diego's major economic engine is the defense industry, manufacturing—particularly high tech research university spillovers such as biotechnology, computer science, electronics, software development, and wireless telecommunications—and agriculture, including agricultural biotech. The latter strength supports California's premier position as a flourishing exporter of fruit and vegetables, along with more value added products based on them such as wine (Troutman 2004) (Fig. 8.8).

Silicon Valley acquired its name in 1971, referring to the location of this area in the Santa Clara Valley (including parts of Alameda, San Mateo and Santa Cruz counties) south of the San Francisco Bay and its early importance as the birthplace of the semiconductor industry, based on the invention of the silicon chip. Major cities in the Valley are San Jose and Mountain View. A long and distinguished list of high tech innovative companies includes Bell Labs, Xerox, and Cisco, feeding off their proximity to the local cluster of research universities including Stanford and various University of California and State University branches, from Berkeley north of San Francisco to San Jose State and UC Davis near Sacremento. Several features sustain this regional economic corridor: (1) an entrepreneurial spirit among well-paid workers that enables them to start their own companies to explore the feasibility of their own ideas, (2) a local pool of venture capital stemming from computer millionaires seeking to invest funds in the Next Big Idea, and (3) the convenience of staying close to universities with facilities and talent to develop new ideas (Saxenian 1994). Almost half of the Valley residents speak a non-English language at home; 60 % of the scientists and engineers were born outside the U.S. (Fig. 8.9).

Fig. 8.9 Sizzling Silicon Valley, north central California innovation habitat

Seattle is a seaport on an isthmus 100 miles south of the border with Canada, stretching over 142.5 sq. miles (369.2 km^2). Its population is 617,000, with a metropolitan population of 4.2 million, similar to Atlanta, Georgia. Seattle took off as a staging area for the trek to the Alaskan gold fields, but blossomed as a transportation corridor (including highways and rail) and port known as "The Gateway to the Pacific". Further recent distinctions include being highest in coffee drinkers, college graduates, and literacy. Seattle's economic base is a diversified blend of old and new economy, from lumber to internet and technology services, design and green technology startup firms. Six young Fortune 500 companies have headquarters in this city: Washington Mutual (banking), Amazon, Starbucks, Nordstrom, Safeco (insurance), and Expeditors (global logistics). Nearby giants include Costco, Microsoft, Nintendo, Weyerhauser (forest products), truck manufacturers, and T-Mobile USA. The city's largest private manufacturing employer remains Boeing, though headquarters moved to Chicago (Morrill 2013) (Fig. 8.10).

8.7.3 Urban-Industrial Transformation Strategy

All three city regions thrive from local university technology transfers, consciously linked to a business community that promotes and privately finances risk-taking entrepreneurship. As one local leader said, "It's not how many times you fall down; it's how fast you get up." A supportive community that seeks to coordinate policies nurturing the local workforce and businesses is also common to all three areas, in the frontier pioneering spirit of the West Coast. They also benefit from a global workforce that attracts bright and energetic minds from many countries, a particularly prominent demographic in the case of Silicon Valley (Kenney 2000).

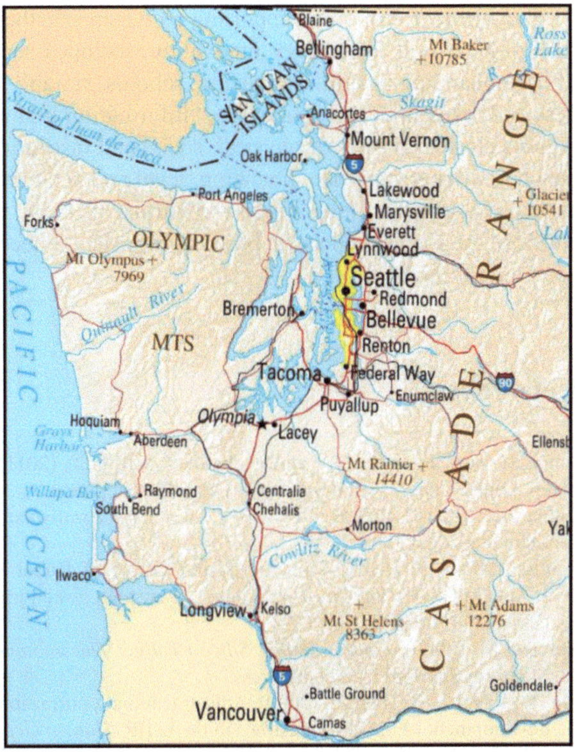

Fig. 8.10 Seattle: Northwest Fortune 500 frontier for entrepreneurial research (www.nationalatlas.gov)

8.8 Conclusions

Despite their different regional locations stretching from the east to west coast and the continental heartland of their country, each urban cluster displays some similar characteristics that have sustained it through economic downturns and transitions. These include:

- A location that benefits from transportation by a combination of road, rail or water;
- A key profitable industry base;
- An amenity environment that creates and sustains a well-educated labor force;
- Good city planning which directs incentives in line with business attraction goals;
- Technological innovation within the cluster;
- Rapid adaptation to changing circumstances due to information networks.

While technology's innovation ability has a significantly positive effect on an industrial cluster's development, leadership that actively works with political and business leaders to promote mutually beneficial policies for forming a coalition "growth engine" is key to bringing all the listed factors together. Adjacent cities that demonstrate cooperation with each other also promote a mutually beneficial urban economic cluster. Despite many intriguing similarities, major differences between the situation in the Chinese and U.S. cities discussed include the relative proportion of government and private involvement in the political economy of each area.

The political economy supporting a cluster, whether based on advanced manufacturing or high technology innovation, includes services and attractive quality of life-enhancing amenities, housing, schools, suitably configured and priced business and residential real estate. The "information highway" along which these city regions cluster consists of physical, social and institutional infrastructure including bench to market expertise, financing, legal and accounting counsel, a favorable tax structure, and community support to nurture the new, which could be quite different from the former economic base but represent a necessary transition to a vibrant economic future.

References

China Research Society of Urban Development (CRSUD). (2011). *The yearbook of China's cities*. Beijing: China Research Society of Urban Development.

China State Statistical Bureau (CSSB) (2002). *China statistical yearbook* 2002. Beijing: China State Statistical Press.

Foreign Affairs (Overseas Chinese Affairs) Office of Hubei Provincial Peoples Government (FAOHPPG). Wuhan City Circle. http://www.fohb.gov.cn/jrhb/whcsq/201112/6117.html. Accessed 20 Aug 2013.

Kenney, M. (2000). *Understanding Silicon Valley: The anatomy of an entrepreneurial region*. Stanford: Stanford University Press.

Ma, Y., Jiang, J., & Tong, L. (2000). Adjustment and optimization of the industrial structure and de-velopment of the key industrials in the IZHR. *Human Gepgraphy, 15*(6), 42–45.

Morrill, R. (2013). The Seattle central district over eighty years. *Geographical Review, 103*(3), 315–335. doi:10.1111/j.1931-0846.2013.00001.x.

Saxenian, A. (1994). *Regional advantage: Culture and competition in Silicon Valley and Route 128*. Cambridge: Harvard University Press.

Sies, M. C., & Silver, C. (1996). *Planning the twentieth-century American City*. Baltimore: Johns Hopkins University Press.

Troutman, P. (2004). A growth machine's plan b: Legitimating development when the value-free growth ideology is under fire. *Journal of Urban Affairs, 26*(5), 611–622. doi:10.1111/j.0735-2166.2004.00217.x.

Wang, J. C., & Wang, J. X. (1998). An analysis of New-tech agglomeration in Beijing: A new industrial district in the making? *Environment and Planning A, 30*(4), 681–701.

Wang, R., & Zhao, L. (2004) Study on urbanization response of Harbin-Dalian traffic economic belt in northeast *China. Scientia Geographica Sinica, 24*(5), 535–541.

Wang, W. (2008). Revitalizing northeast industrial base in an all-round way and recreating the fourth growth pole of China's economy. *Science-Technology and. Management, 10*(2), 84–86.

Wilson, J., & Lindsey, G. (2005) Socioeconomic correlates and environmental impacts of urban development in a central Indiana landscape. *Journal of Urban Planning & Development, 131*(3), 159–169. doi:10.1061/(ASCE)0733-9488 (2005) 131:3(159).

Zhao, J., & Chen C. (1999). *China geography*. Beijing: Higher Education Press.

Zhongguancun Science Park (ZCP). Introduction of Z-park. http://www.zgc.gov.cn/sfqgk/56261.htm. Accessed 20 Aug 2013.

Chapter 9
Urbanization, Urban Form and Adaptation: A Comparison of Four U.S. and Chinese Cities

Tao Ye, Brian Muller and Peijun Shi

9.1 Introduction

Urbanization has been a fundamental dynamic over the past millennium in the social and economic history of most regions across the globe. There are four processes by which human environments and economic livelihoods have changed through urbanization and modernization. First, urbanization processes result in the spatial concentration of populations, occurring through the migration of rural populations to urban centers. Second, urbanization is associated with land use changes including the shift from agricultural to urban activities. Third, urbanization modifies natural landscapes to support expanding urban footprints. Finally, urbanization is associated with a shift of dominant economic activities from agricultural production to manufacturing, service and tertiary activities. In addition to the above changes, urbanization has also been accompanied by shifts in culture and life style. The proliferation and extension of human settlements, through which relatively dense settlement patterns are extended into agricultural and natural lands and rural populations are incorporated into urban economies and cultural modes, is an important spatial expression of urbanization. The term "sprawl" is frequently used to describe dispersed and fragmented urban expansion patterns typically associated with weak physical and environmental planning.

The industrial revolution of the nineteenth century stimulated rapid urbanization processes. In the United Kingdom, urban population exceeded rural population for the first time in 1851. Worldwide, urbanization processes occurred relatively slowly

T. Ye (✉) · P. Shi
State Key Laboratory of Earth Surface Processes and Resource Ecology,
Beijing Normal University, Beijing, China
e-mail: yetao@bnu.edu.cn

P. Shi
e-mail: spj@bnu.edu.cn

B. Muller
Program in Environmental Design, University of Colorado Boulder, Boulder, USA
e-mail: brian.h.muller@colorado.edu

R. Hartmann, J. Wang (eds.), *A Comparative Geography of China and the U.S.*,
GeoJournal Library 109, DOI 10.1007/978-94-017-8792-5_9,

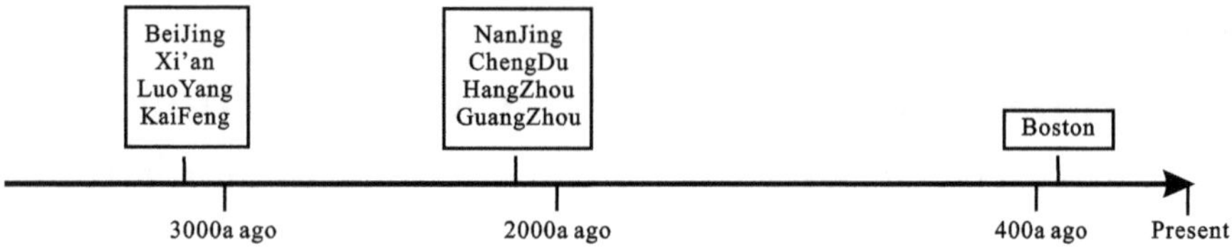

Fig. 9.1 The emergence of cities in China and the U.S. in the historical perspective

during the early nineteenth century to the early twentieth century then accelerated from about 1920 to 2000 as the annual average urban population growth rate exceeded above 0.3 %. In twenty-first century, researchers anticipate that the growth rate will return to a slower rate. Due to the differences in industrialization and economic development as well as social and cultural factors, urbanization in different countries and regions shows considerable disparity in its form and characteristics.

China has a long history of urbanization (Fig. 9.1). The four major ancient capital cities of China (*Beijing, Xi'an*, *Luoyang* and *Kaifeng*) are all more than 3,000 years old. Several other major cities including *Nanjing*, *Hangzhou*, *Chengdu* and *Guangzhou*, have a history extending more than 2,000 years. Until recently, however, the growth of cities in China has been relatively slow, only heating up after World War II. In 1949, Mainland China had only 132 cities, with an urbanization rate of only 7.3 %. By 2009, China had 655 cities, with increases of 46.6 % of the population living in urban areas.

Compared to China, America is a relatively new country with only about 400 years of history. Urbanization in general has been strongly influenced by immigration, which in the U.S. has had both internal and external phases. As late as 1870, the proportion of urban population in the U.S. was only about 5.1 %. With the speed-up of industrialization in the later decades of the nineteenth century, the urbanization process accelerated. The proportion of urban population increased to about 40 % in 1900 and 45 % in 1910. Around 1920, the urban population proportion reached 50 %, completing what researchers describe as the "primary stage" of urban growth. Since the 1920s, the pace of urbanization has slowed. By 2002, the urban population had increased from 200 to 250 million, accounting for about 80 % of the total U.S. population.

Urbanization has been a driving force in human-environment systems, and the pattern by which urbanization occurs has important implications for the capacity of nations and societies to sustain themselves over time. Urbanization and related problems of sustainable development have been a focus of research by among geographers and planners for more than a century. The S-curve is one characterization of the relationship between urbanization and sustainability. As cities mature through prolonged urbanization, urban and natural management systems are stressed, disamenities emerge and cities tend to become less sustainable (and livable). Along with urban maturity, a multitude of other factors that may also influence the process by which cities adapt to problems of sustainability, however, including local geography, land use planning, regional economic drivers, and macro-economic conditions.

Fig. 9.2 Illustration for geographic location of sample cities

In this chapter, we present case studies of four cities in China and the U.S. as a framework for comparing patterns of urban growth and sprawl across the two countries and different types of cities. Our case study cities are Chengdu (Sichuan Province) and Phoenix (State of Arizona), both of which were the historical centers of inland resource regions, and grew rapidly after World War II, as well as Shenzhen (Guangdong Province) and Las Vegas (State of Nevada), which have short histories of urbanization, with explosive growth rates after 1970 (Fig. 9.2).

This chapter is organized as follows. First, we provide background on each of the cities including their geographic location and historical development. Second, we reconstruct the urbanization processes in the four cities, focusing on population and landscape change. Third, the key factors or characteristic drivers behind the growth of these cities are reviewed as a basis for conceptualizing differences of urban sprawl in China and the U.S. Finally, we discuss sustainability challenges in the four cities as a framework for exploring issues of changing human-environment relationships during the processes of rapid urban expansion.

9.2 Inland Centers: Chengdu and Phoenix

Chengdu and Phoenix are well-known cities, with strong cultural identities and relatively long histories. Chengdu was founded in 611 BC. During the mid-fifth century, B.C., the Liberal Kingdom named *Shu* shifted its capital city to Chengdu and built the city walls and moats. For 37 years during the late *Ming* and early *Qing* Dynasty Sichuan was in a state of almost constant war. Chengdu was largely destroyed. In the *Qing* Dynasty, Chengdu was rebuilt and immigration encouraged. Local officials opened up common lands to settlement and expanded opportunities for migration to Sichuan. Chengdu is now the capital city of Sichuan Province, an important central city in the central and western region of China, and a major megalopolis of China.

The Native American Hohokam Civilization emerged in the region where Phoenix is now located around 700 AD. The Hohokam people dug an irrigation canal 135 miles (217 km) long and developed an extensive system of agricultural cultivation in the area. During the period from 1300 to 1450 AD, possibly due to droughts and fierce floods, the Indian Hohokam civilization disappeared from the Phoenix region, leaving behind cultural relics as a foundation for further growth. In 1881, Phoenix became a city. In 1891, the railway was constructed. In 1912, Phoenix was designated the capital of Arizona. In 1911, Roosevelt Dam was built and irrigated agriculture emerged as an economic driver in the area. After the Second World War, the city developed rapidly, building on its agriculture and mining industries, but also emerging as a regional distribution and manufacturing and a center for retirement and recreation. At present, Phoenix is the capital and the largest city in the state of Arizona.

9.2.1 Urbanization Overview

9.2.1.1 Geography

The geographic location and climate characteristics of Chengdu and Phoenix are very different. Chengdu is situated between 102°54’ and 104°53’ E and between 30°05’ and 31°26’ N, in the western part of the Sichuan Basin and the eastern part of the Chengdu Plain in Southwest China, with an average altitude of about 500 m.

The Chengdu Plain is the biggest plain in the southwestern region of China, extending east to *Longquan* Mountain Rang and west to *Longmen* Mountain Range, on the West Sichuan Fracture Zone. The plain region is high in the northwest and low in the southeast, with an average gradient of 0.3 %. The city area of Chengdu is 192 km long from east to west and 166 km wide from south to north, with a total land area of 12,390 km^2, mostly on the plain, with *Jinjiang* River, *Fuhe* River and *Shahe* River running through the city area. Chengdu is a subtropical, humid monsoon region, with a mild climate, distinct seasons, a long frost-free period, abundant rainfall, and a limited number of sunny days. The average annual temperature is 16.2 °C, the total annual rainfall is 918.2 mm and the average annual sunshine duration is 1,071 h.

Phoenix is situated in the central part of the Arizona State, between 33°31' N and 112°4' W. The region has become known as the Salt River Valley or Sun Valley. Phoenix is bounded by the McDowell Mountains in the north, the White Tank Mountains in the west, Superstition Mountains in the east and the Sierra Estrella Mountains in the south. Two mountains are located inside the city: Phoenix Mountain and South Mountain. Phoenix has an average altitude of 1,117 ft (340 m) and borders the Sonoran Desert. Phoenix has a dry and hot climate, ranking first among major U.S. cities in average annual temperature. During an average year, there are 89 days that the temperature of the city exceeds 100°F (38°C), typically during June to September. Phoenix is sunny for 85 % of the daytime hours during an average year. The city's average rainfall is only 8.3 in (210 mm) at Phoenix Sky Harbor International Airport.

9.2.1.2 Patterns of Population Growth

Urbanization is a self-organized system with positive feedbacks, where urban functions attract population while urban population contributes to promote urban functions. Chengdu and Phoenix have a long history. The population size of Chengdu is greater than Phoenix. In the process of urban population growth, both of them experienced the assimilation of mainstream nation, as well as rapid population growth driven by industrial development.

Chengdu has experienced dramatic population shifts over its long history (Fig. 9.3). After destroying *Bashu*, *Qin* Kingdom "shifted ten thousand of its households into *Shu*" and established the *Shu* County at Chengdu, with the population of Chengdu increasing sharply. By the *Han* Dynasty, the economy of Chengdu was flourishing. During the *Western-Han* Dynasty, its population reached 76,000 households (about 400,000 people) and it became one of the six major cities (the other five of which were *Chang'an*, *Luoyang*, *Handan*, *Lintao* and *Wan*) in the *Han* Dynasty. During the *Western-Jin* Dynasty, the minorities re-occupied Chengdu and much of the *Han* population fled the city. During the periods of the *Sui* Dynasty, *Tang* Dynasty and *Song* Dynasty, the minorities were suppressed on the one hand and assimilated on the other hand, with more and more people migrating to Sichuan from Central China. During the period of *Sui* and *Tang*, the economy of Chengdu

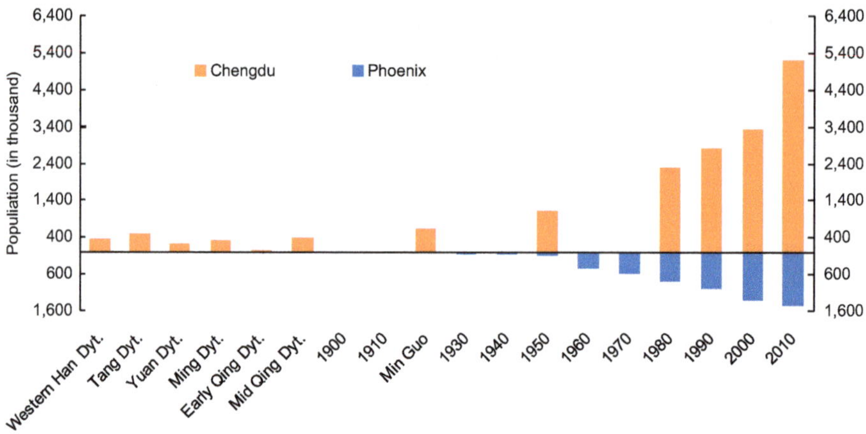

Fig. 9.3 Population change in Chengdu City and Phoenix City. (Note: the time scale is inconsistent and discontinued)

expanded rapidly, and the population grew to over 500,000 people, ranking second in population only to *Chang'an*. In this period Chengu was considered one of the four primary cities in China, alongside *Chang'an*, *Yangzhou* and *Dunhuang*. During the period of late *Song* Dynasty and early *Yuan* Dynasty, the population of Chengdu again decreased massively due to wars, diminishing to about 200,000 residents during the period of *Kublai Khan* (*Yuan* Dynasty). Additional wars during the period of late *Ming* Dynasty and early *Qing* Dynasty were responsible for extreme suffering in Chengdu. At the end of this period less than one hundred households remained in the city and its population dropped down below 10,000. In the 3rd year of *Qing*'s *Shunzhi* Emperor (1646 AD), the entire city of Chengdu was destroyed in the war and nothing remained. The capital city of Sichuan was shifted to *Langzhong* in *Baoning* Prefecture. During the period of *Kangxi* Emperor, the government started to implement the great "Migration from *Hu* (Hubei Province) *Guang* (Guangdong Province) to Sichuan" which lasted nearly one hundred years. Nearly one million people migrated into Sichuan Province from other provinces. Because of the settlement of immigrants, Chengdu was gradually revitalized and its population increased gradually. During the period of *Qing*'s Emperor *Jiaqing*, Chengdu's population grew rapidly to about 386,000. Afterwards, its population grew more slowly, reaching about 650,000 during the Republic Period. After the People's Republic of China was founded, as the capital city of Sichuan Province and the political, economic and cultural center of Southwest China, Chengdu experienced a sharp increase of population, growing from 1.13 million in 1949 to 5.10 million in 2008, an increase of 3.53 times at an annual growth of 7.56%. With the Chinese governmental reforms beginning in 1978, the proportion of population living in the Chengdu metropolitan area increased sharply, from 28.38% in 1978 to 45.35% 2008. The population in the central city area of Chengdu was 1.61 million in 1990, increasing to 2.05 million people in 2000 and over 3 million in 2006. In 2008, with

Fig. 9.4 Changes in the urban landscape of Chengdu. **a** Chengdu during Minguo period (http://www.lishi.net/html/laozhaopian/2011-12/2900.html). **b** Chengdu at present (http://www.lvxiaobao.com/index.php?doc-view-1423.html)

a permanent population of 4.41 million in the central city area, Chengdu could be considered a megacity.

Phoenix was formed on in February 25, 1881 through adoption of the "Phoenix Charter", which established the foundations of a governmental system for the city. The city and region of Phoenix experienced rapid population growth after World War II. The population of the city of Phoenix was 106,818 in 1950, 789,704 in 1980, 1,321,045 in 2000, and 1,475,834 in 2005. The Phoenix region had a population of 4,281,899 in 2008, making it the 12th largest metropolitan area in the United States.

9.2.1.3 Patterns of Spatial Expansion

Agriculture laid the foundation for the development of both Chengdu and Phoenix. In both regions, the expansion of agriculture was associated with the development of a reliable supply of water and with the grading of agricultural sites which also served to prepare land for urban development. Moreover, agriculture created pastoral scenery which played a key role in attracting immigrants. The built-up area of the central city area of Chengdu was 60 km^2 in 1980, 74.4 km^2 in 1990, 129 km^2 in 1995, more than 200 km^2 in 1999, and up to 427 km^2 in 2008, 7.12 times of that in 1980 (Sichuan Statistical Bureau 2011). Chengdu's urban landscape has changed dramatically, associated with fundamental shifts in architecture, infrastructure and all other dimensions of the built environment (Fig. 9.4). Figure 9.5 provides a vivid presentation of the city's expansion, in which the built-up area of Chengdu gradually extends across nearby lands. Because Chengdu city is surrounded by a developed agricultural area, the expansion of the city has been associated with the conversion of a significant amount of land from agricultural to urban uses, which makes the city suffering from such conflicting uses of conversion (Peng et al. 2008).

In 1950, Phoenix extended over 17 square miles, ranking 99th in the United States in area (Table 9.1). By 1980, the area of the city had increased to 330.59 square

Fig. 9.5 Urban sprawl and land use change of Chengdu City from 1992 to 2006. (Peng et al. 2008)

Table 9.1 City area and rank in the U.S. cities of Phoenix

Year	City area (km^2)	Rank in the U.S. cities
1950	17	99
1980	331	9
1998	470	6
2008	839	3

miles, ranking 9th among U.S. cities in area. By 1998, with a land area of 470 square miles, it became the sixth largest city in the United States in terms of area.

Figure 9.6 illustrates land uses in the Phoenix region over five periods: 1912, 1934, 1955, 1975 and 1995. The light gray indicates desert, the green indicates agricultural areas and yellow indicates built-up areas. These maps vividly display urban expansion in the Phoenix region. In general, the growth of the metropolitan area occurs through the conversion of desert and agricultural land to urban uses. As we see, before 1955, the city of Phoenix was surrounded by a large agricultural region. Simultaneously, more and more farmland was reclaimed due to the contributions of the Salt River Project and Central Arizona Project, which brought water resources to the valley. Beginning in about 1955 or so, a significant change occurred in development pattern as the city extended into desert lands. From 1955 to 1995, the metropolitan region expanded significantly, mainly towards the north and west. Considerable amounts of farmland and desert were occupied and turned to urban land use.

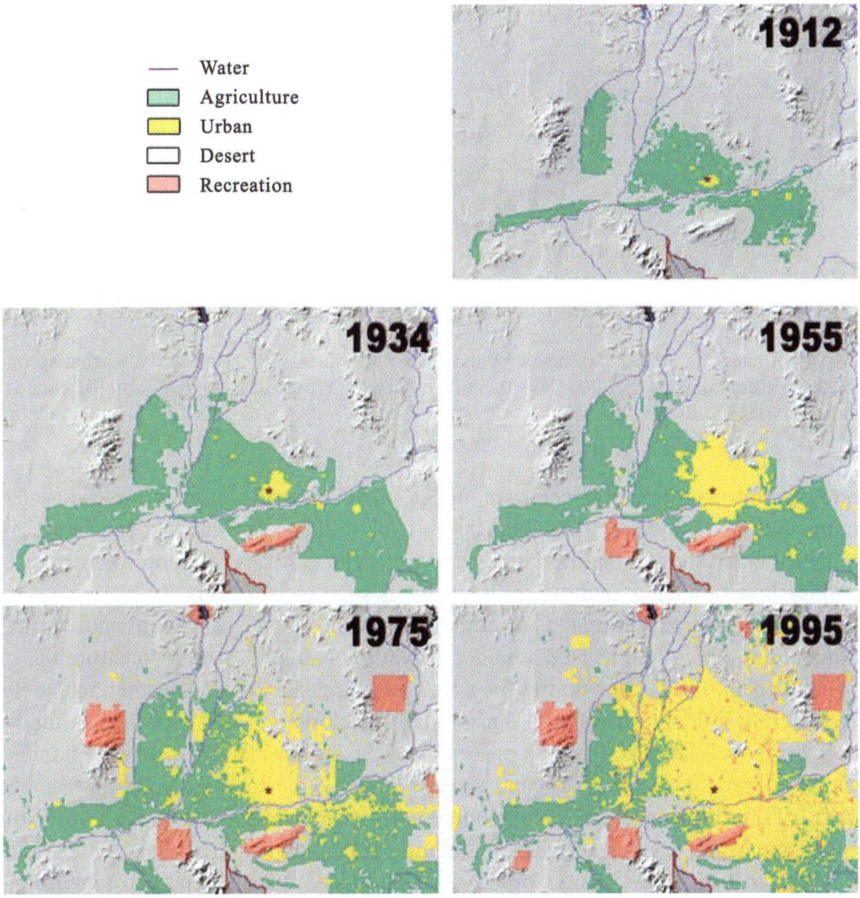

Fig. 9.6 Landscape urbanization of Phoenix. (Wu et al. 2011)

In addition to the continual expansion of city area, the urban landscape of Phoenix City also changed significantly, from interspersed urban and agricultural lands to a modern metropolis intensively developed with high-rise buildings (Fig. 9.7).

Agriculture laid the foundation for urban development in the region. It played a key role in securing water supply, preparing development sites, and creating pastoral landscapes that attracted immigrants from other parts of the country. Agriculture in central Arizona developed with water-intensive practices including flood irrigation of crops such as citrus. In many cases water rights were passed to the subdivision developments that succeeded them. In the post-World War II era, subdivisions were built in the citrus groves and marketed as garden suburbs. At least for a time, citrus trees and flood irrigation practices were maintained in the new suburbs. As across much of the Western U.S., urban growth in Phoenix has been heavily promoted

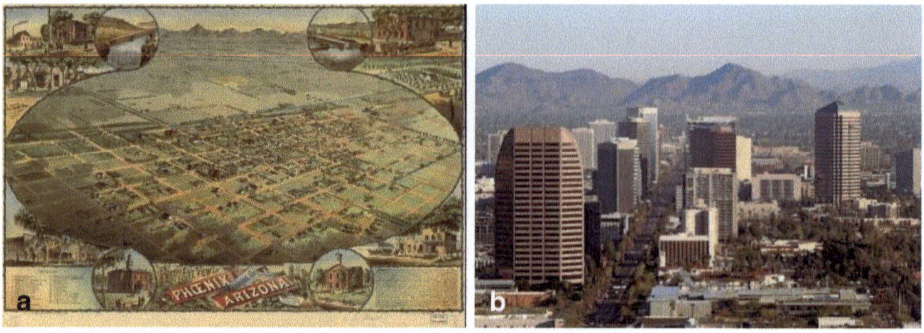

Phoenix in 1885 Phoenix at present

Fig. 9.7 Change in urban landscape of Phoenix. **a** Phoenix in 1885 (http://www.worldmapsonline.com/historicalmaps/kr-1885-phoenix.htm). **b** Phoenix at present (http://www.mplife.com/us/inquire/120828/76833009901.shtml)

through various forms of boosterism, including advertising campaigns by railroads, municipalities and developers themselves (Sokol 2005). In both Phoenix and Las Vegas, this boosterism featured images of well-watered oases and low-density residential development associated with amenities such as golf courses.

In the non-agricultural areas of both Phoenix and Las Vegas, subdivision development has typically followed a model of grading and landscape reconstruction, in which vegetation is stripped and the contours of the site are transformed to satisfy engineering guidelines. As with agricultural development, this process has modified the ecological character and processes of the region including wildlife habitat, species diversity, and hydrological processes including drainage patterns (Steiner et al. 1999). In general, urban development has also been associated with dramatic expansion of impervious surfaces, including both concrete and asphalt, which have further transformed hydrologic processes and reduced space available for landscaping. These development patterns are likely associated with urban heat island effects. Temperature in Phoenix has increased at a rate of about 1 °C a year. Indeed, both land use patterns and global warming may contribute to heat increases. More water efficient landscaping concepts such as xeriscaping have not taken hold until the last couple of decades, and water use in Phoenix is not sustainable over the longer term (Grimm et al. 2008)

9.2.2 Drivers of Urban Growth

Though Chengdu and Phoenix have considerable differences, there were two basic similarities in the process of urban expansion between the two cities. On the one hand, local geographical conditions supported high agricultural yields. Efficient productivity allowed a part of the population to leave the primary production and move to secondary and tertiary industries. This in turn created market and population agglomerations, and trigged the early stages of urban development. On the other hand, the advantage of geographical location attracted heavy investment, and promotes the rapid development and promotion of the industry. The confluence of these two forces, generated a powerful urbanization dynamic that turned Chengdu and Phoenix into modern cities.

9.2.2.1 Water Supply and Flood Management

Both the Phoenix and Chengdu regions have struggled for centuries with management of water supply and flooding. Chengdu is situated near the mountain pass of the *Minjiang* River, a major branch of the Yangtze River, running out of the *Minshan* Mountain Range into the Chengdu Plain. From *Yulei* Mountain, the mountain pass of the *Minjiang* River, the Chengdu Plain inclines southeastward, with a sharp gradient. In ancient times, when there were major floods on the *Minjiang* River flooded the Chengdu Plain became an inland sea. When there was drought, the area became a barren landscape. Phoenix is also situated in relatively flat valley surrounded by mountain ranges. While the Phoenix valley itself is arid, the higher elevations accumulate moisture both in the form of snow and storm events. Both drought and flooding have been obstacles to the urban and agricultural development of Phoenix over much of its history. One of the primary hypotheses for the disappearance of the Hohohkham civilization is long-term drought. While the group constructed sophisticated irrigation works, these may not have been sufficient to sustain the society during periods of severe drought.

Both Phoenix and Chengdu have addressed both water supply and flood issues through a succession of public infrastructure projects. In 256 BC the procurator of *Qing* Kingdom's *Shu* Count, *Li Bin* organized the construction of the well-known *Dujiangyan* Weir project. This project split the water flow of the *Minjiang* River into two parts for the purpose both of reducing flood risk and extending irrigation into the Chengdu Plain. As a result of the *Dujiangyan* Project, the Chengdu Plain was an important economic center for more than 2000 years. The Chengdu Plain provided an economic foundation and material reserve for significant military investment and aggressive diplomacy. For instance, during the period of *Three-Kingdom*-dynasty, relying on *Dujiangyan* Irrigation Area, *Shu-Han* dynasty fought with the two other nations and sent its forces to Central China several times. *Zhuge Liang,* (181–234 AD, celebrated adviser to the empire, and the founder of the *Shu-Han* dynasty (221–263/264) "took the weir as the base of agriculture and resources for the kingdom". He attached great importance to the management of the project and deployed 1,200 workers for its maintenance. During the *Tang* Dynasty, *Yizhou* (name of Chengdu in the ancient times) was flourishing and as famous as the economic capital of China, *Yangzhou*. At present, the *Dujiangyan* Irrigation Project with the *Dujiangyan* Weir as the canal pivot is responsible for irrigating more than 68.4 million hectares of farm land in seven cities (prefectures) and 37 counties (districts) in the central and west region of Sichuan Basin (Fig. 9.8). It supplies water for the enterprises and urban residential areas of Chengdu and offers various integrated services including flood control, power generation, agricultural and aquatic production, tourism, and

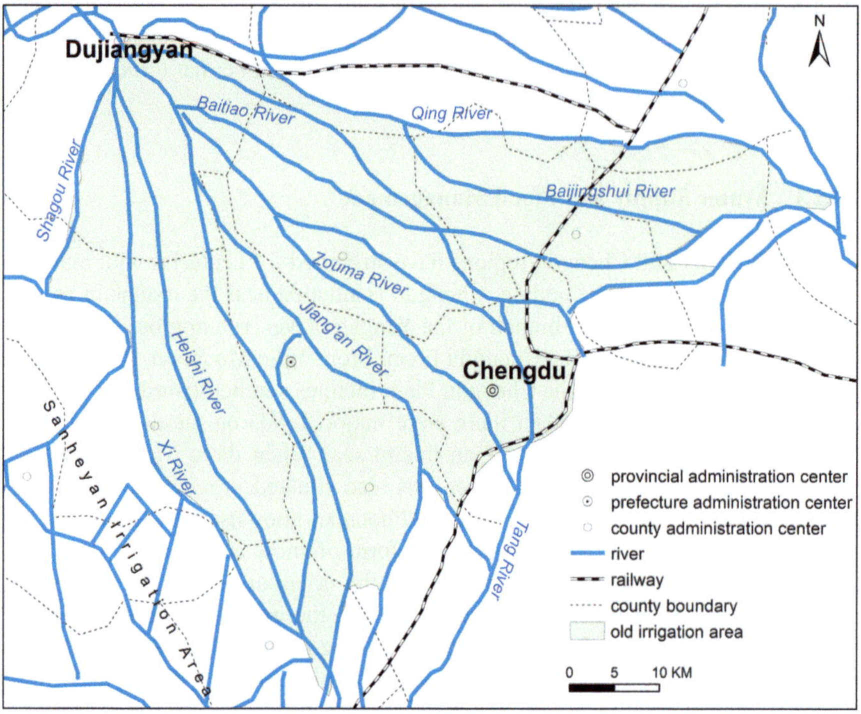

Fig. 9.8 Illustration of *Dujiangyan* irrigation area

environment protection. It is the largest irrigation project in China and necessary for continued support of the economic vitality of Sichuan Province.

The early development of Phoenix started with water diversion for irrigation. Beginning with Jack Swilling, who introduced farming and irrigation to the Salt River Valley after the American settlement, many of the early farmers built canals in the area. Later the Salt River Project was developed to support agriculture and urban development in the region. The demand for water exceeded the supply in the local rivers, however, and Phoenix turned its attention to the primary water source in the Intermountain West, the Colorado River (Fig. 9.9).

The development of water supply in Phoenix follows a pattern characteristic of many cities in the western United States. Construction of irrigation canals and ditches began in the 1880s. During the previous century civil engineers had acquired considerable experience in water transportation through development of canal projects such as the Baltimore and Ohio the Erie Canals, urban drinking water systems such as the Croton Aqueduct in New York City, and the engineering of flumes and other hydrologic systems for mining. In Arizona, the irrigation canals of the 1880s were constructed on the remnants of an historical canal system constructed by the Hohokam Native American group and abandoned in the sixteenth century. Use of historic canals was also a characteristic pattern in development of

Fig. 9.9 **a** Water resources project for Phoenix (Source:http://brazilbrazil.com/urban.html), **b** a part of the central Arizona project system map, by Central Arizona project (Source: http://www.cap-az.com/AboutUs/SystemMap.aspx)

other areas of the U.S. West such as the Rio Grande Valley of New Mexico. This phase of water supply development was organized largely by local government jurisdictions and private entities including municipalities, ditch companies, cooperatives, and developers.

The federal Bureau of Reclamation, organized through the *National Reclamation Act* signed by President Theodore Roosevelt in 1903, strongly influenced urban development in Arizona. The Bureau of Reclamation grew in part out of Progressive Movement reforms calling for professionalization (and expansion) of the federal government. The Bureau of Reclamation focused initially on development of diversion dams; farmers were responsible for the construction and maintenance of ditches that transported water from the dams for the flood irrigation of crops such as including citrus, melons and cotton. Dams were also constructed in the Phoenix area to manage intermittent floods resulting from storm water accumulation across the Salt and Verde River watersheds. This system of agricultural irrigation and flood management survived the great depression and became the foundation for urban development in the great development boom after World War II. In the mid-to-late 1950s, water planners argued for another phase of investment and system expansion. The seven dams and other system elements managed by the Bureau of Reclamation were consolidated into the Salt River project. This initiative was originally designed to support agriculture but its focus later shifted to urban uses. The Salt River project was designed as a comprehensive and integrated system for transportation and management of surface water in the Phoenix region.

During the 1990s, groundwater depletion became evident. Rates of withdrawal from groundwater sources continued to increase with the growth in irrigation and development. By the 1990s, groundwater depletion was recognized as a major water supply issue, and in 1977, the federal government required that Arizona address the problem or lose federal funding for water development. Arizona responded with state legislation that established a state regulatory presence over land and water development. This state activity has expanded over subsequent decades. State regulation may be constrained, however, by recent state ballot initiatives in Arizona designed to protect private property rights (Sparks 2009). These limit a regulatory 'taking' of land value without reimbursement to landowners and other affected parties.

Another phase of water supply development in the Phoenix area is the transfer of water from the Colorado River. When complete, this effort will divert up to 1.7 billion m^3 of water from Lake Havasu and move it more than 450 km to the Phoenix region. The Colorado River is the primary drainage system in the intermountain region of the Western United States between the Rocky and Sierra Nevada mountain ranges. The diversion of water to Phoenix has implications for both upstream and downstream users of the system, and potential ecological effects throughout the region. Among these effects, the project may limit opportunities for modification of the dam and reservoir system along the Colorado River, which has been proposed both to address siltation of reservoirs and recreate historical bottom land ecologies along the river. Likewise, urban water demands both in Phoenix as well as other cities such as Las Vegas may influence opportunities for restoration of the historically important wetlands at the mouth of the Colorado River, located in Mexico. Over the past decade or so these wetlands have become an increasingly sensitive concern among environmental organizations and in U.S.-Mexico diplomacy. The recent legal settlement with the Gila Indians will also have an important effect on water supply for urbanization. It allocates about 800 million m^3 for specific uses designated by the tribe, and is expected to support an expansion of the Phoenix metropolitan area in new directions. In this sense, decisions about water to support future growth in Phoenix have implications across a variety of boundaries: aquifers, watersheds, states, sovereign entities such as tribes, and nations.

City and state governments in Arizona have used a variety of regulatory tools and incentives to manage the allocation of water sources and rationalize water use.

Farming practices have been an important focus for water conservation efforts. Most farming around Phoenix is dependent on irrigation and historically the agricultural community has competed with urban users for access to water. Federal and state governments encouraged farmers to systematically evaluate water consumption and apply water-efficient farming methods. Another emphasis of government programs has been education of urban users to conserve water and invest in water-saving technologies. Although local governments in the Phoenix area have dedicated significant effort to water conservation, there are still significant barriers to reduction of water use to sustainable levels. (White et al. 2008; Larson et al. 2005).

9.2.2.2 Regional transport development

The development of both Chengdu and Phoenix is closely associated with the regions where they are located. Chengdu is situated in the central part of the Chengdu Plain and thus is in a favorable urban position with respect to Western Sichuan. However, as Chengdu Plain itself is located in the western part of Sichuan Basin and is in poor connection with other regions outside the basin, the development of the entire Chengdu region was constrained until fairly recently. Due to the rough roads and dangerous water crossings, external communication has been difficult and hazardous since ancient times. Historically, the *Yangtze* River was the primary transportation route out of the Sichuan Basin, but formed a series of deep valleys in eastern Sichuan (the Three Gorges in Mid-Yangtze River), where the river goes through rapids that are unfavorable for navigation.

The rapid development of Chengdu after the formation of the People's Republic of China is closely associated with improved regional transportation achieved by breaking through these natural obstacles. In the early 1980s, Chengdu was connected with areas outside Sichuan Province through three major railway lines: Cheng-Yu Railway, Cheng-Kun Railway and Bao-Cheng Railway. Cheng-Yu Railway built during 1950–1952 was the first major railway line in Southwest China. It crossed the Sichuan Basin, promoting the transportation of materials throughout the southwestern region and playing an important role in developing production and flourishing economic construction. The eastern terminal of the Cheng-Yu Railway is Chongqing City, which was connected to the line in 1965. In 1979 the line was connected to Guiyang of Guizhou Province and Xiangfan City of Hubei Province. The Bao-Cheng Railway began operation in1958, extending north from Baoji in Shaanxi Province and south to Chengdu. The Cheng-Kun Railway opened in 1970 and extends across 130,000 km^2 including seven prefectures and cities of Sichuan and Yunnan. Thus, Chengdu is a pivot point for four railway lines (the above three lines and the railway from Chengdu to Dazhou) and a primary route for movement out of Sichuan.

Phoenix has a different, although also significant, transportation history. It sits astride one of the major routes between Southern California and Texas, and as such has become an important regional transportation center. The Southern Pacific railroad completed its line through Maricopa (near Phoenix) in 1883, and the Santa Fe railroad entered the city in 1890. A number of other railroads were constructed in the state primarily for transportation of minerals. Interstate 50 was completed through the city in 1958 connecting Los Angeles and Texas. However, unlike Chengdu, Phoenix historically did not function as a major transfer point nor as a gateway to a major region. The Southwestern U.S. was sparsely populated until recently, and the major ports of Los Angeles, Houston and New Orleans focused on population centers in California, Texas and elsewhere. As the population of Arizona and surrounding states have grown over the past few decades, the role of Phoenix as a transfer point and logistics center has become more important. Moreover, Phoenix has become a primary gateway for immigration from Mexico to the United States, and along with this population shift, immigration has become a heated issue in Arizona politics over the past couple of decades. A limited tonnage of goods from Mexico is also shipped through Phoenix although this traffic is dwarfed by trade through Texas and California. In these respects transportation links to the south of Phoenix have become salient. However, Phoenix is competing in all these dimensions with other cities in the region including Salt Lake City, Las Vegas, Denver, El Paso, Laredo, San Diego and Los Angeles.

9.2.2.3 National Policy for Inland Cities

In late 1950s, due to fierce international conflicts and strategic considerations, China shifted its heavy industries from Northeast China and East China to the inland western region. This shift was known as the "third-line construction" period. During this period, Sichuan was one of the key receiving regions for the industrial migration. Chengdu was the command center of the southwestern third-line construction. From 1965 to 1976, a total of 33.5 billion Yuan (equal to 500 billion Yuan at present) was invested in construction activities in Sichuan, about 80 % of the total national investment in Sichuan from 1949 to 1976.

During this period, important enterprises of the "first-line" region were shifted to Chengdu including the Sichuan Gear Factory. Additionally, the central government invested in the establishment, expansion and rehabilitation of large and medium military enterprises including Chengdu Aircraft Company and Chengdu Engine Company. The relocation, construction and operation of these enterprises strengthened the industrial base of Chengdu and massively enhanced its capacity to manufacture machinery, electronic elements and military equipment. Other types of manufacturing were also being promoted in the region, and by the 1970s Chengdu could produce more than 100 types of industrial products such as seamless steel pipes, automobiles, mechanical equipment, measuring and cutting tools, fertilizers, and basic chemical materials and antibiotics. Additionally, Chengdu became one of the four centers of the electronics industry in China and a base of fighter plane and radar production. By 1977, the city had more than 4,010 enterprises with an annual industrial output of RMB 3.65 billion Yuan.

The "Third-line Construction" policy played a strong role in promoting the urban growth of Chengdu. Before 1949, Chengdu's total industrial output was only RMB

108 million Yuan, with 14,000 small businesses in the handicraft industry; factories and businesses of mechanical and semi-mechanical production including small-size textile mills and power plants, machinery plants and workshops providing repair services, processing factories for agricultural products and by-products; and brick and tile factories. As a result of public investment during the "1st five-year" period (1953–1957) and "2nd five-year" period (1958–1962), Chengdu's industries were strengthened The growth of industries such as electronics, machinery, metallurgy, chemical, textile, light industry, building materials and foodstuffs laid a foundation for the further growth of the Chengdu region. By 1965, the city had 2,270 industrial enterprises, with a total industrial output of RMB 1.05 billion Yuan.

The central government's preferential investment in Chengdu during the "Third-line Construction" period was a primary driver for the transformation of the city's economy. Chengdu changed from a city of consumption into a city of modern industry, represented by a significant presence in electronics and electromechanical industries rather than heavy manufacturing.

Economic growth policies in the Southwestern U.S. had their roots in post-civil war reconstruction and the shift of textiles and other industries to the Southeast. This industrial migration was accompanied by the emergence of a new economic development strategy. Cities across the U.S. South reorganized local and state tax policy and constructed public-private investment partnerships to enhance their competitive advantage in wage rates and land costs and more relaxed regulatory environments. This urban growth strategy—competition over wages, land development cost and regulatory liberalization—was transferred across the western United States after World War II. Moreover, federal and state governments made significant public investments in the Southwestern U.S. including construction of the interstate highway system and development of electrical power generation and distribution systems. Finally, military investment and retirement played a role in the growth of the Southwest (Abbott 1998). The army had a considerable presence in the Southwest during the Indian wars, which ended late in Arizona with the capture of Geronimo in 1883. During World War II, The U.S. military shifted the geography of defense industries and facilities to the southern and western parts of the United States, creating huge new regional centers of military training and supply. Both Phoenix and Las Vegas benefitted from this shift in military investment. Low population densities, land costs and sunshine made the Phoenix region particularly attractive to the air force. In the early years of World War II, the government established Lockland Air Force Base, the Thunderbird Training Grounds, the Garrett Research Center, three army camps and six air force bases in the Phoenix area. The presence of military facilities and defense contractors provided a primary stimulus to the emergence of technology industries, which have been a cornerstone in the growth of the Phoenix economy after the 1960s. State and federal government also emerged as an important employer in Phoenix as state government grew after the 1950s and federal agencies established regional and state offices in the city.

Finally, Phoenix became a magnet for retirement very early in the twentieth century, in part as a result of tourism and booster advertising campaigns mentioned earlier. For example, the Santa Fe railroad organized a well-known campaign featuring native American art, Spanish cultural influence, cowboys, sunny days, warm winter climate and stark landscape. Later, Phoenix was marketed to retirees as a specialized and modern suburb, with age-appropriate housing convenient shopping, contemporary architecture, access to amenities such as golf courses, and availability of high-quality health care. The federal social security system and public and private retirement benefits provided the income to support this retirement migration. The military interacted with these other elements in the regional development process by creating a pool of retirees, at least some of who may have had second careers in defense industries.

9.3 Institutional Innovators: Shenzhen and Las Vegas

Our second set of cases is "young" cities that emerged not primarily as capitals of resource regions but out of a combination of institutional innovations and spill-over effect. We evaluate Shenzhen and Las Vegas as case studies of growth processes of this kind. These cities are the product of administrative innovations as national, state and local governments created regulatory experiments in what were then small, peripheral towns. Moreover, these cities developed as a result of spillover effects as mature cities spun off economic activities to low-cost, proximate areas. In both places, regulatory regimes that were dominant elsewhere in the country were relaxed. Regional economic development around both cities was stimulated by their regulatory uniqueness. Regulatory innovations were also accompanied by a specialized geographical advantage. Each city was located near an established metropolis, and derived further advantage from interacting with its mature sibling: Hong Kong for Shenzhen, and Los Angeles for Las Vegas. Finally, both cities developed strong investment and growth strategies to take advantage of their comparative strengths. The combination of these factors has created enormous economic momentum in both places (although the economy of Las Vegas has suffered over the past few years from the collapse of the housing boom).

Shenzhen became an administrative establishment first in the 8th year of *Qing's Kangxi* Emperor (1669 AD), when the *Qing* government set up *Xin'an* County for jurisdiction of the current Shenzhen and Hong Kong region, with an area of 3,076.00 km^2. After the Opium War, Hong Kong was functionally independent of Beijing. After P.R. China was founded in 1949, the Shenzhen region set up Baoan County, with the county center located at Shenzhen Town. In 1979, Baoan County was changed into Shenzhen City, with the municipal government located in Shenzhen. In November of the same year, Shenzhen was designated a city under the provincial administration. In 1980, the "Shenzhen Special Economic Zone" was established. In 1988, the State Council approved the designation of Shenzhen City in the state plan and granted economic management authority to the city. During this period of about three decades Shenzhen grew from a small fishing village to a global capital with GNP ranking fourth in China and a population of nearly 10 million. Shenzhen has observed great changes during this period, and its urbanization rate is known as "Shenzhen Speed".

At present, Shenzhen has one of the strongest export economies in mainland China with strengths in high technology, logistics and finance. In 2009, Shenzhen's GDP reached RMB 820,123 billion Yuan, per capita GDP was RMB 92,771 Yuan, social fixed-asset investment amounted to RMB 170,915 million Yuan, and social consumable retail turnover amounted to RMB 259.868 billion Yuan. With the total export volume of USD 270,155 billion, Shenzhen has ranked the first among the large and medium cities in China for 20 consecutive years (by the end of year 2012).

Las Vegas was settled in 1854 by a small group of Mormon settlers, who located near a spring in an area inhabited historically by native-American populations. Soon after the Mormon settlement, the U.S. military established an army post in the area for the purpose of governing native American populations and managing border disputes with Mexico. The city of Las Vegas was created in 1905. Through the early 1930s Las Vegas flourished as a small mining and ranching service center, stimulated by the discovery of gold and silver deposits in the region and resulting population immigration. The Nevada legislature legalized gambling in 1931, apparently with little discussion of possible long-term effects. Las Vegas slowly diversified into a gambling capital during the 1900s and 1940s. With gambling as the initial economic driver, communications and other industries also emerged. Tourism continued as a pillar of the city's economy, however, and by 2000 it annually attracted close to 40 million tourists and provided about 300,000 jobs. The city's entertainment and consumer markets have attracted entrepreneurs and capital from other parts of the U.S. and the world. With rapid investment in casinos, hotels and other amenities, Las Vegas experienced rapid population growth—at times it was the fastest growing city in the United States—and its landscape underwent multiple transformations.

9.3.1 *Patterns of Urban Expansion*

9.3.1.1 Geography

Shenzhen City neighbors Hong Kong in the south, at 113°46'–114°37'E and 22°27'–22°52' N. Its total area is 1,952.84 km^2 including an area of 391.71 km^2 for Special Economic Zone. Shenzhen has a tropical ocean monsoon climate, with the constant annual average temperature of 24.0 ℃, extreme maximum temperature of 38.7 ℃, minimum temperature of 0.2 ℃, frost-free period of 355 days, average annual rainfall of 1933.3 mm, and sunshine of 2120.5 h. Topographically, Shenzhen is high in the southeast and low in the northwest, comprised of both mostly hilly land and gentle terraces. Its western coast is the beach plain. The main river is Shenzhen River with a total length of 35 km.

Las Vegas is located at 36°10' N and 115°08' W, with an area of 340.0 km^2, including a land area of 339.8 km^2 and the water area of 0.16 km^2. The metropolitan

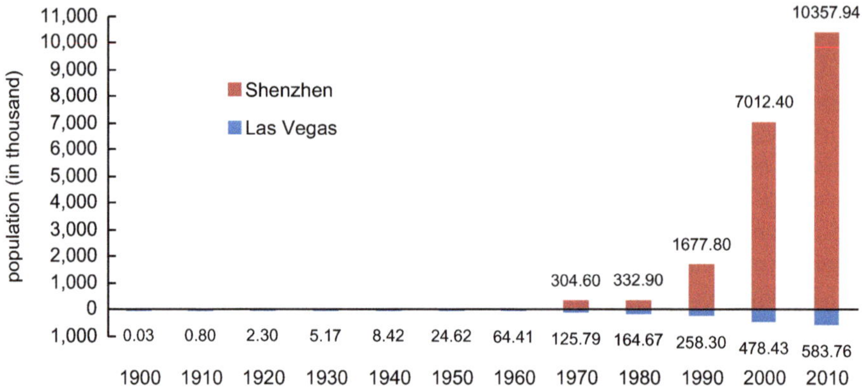

Fig. 9.10 Urbanization of population in Shenzhen and Las Vegas

area is 712 km². Its average altitude is 610 m. Las Vegas is of distinct seasons. In summer, it is of typical desert climate, with the temperature often up to about 38 ℃ at noon and a relatively cool temperature at night. In winter, the temperature is mild and comfortable as a whole, with an average temperature of about 15 ℃ at daytime. Las Vegas has the lowest relative humidity and perhaps the lowest annual rainfall of any major city in the United States (about 100 mm a year).

9.3.1.2 Patterns of Population Growth

Although both cities are young compared to Chengdu and Phoenix, Shenzhen is even younger than Las Vegas. As Fig. 9.10 shows, the population growth as a part of the urbanization process in Las Vegas began in 1900s with rapid increases since the 1980s. Shenzhen's rapid population growth began as it was established as a Special Economic Zone in 1978. The urbanization process in the two cities is closely related to the influx of large populations.

Shenzhen was the original Baoan County of Guangdong Province. The population of Baoan County was 304,600 in 1970 and 312,600 in 1979, with an increase of only 8,000 over an entire decade. After 1979, Shenzhen City experienced a process of rapid population urbanization, with its population growing sharply (Fig. 9.11). From 1979 to the end of 1990, the permanent population grew at an average rate of 127,000 persons per annum. During the period from 1991 to 2000, the permanent population grew at an average rate of 533,000 persons per annum. From 2001 to the end of 2008, the permanent population grew at an average rate of 220,000 persons per annum. The average growth rate of permanent population was about 293,000 per annum in three decades, equal to population of a medium-size city. The pattern of its population growth was slow, fast and slow again. At present, Shenzhen's population density is up to 24,564 persons per km², far higher than the provincial and national standard. Shenzhen is now the densest city in China.

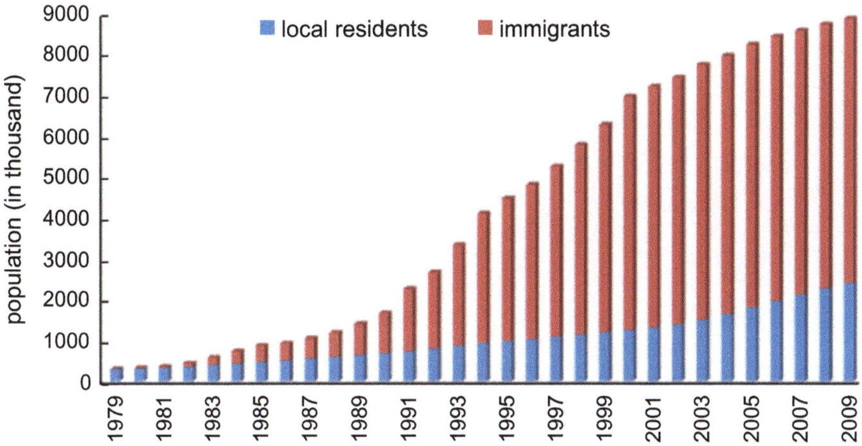

Fig. 9.11 Changes in number of Shenzhen's permanent population and temporary population (1979–2009). (Shenzhen Manicpal Statistics Bureau 2010)

In 1900s, there was only 25 people in Las Vegas region (Fig. 9.11). Nevertheless, the number reached several thousand since 1920s, and became tens of thousands in 1950s. During much of the 1990s, Las Vegas is the city with fastest population growth rates in the U.S., with annual average growth rate of 85.2 %.

9.3.1.3 Process of Landscape Transformation

Rapid urbanization of Shenzhen is not only represented in its social and economic development, but also in significant landscape change. Before 1979 Shenzhen was only a small frontier town (Fig. 9.12) with GDP of only 196 million Yuan in an area of only 3 km^2 Shenzhen has now become a modern international metropolis with high-rise buildings. Beginning largely as barren desert, Las Vegas evolved over just a few decades into a global leisure and entertainment capital.

The dynamic changes in Shenzhen land use can be observed by classifying and assessing the Shenzhen region using remote sensing imagery (Fig. 9.13). We see that the built up land uses in Shenzhen expanded from a scattered distribution in 1979 to a dense metropolis in 2010. In the early period of urbanization, development sprawled gradually outward around the already built-up area and traffic corridors. In recent years, the city's spatial development pattern has been affected by topographic factors. The growth rate of the established area increased from 6.3 % in 1980 to 33.5 % in 2005. At the same time, the area of built up land in the Shenzhen region expanded from 2.9 km^2 in 1979 to 813 km^2 in 2009, with the proportion increasing from 0.15 to 41.63 % in three decades, at an annual average growth of 20.67 %.

Historical satellite imagery shows the rapid urban expansion in Las Vegas (Fig. 9.14). The extent of the build-up area in Las Vegas grew by three times during

Fig. 9.12 Changes in urban landscape of Shenzhen and Las Vegas. **a** Shenzhen 1980s (http://ly.sz.bendibao.com/tour/200986/ly119668.html). **b** Shenzhen 2000s (http://citylife.house.sina.com.cn/detail.php?gid=49738). **c** Las Vegas 1900s (http://www.gamblingweblog.com/tag/origins/). **d** Las Vegas 2000s (http://www.realbollywood.com/2012/09/spending-million-single-night-sin-city.html)

this period (UNEP 2005; Wu et al. 2011). Visual browsing of the satellite imagery also suggests that the build-up area became denser during this period. Las Vegas developed in a desert without significant agricultural cultivation, which is one of the prominent ecological factors distinguishing Las Vegas from our other three case studies. In Las Vegas, the low level of land utilization around the city may have helped keep land costs low, in turn lowering overall development costs relative to competing regions such as southern California.

The expansion of Las Vegas has occurred largely in the western and southern directions. The old downtown, once close to the center of the metropolitan area, has been supplanted as the city's primary economic center by the Las Vegas 'Strip', which experienced an enormous influx of investment after 1990 by a new generation of casino corporations. The locations of the strip, airport, I-15 and other infrastructure have tended to pull new development investments toward the southern sector of the metropolis. Much of the new development has also occurred in independent jurisdictions such as Henderson, and the patterns of metropolitan expansion in Las

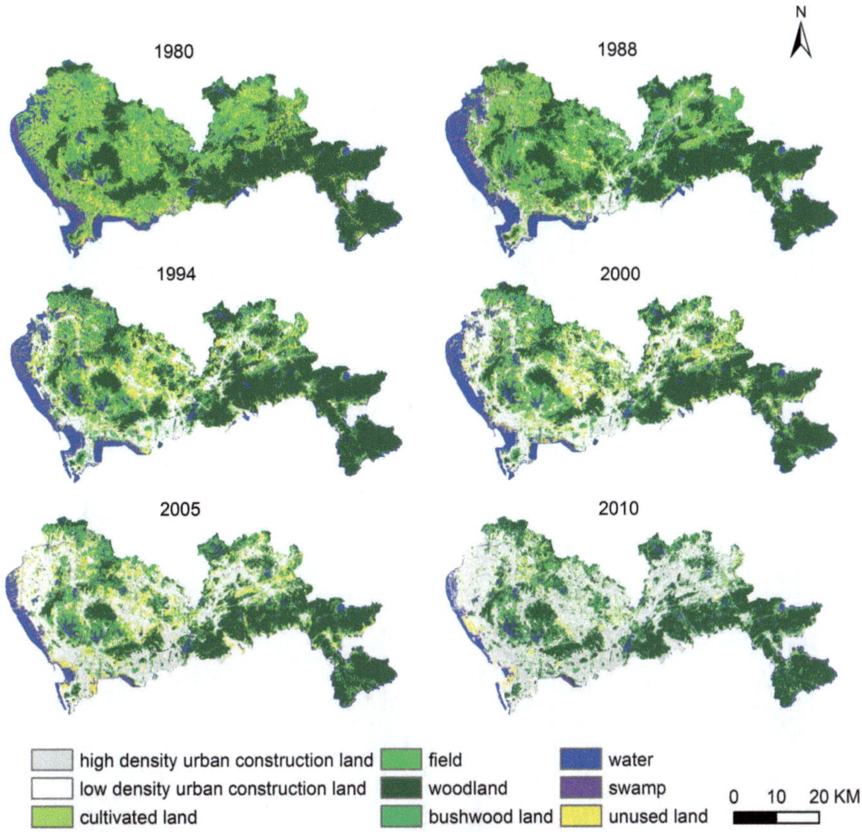

Fig. 9.13 Expansion of Shenzhen's urban and built-up lands. (Yuan 2003; Zheng 2007; Xu 2004)

Vegas can be written in part as a story of competition among small and mid-sized jurisdictions. Amenities such as parks, golf courses and open spaces have also been concentrated in the south and west. Finally, development in the east has been limited in some areas by geographical factors including rough terrain.

9.3.2 *Drivers of Urban Growth*

The urbanization processes in our four case studies have important differences but also common elements. In the cases of Chengdu and Phoenix, cities emerged from advanced agriculture, post-World War II defense policy and as capitals of inland resource empires. Shenzhen and Las Vegas are a rather different type of case: they took off in a post 1980 growth boom as a result specialized interactions between economic dynamic and institutional innovations. We shall see in the following

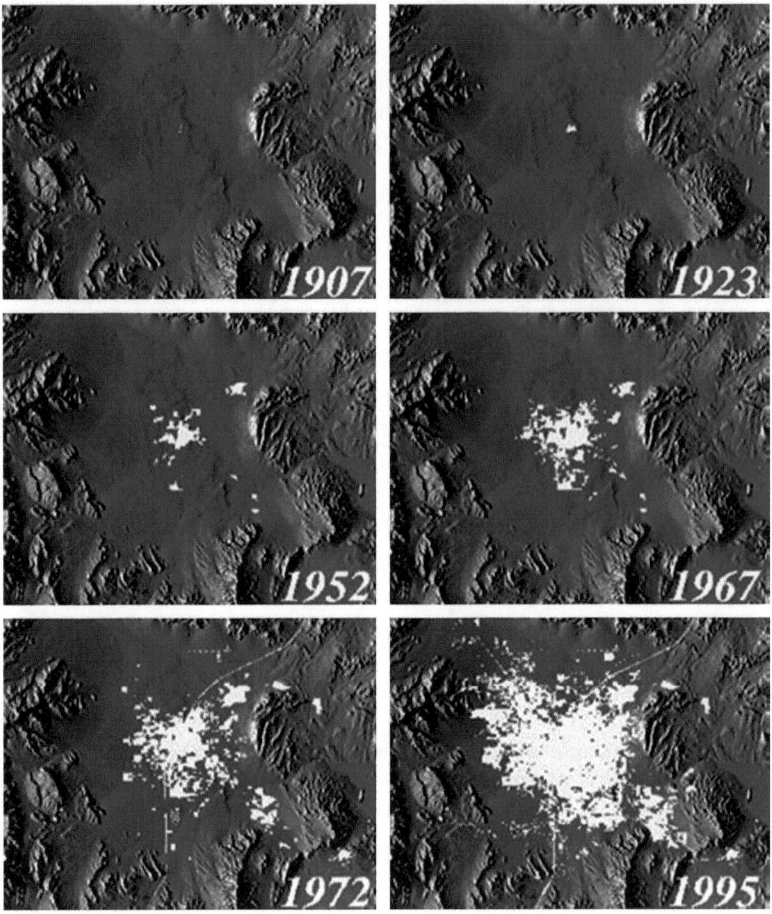

Fig. 9.14 Landscape urbanization of Las Vegas (1907–1995). (Wu et al. 2011)

sections the critical issues in the expansion of Shenzhen and Las Vegas and how policy design successfully create attractiveness to immigrants.

9.3.2.1 Regulatory Innovators

Without external incentives and economic reform, Shenzhen and Las Vegas probably would not have experienced rapid growth rates. In Shenzhen a special economic zone was established as an "experimental field" of the national economic reform and a "window" for the outside world. In conjunction with this zone the central government granted Shenzhen a series of preferential policies. With the support of such policies, Shenzhen became the pioneer in the reform of China, innovating successively in various aspects such as land use system, taxation system, employment system and administrative system and laying a foundation for the rapid development of Shenzhen.

The most outstanding systemic innovation of Shenzhen was in the land use system. Before the reform the three fundamental problems with the policy of China for urban state-owned land were administrative allocation, charge-free and limitless use and prohibited transfer by users. In view of the above problems, Shenzhen Special Economic Zone carried out the reform of land policy mainly in two aspects: reform of land property system and land lease and pricing system. In November 1981, the standing committee of Guangdong's 5th Provincial People's Congress adopted *the Provisional Regulations for Land Administration of Shenzhen Special Economic Zone*, indicating that the land of Shenzhen entered the period of charged use. In 1987, Shenzhen Municipal Government promulgated the *Administrative Regulation for Land of Shenzhen Special Economic Zone*, indicating the official establishment of charged use and transfer system of state-owned land and the gradual formation of a land market in the special economic zone. On December 1, 1987, for the first time, Shenzhen auctioned the land, with a land lot of 8588.25 m^2 acquired by Shenzhen Special Economic Zone Real Estate Corporation at a cost of 5.25 million Yuan, with an average land price of 611.3 Yuan/m^2 and a use period 50 years. This auction was called "*a milestone for the critical breakthrough in the land reform of China*". On the premise of the experiment in Shenzhen Special Economic Zone, the Amendment to the Constitution of the People's Republic of China (1988) specified that "the land use right can be transferred according to law". In the same period, the Land Administration Law was promulgated, further specifying the charged use system of land. With such relaxation in land use system, land can be transferred (traded or leased) by its "use-right", which allows the emergence of a land market as well as the price system, leading to more efficient allocation of land resources. Availability of land is important to urban expansion of Shenzhen. In addition to the land use system, the Shenzhen Special Economic Zone also developed other economic innovations in areas such as taxation and finance. In general, the area became a pioneer of reform for the joint objectives of industrial promotion and urban development. In the special economic zone, preferential tax rates were implemented to attract domestic and foreign capital into the zone. During the early period, Shenzhen Special Economic Zone lowered the corporate income tax to 15 %, which was the same as that in Hong Kong. In contrast the inland enterprises paid the tax at a progressive tax rate of 55 %. In terms of finance, the first joint-stock commercial bank of China was established in Shenzhen in 1987, breaking through the conventional theory of state-owned banks. The first securities company also emerged in Shenzhen in the same year. Likewise, the Shenzhen Stock Exchange was founded in 1992 and became one of the two major stock exchanges in China. Under a regime of strict foreign exchange control, Shenzhen also took the lead in establishing the foreign exchange center in 1985. Other financial innovations such as mortgage and automobile loans and offshore financial services were all exported to the inland areas of China after being developed in Shenzhen.

With its policy advantages and bold innovations, Shenzhen provides a better investment environment than other regions and thus has created relatively strong attraction for foreign and domestic funds, technology, talents and other production factors. The area has organized itself as a conglomeration of modern production factors in a short period of time. Because of such open policies, in contrast to inland areas of China, Shenzhen became a national "policy control point", forming agglomeration effects and becoming the source power for urban development.

The ban on gambling was repealed by the Nevada state legislature in 1931. At the same time, the Nevada legislature adopted liberal legislation governing marriage and divorce, which became another attractor for people fleeing the strict behavioral codes present across much of the United States. The first casino, the Pair-O-Dice Club, opened in 1931, with four more casinos opening up over the next 15 years. In 1946, the Flamingo Club opened its doors, establishing a new Las Vegas standard for clubs offering a 'complete experience', with luxurious hotel rooms, gardens, pools and entertainments. The industry has gone through another major upheaval over the past two decades or so, resulting in development of ever-larger resorts and themed experiences such as an emphasis on classical art and architecture at Belaggio. In alliance with the gaming industry, local and state governments have worked actively to stimulate further growth through marketing of the region; encouragement of public investment in water supply, electrical generation and defense industries; and regulation to strengthen the gaming industry and improve its image among policymakers and the public. Las Vegas has marketed itself through sophisticated cultivation of its image, beginning at least in 1945 when the city hired J. Walter Thompson to develop an advertising campaign promoting tourism. The city of Las Vegas has also been successful in securing infrastructure to support growth. Boulder dam, built with federal monies during the 1930s, has provided a relatively low-cost source of electricity and the first substantial growth spurt in the region. The city has also been successful in capturing water supply to support urban development, in part through the Southern Nevada Water Authority. Although Las Vegas has successfully implemented some conservation measures, its demand for water use is expected to stress regional water and other resources (Cooley et al. 2007). Finally, the city has benefitted from federal defense investments such as Nellis Airforce Base. With respect to regulation, Las Vegas developed an early reputation for lawlessness even during the 1930s when it provided entertainment to workers building Boulder Dam. By the mid-1940s, city leaders worked actively to discourage organized crime involvement in the casinos, and national policymakers put further pressure on the city and state through a Congressional investigation of organized crime in 1950. The state of Nevada adopted an important set of rules in the late 1950s designed to constrain the reach of organized crime in the gaming industry.

More recently, the gaming industry in Las Vegas has adapted to global competition by developing specialized marketing strategies and constructing more extensive infrastructure. Over the past 20 years, casinos have been designed with iconic architecture to attract visitors. They are open for business around the clock. Meals are ordered by a wave of the hand. The airport provides access to destinations worldwide and supports a variety of private craft. Other cities in Nevada and elsewhere in the country have also emerged as gambling centers, but Las Vegas still represents the largest spatial concentration in the U.S. gaming industry because of its convenient transportation links and other historical advantages.

9.3.2.2 Advantageous Regional Conditions

In our two cases, the advantages of regulatory innovation are associated with two factors: proximity to a global city, and opportunity to exploit comparative advantage by attracting economic activity from places under a traditional regulatory regime. The rapid development of Shenzhen and Las Vegas are closely associated with these advantageous regional conditions.

The regional advantage of proximity to Hong Kong is a fundamental advantage that enabled Shenzhen to grow at a rate faster than that of other cities in China. The geography of Shenzhen is intertwined with that of Hong Kong. They have a common boundary of 27.5 km. Shenzen is astride the only land route by which the inland can communicate with Hong Kong, and by which Hong Kong can enter the *Pearl* River Delta and interact with the inland. Due to the long-term operation of global regional political and economic pattern, Hong Kong became a central node of the world economy. The economic and social differences between Hong Kong and Shenzen acted as a mutual attractor. Hong Kong was a "free port" with a highly-developed economy, differing from its neighboring Shenzhen in social system, institutions, economic strength, urban position and social form. The development of Shenzhen, especially its initial stages, is "Hong Kong-oriented", and its urban form had its start by sprawling outward from *Luohu*, *Shatoujiao* and *Shekou*, all of which lie in close contact with Hong Kong.

Hong Kong has influenced the urban economic development of Shenzhen in several dimensions. First, the large-scale industrial shift from Hong Kong was the fundamental driver for the takeoff of industry outside the special economic zone in Shenzhen. After the late 1970s, the tertiary industries of Hong Kong had developed rapidly, but its export labor-intensive industries were constrained by the increasing land price and high labor cost. Thus, Hong Kong needed access to a development region with lower total costs. When China broke through the political boundaries separating Shenzhen and Hong Kong with the unique policy of the special economic zone, Hong Kong moved quickly to take advantage of the availability of land and low labor costs in Shenzhen. Second, Hong Kong moved capital as well as industry to Shenzhen. The large-scale investment from Hong Kong stimulated the process of urban construction in Shenzhen, and the rapid development of the built-up area. Hong Kong capital has continued as the major source of foreign investment in Shenzhen. Third, the regional advantage of neighboring with Hong Kong has inspired Hong Kong people to consume in Shenzhen, driving the rapid development of tourism, commerce, housing and other material forms in Shenzhen.

The competitive advantage of Las Vegas derives substantially from its location near the population centers of southern California, which provided investment capital, industry expertise, a customer base and access to the entertainment industry that were a critical ingredient in the expansion and marketing of Las Vegas. In 1938, California began an aggressive campaign against gambling, which created a sharply different regulatory environment between the two states, Parts of the gaming and entertainment industries moved to Nevada as the regulatory environment for gambling in California became less permissive. For many years, southern California remained an important market for the Las Vegas gaming and entertainment industries. Beginning in the 1940s, Las Vegas developed a glamorous cachet based on Hollywood connections, particularly association with stars such as Liberace, Frank Sinatra and Sammy Davis Jr. This relationship was a competitive advantage of Las Vegas during this period, differentiating it from other historical centers of gambling and entertainment in the country such as Galveston Island, Texas and New Orleans, Louisiana. The growth spurt of Las Vegas after 1980 is in large part a product of its low taxes, housing and living costs relative to California. As California's urban population growth took off in the post-World War II period, costs of land, construction, retail products, public services, labor and other factors escalated, and home ownership was no longer feasible to many residents of the state. Las Vegas emerged as a primary low-cost alternative, and attracted migration both by individual households as well as employers.

Las Vegas has also grown as a result of multi-modal transportation links that provided convenient access to first domestic and now international markets. McCarran International Airport has become a significant hub for air traffic. The Union Pacific Railroad extends through Las Vegas and at least earlier in the century was an important mode for passenger traffic in the region. Interstate I-15 is an increasingly important north-south route through the Intermountain West.

9.3.2.3 Industrial strategy and competitive advantage

In the early period of establishing the special economic zone, Shenzhen developed the economic model of "processing with supplied materials, processing as per sample, processing with samples supplied, assembling parts supplied and compensated trade" which largely simulated the growth of secondary industry. After the 1990s, seizing the opportunity afforded by the rise of global information and distribution systems; Shenzhen diversified its industrial structure. Tertiary industries starting to develop and the proportion of the high-tech sector in the secondary industries gradually increased. More recently, the modern tertiary industry with finance and logistics as the core developed rapidly. Business, trade, tourism and real estate also have emerged with a major share in the economic activities. At present, Shenzhen has established a new industrial system of "adapting to the functions of modern central city with the high-tech industry and advanced manufacturing as the base and with the modern tertiary industry as the support".

Gambling is the best-known industry in Las Vegas but the city's economy is not limited to it. The rapid development of gambling industry in Las Vegas stimulated the rise of general tourism and other industries. After the 1980s, Las Vegas built a broader range of service and manufacturing industries and developed an increasingly diverse employment base, with an important concentration in housing and land development. Because of the city's rapid population growth since the early 1980s, it was deeply exposed to the national recession and decline in housing markets after about 2007. The city proved to be particularly vulnerable to weakness in the construction industry and highly-leveraged housing credit markets. The contraction and troubles in these industries have led to high unemployment rates and thrown into question the recent Las Vegas model of economic development.

9.4 Conclusions

Urbanization and urban sprawl are the produce of human society and economy developing to a certain extent, but still a kind of result and representation of the interactions of man-land system on the earth surface. These four cities have different origins and influences, but many areas of commonality. From the perspective of urban geography, the development of a city is constrained or stimulated by three factors: natural conditions; regional and national economic context; and growth strategy. In this chapter we compare four Chinese and U.S. case study cities across these three dimensions, contrasting two U.S. and Chinese cities with a relatively long urban history and two U.S. and Chinese cities with a relatively short history.

Topographically, the four cities are all situated in a relatively flat region, notably the Chengdu Plain where Chengdu is located. They are all located in the northern temperate zone. Water supply is a critical natural resource for all four cities. Chengdu, Phoenix and Las Vegas have secured water resources necessary to support urban growth through large-scale water supply and conservation projects. Shenzhen mainly depends on large reservoirs in the city to assure the supply of fresh water.

In terms of regional geography and investment strategy, all four cities had considerable intrinsic advantage which they were able to build through policy and investment. The young cities of Shenzhen and Las Vegas had the advantage of location at the periphery of a global megalopolis, which they exploited through regulatory and administrative innovations. Chengdu and Phoenix had the advantage of location in the center of inland resource regions, which they cultivated through military and infrastructure investment. Chengdu, for example, is well-situated within Sichuan Province, but the establishment of its economic position with respect to Southwest China is closely associated with development of several important railways. All four cities used their geographical and policy advantages and early-stage investments as a foundation for development of industry specializations including technology industries, construction and services in Phoenix; gaming, tourism and construction in Las Vegas; and technology, logistics and finance in Shenzhen.

Growth processes in all four cities are associated with patterns of spatial dispersion and fragmentation typically characterized as sprawl. In both our U.S. and Chinese case studies, regional and global immigration was a primary driver of urbanization process, and the process of immigration in itself may have led to dispersed and fragmented development. For example, in our U.S. cases, cultural images including the oasis, the retirement suburb and Los Angeles-style modernism attracted immigrants to new, low-density subdivisions. The suburban morphologies characteristic of postwar residential development in the U.S. have more recently been adopted in China. Immigration has also stimulated development of specialized districts such as low-cost housing areas and ignited political and social tensions that may have tended to fragment overall urban patterns. In all of our case studies secondary industries emerged including public employment, distribution and logistics, and service industries such as health care. These generated additional growth dynamics around development of dispersed sub-centers. Military and government facilities and technology industries have also, at least until recently, tended to choose low-cost locations on the urban periphery. Demands for semi-natural amenities such as golf courses have also stimulated development at the urban periphery. In general, these pressures over time have led to development of a dispersed metropolitan geography, conspicuous in our U.S. cases and emerging in the two Chinese cities.

This chapter also suggests that even with urban applications of advanced science and technology, the viability and further growth of cities remains sharply constrained by natural conditions and availability of resources. Current development patterns in all four cities tend to be associated with high rates of energy, land and water consumption and depletion of ecological services such as habitat and hydrological function. All four cities have to a greater or lesser extend modified their environmental conditions through infrastructure development and adoption of conserving behaviors, but they still appear vulnerable to long-term problems of resource depletion, ecological disruption or natural disaster. The urban sustainability movement, focusing on compact cities alternative travel systems and resource-efficient building systems, offers one strategy to reduce resource consumption in both U.S. and Chinese cities. It is unclear whether the ideals of this movement can be engineered and implemented. The history of the *Dujiangyan* Water Irrigation Project near Chengdu is uplifting, however: it suggests that long-term environmental adaptation—sensitive to or even in harmony with local conditions—is achievable.

In discussions of global climate change, capacity for adaptation has become an important measure of the success of human systems at a global, national or urban scale. Cities are able to adapt if they have the institutions and abilities in place to manage complex and changing relationships between human and natural factors such as water use, hydrologic and ecological process, heat, landscaping, agriculture and economic development strategy. Building the knowledge, technology and institutions to manage these relationships is the fundamental challenge for the next generation of cities. Over their recent histories, Phoenix, Las Vegas, Shenzhen and Chengdu have been successful in building urban development systems that emphasize promotion of economic growth while also managing environmental resources. In the past decade or so, these cities have begun to bump up against less tractable resource constraints and environmental risks. Their challenge for the next decade is to rethink urban form around new relationships between growth strategy and resources, and rebuild the adaptive systems to manage it.

References

Abbott, C. (1998). *The metropolitan frontier: Cities in the modern American West*. Tucson: The University of Arizona Press.

Cooley, H., Hutchins-Cabibi, T., Cohen, M., et al. (2007) *Hidden Oasis: Water conservation and efficiency in las vegas*. Pacific Institute/Western Resource Advocates. http://pacinst.org//wp-content/uploads/sites/21/2013/02/hidden_oasis3.pdf. Accessed 15 Mar 2014.

Grimm, N. B., Faeth, S. H., Golubiewski, N. E., et al. (2008). Global change and the ecology of cities. *Science, 319*(5864), 756–760. doi:10.1126/science.1150195.

Larson, E. K., Grimm, N. B., Gober, P., et al. (2005). The paradoxical ecology and management of water in the Phoenix. USA metropolitan area. *Ecohydrology & Hydrobiology, 5*(4), 287–296.

Peng, W. F., Zhao, J. F., Zhou, J. M., et al. (2008). *Dynamic detection on urban sprawl based on RS and GIS: A case study of Chengdu city*. 2008 International workshop on education technology and training & 2008 international workshop on geoscience and remote sensing, Shanghai, China.

Sichuan Statistical Bureau. (2011). *Sichuan statistical yearbook 2011*. Beijing: China Statistics Press.

Sokol, M. (2005). Reclaiming the city water and the urban landscape in Phoenix and Las Vegas. *Journal of the West, 44*(3), 52–52.

Sparks, J. L. (2009). Land use regulation in Arizona after the private property rights protection act. *Arizona Law Review, 51*(1), 211–234.

Steiner, F., Sherry, L., Brennan, D., et al. (1999). Concepts for alternative suburban planning in the Northern Phoenix area. *Journal of the American Planning Association, 65*(2), 207–222. doi:10.1080/01944369908976048.

UNEP/GRID. (2005). Arendal, urban sprawl, Las Vegas, UNEP/GRID-Arendal maps and graphics library. http://maps.grida.no/go/graphic/urban-sprawl-las-vegas. Accessed 6 Feb 2012.

White, D. D., Corley, E. A., & White, M. S. (2008). Water managers' perceptions of the science–policy interface in Phoenix, Arizona: Implications for an emerging boundary organization. *Society and Natural Resources, 21*(3), 230–243. doi:10.1080/08941920701329678.

Wu, J. G., Jenerette, G. D., Buyantuyev, A., et al. (2011). Quantifying spatiotemporal patterns of urbanization: The case of the two fastest growing metropolitan regions in the United States. *Ecological Complexity, 8*(1), 1–8.

Xu, X. L. (2004). *A study on regional urbanization and its impact on environment: An example of Shenzhen City*. A doctoral thesis at Beijing Normal University.

Yuan, Y. (2003). *Changes in land use of Shenzhen City and ecological safety response mechanism*. A doctoral thesis at Beijing Normal University.

Zheng, J. (2007). *Modeling the hydrological response to land use change in rapid urbanization areas—a case study from Buji River Basin, Shenzhen*. A master's thesis at Beijing Normal University.

Chapter 10
The Main Agricultural Regions of China and the U.S.

Mark Leipnik, Yun Su and Xinyue Ye

10.1 Introduction

Agriculture defines the way of life of a majority of the residents of the People's Republic of China (China). Over 300 million of the total population of 1.3 billion Chinese are farmers and in total 51.3% of Chinese lived in rural areas in 2010. Thus a majority of Chinese are impacted by agriculture on a daily basis. China has made great strides in becoming self-sufficient in food and this is particularly true for the three great cereal producing regions considered in this chapter. These are the Northeastern Plain, the North China Plain and the Middle-Lower Yangtze Plain. Benefiting from climate, topography, economic and historical factors, eastern China has become the major agricultural production area in the country. This chapter will focus on the agricultural areas of Eastern China specifically the Northeastern China Plain, Northern China Plain and the Middle-Lower Yangtze Plain. Analyzing the various features of the geographical environment in each area will illustrate the influence of natural conditions on the formation and characteristics of agriculture, of which climate and soil, which are related to each other, are most important. Also included is an analysis of the importance and role of these different agricultural regions in the overall pattern of Chinese agricultural production. The surplus production of grains from these "three storehouses" as well as production of other crops like cotton grown there, provide for the stability of prices and supply in the whole of China and thus are crucial on a global basis.

Y. Su (✉)
School of Geography, Beijing Normal University, Beijing, China
e-mail: suyun@bnu.edu.cn

M. Leipnik
Department of Geography and Geology, Sam Houston State University, Huntsville, USA
e-mail: geo_mrl@shsu.edu

X. Ye
Department of Geography, Kent State University, Kent, USA
e-mail: xye5@kent.edu

R. Hartmann, J. Wang (eds.), *A Comparative Geography of China and the U.S.*,
GeoJournal Library 109, DOI 10.1007/978-94-017-8792-5_10,

While relatively few residents of America (the United States of America, USA or just the U.S.) are farmers they conduct operations on such an extensive and capital and technology intensive scale that the surplus they produce is also a key element in global food security. Only about 960,000 Americans are full time farmers and perhaps another million are farmers, ranchers or fishermen on a part time basis. In contrast to China, only 16 % of Americans lived in rural areas in 2010. Chapter 4 discusses the great diversity in agricultural in the U.S. While there are many areas where agricultural surpluses are produced, two stand out in particular. The two great "bread baskets" of the U.S. are the Corn Belt and the Great Plains regions. These two regions that have been blessed by nature with rich soils and adequate water and extensive arable lands and thus produce a great surplus of cereals that in turn are often used to feed animals. Much of the production of both cereals and livestock is exported and these exports are absolutely crucial in global food supply and price stability.

To make full use of the different natural conditions and conform to the development trend of modern agriculture, the production regions in China and the United States both have become specialized. Both of them have successfully built several specialized agricultural zones (districts). China has 16 advantage districts, which include food, economic crops and aquatic products, etc. The Northeast Plain, the North China Plain, and the Middle and Lower Yangtze Plain are respectively production center of corn, wheat and rice. The United States also has several agricultural areas discussed in Chap. 4. These areas include those specializing in corn, wheat, cotton etc. Subject to agricultural production conditions, there are differences, between China and the United States in the key food production regions. The main agricultural production areas of the United States are mainly concentrated in the Middle West, central plains, and west coast states such as California and the Mississippi river basin and southeast coastal region. The major crops are corn, soybeans, wheat, cotton, rice, peanuts and so on. The U.S. agricultural production regions hold a leading position in scale, mechanization, specialization and market-oriented production, and also have high production efficiency. The key food production regions of China are concentrated in the three eastern plain areas. The Northeast China Plain and the North China Plain have been greatly improved in terms of production scale, mechanization and commercialization. Integrated production is the advantage of the Middle and Lower Yangtze Plain. China has comparative advantage in the production and export of labor-intensive agricultural products, such as fruits, vegetables, tea, honey, silk, etc. The United States has a comparative advantage in production and export of land intensive and capital intensive agricultural products, such as soybeans, cotton, corn, wool, chickens, pork, etc. These products all come from the major agricultural production areas discussed in the remainder of the chapter.

10.2 Three Main Agricultural Regions of China

10.2.1 The Northeast Plain

10.2.1.1 Geographical Environment and Agricultural Development

The Northeast Plain is located between 38°46' north latitude to 53°23' north latitude and 115°35′ east longitude 134°33' to east longitude; it mainly consists of three flood plains. In the northeast is the Sanjiang Plain, an alluvial flood plain of the Heilongjiang, Songhuajiang and Ussuri Rivers. South of this is the Liaohe River alluvial plain. In the middle is the Song-Nen plain, an alluvial plain of Songhuajiang River and the Nenjiang River. This agricultural region is located within the boundaries of the provinces of Heilongjiang, Jilin, Liaoning and the eastern portion of the Inner Mongolian Autonomous Region. Its area is approximately 35 million km^2 in extent making this area China's largest collection of alluvial plains. (Cheng 1984; Zhou 2000). The Northeast Plain has only a short history of large-scale agricultural cultivation. Before the nineteenth century, only the Liao River Plain had scattered development. After 1860, the Qing government adopted an "open door" policy which encouraged immigration, resulting in a more rapid development of land in the Northeast. Compared to Liaoning, and Jilin provinces, the development of Heilongjiang province came even later. Before 1904, there were prohibitions on land development. The cold and wet climate is the main reason for the later development of this region. The Northeastern winter is very cold and snow remains late into the spring. The summer is mild and humid, but very short. There is a short growing season and high potential for frost damage to crops. However, the climate also brings adequate precipitation and there is an extensive river system with an abundant supply of water. The cold and wet climate means that the area has nearly 3–6 months of freezing weather and widely distributed frozen soil. Saturated soils and wetlands are very common. There are reed swamps in the Songnen Plain, and seasonal marshes in the Song-Nen Plain. Until the 1950s, the Songnen Plain and the Northern Song-Nen Plain were a "Great Northern Wilderness". (Hao 2010).

However, the area has many advantages for agricultural development. Climatic conditions can produce a one-year maturity, and summer precipitation is concentrated during the growing season. The flat and extensive land area is very suitable for agricultural mechanization. Moreover, the soil is very fertile and distributed widely. The widespread black land soils, also known as Chernozems (Russian) or Mollisols (U.S. soil order) and meadow soils, are rich in organic matter (humus) and have a deep, dark colored topsoil horizon. In the middle of the Song-Nen Plain are the most fertile black land soils of the northeastern zone (Fig. 10.1), which is also the only large black land soil belt in Eastern Asia, with a total area of about 11 million hectare. In natural conditions, the black land soil has 1 m humus layer and rich nutrient availability (high cation exchange capacity). Therefore, after the 1950s, the government invested the labor of hundreds of millions of person years of work mostly by soldiers and urban young intellectuals to reclaim the "waste land" in

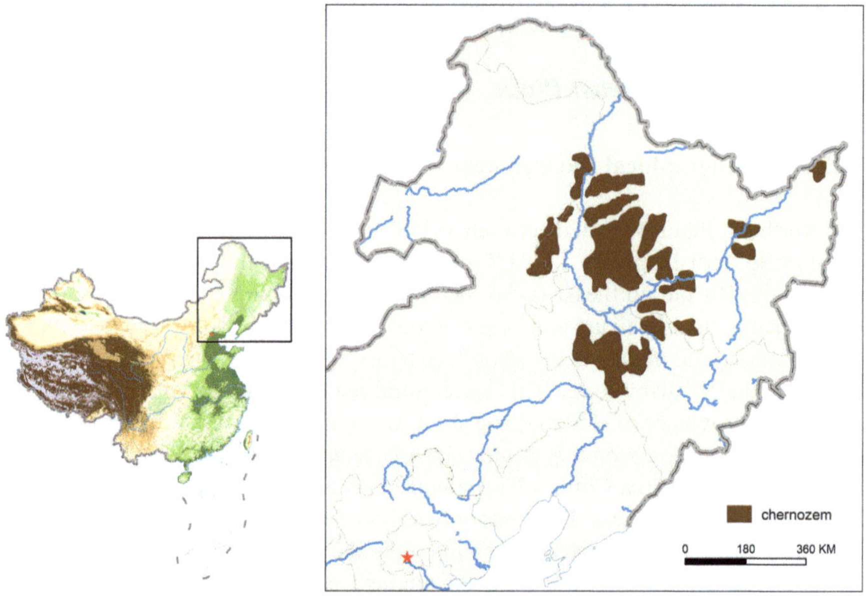

Fig. 10.1 The extent of blackland soils in the Northeastern Region

the Northeastern region and make it suitable for farming. With the development of industry and a transportation network in the Northeast, the level of agricultural mechanization advanced quickly. The cultivated land area and the grain output increased rapidly and the whole region turned into a "big barn". (Li 1993; Chinese Academy of Sciences, Institute of Soil and Water Conservation 1961).

In recent years, global climate change and warming has become a very important factor in agricultural production because agriculture is very sensitive to climatic factors especially water availability and temperature conditions. The extent of the current climatic warming is not beyond the scope of human adaptability, so global warming, especially the increase in temperatures has produced certain positive impacts on grain production in China, especially in the Northeastern area which in the past at least has been severely affected by freeze damage. The rice yield has increased by 2,574 kg/ha due to the warming trend from the 1970s to the 1990s in Heilongjiang province, the estimated contribution of warming to this increase in yield is 19.5–24.3 %. With the agricultural production mechanisms well developed there, the "big barn" status of the Northeastern plain has been consolidated and has improved constantly, but at the same time, a series of ecological problems have been generated, including water consumption, contamination and soil loss, especially the erosion of blackland soils which is mostly due to water.

After being reclaimed, the black land soils became more vulnerable to water and wind erosion, resulting in the thinning of black land soils and a decrease in humus content. The annual loss of black land soils is 0.4–0.5 cm annually in areas with serious soil erosion. In these areas, about two fifths of the black land soils humus layer is less than 30 cm thick at present. Thus, concurrent with the development of agriculture in this area, a black land soil protection project has been actively carried out. Major measures include: change of the traditional farming methods, plowing along the contour; using no-till practices and returning crop stubble to the soil. Also establishing a scientific crop rotation system in the black land soils; and implementing a system of repair and conservation of forest belts. In addition, the practice of conversion of wetlands into farmland by bog and swamp drainage has also been curbed. Starting in the 1990s, these conservation measures have shifted the focus from maximum output of crops to the conservation of biodiversity and wetland ecology.

10.2.1.2 The Characteristics and Location of the Main Grain Producing Areas in the Northeast

Based on the combination of sunlight, warmth and moisture conditions in the Northeast Plain, most of region can meet the needs of late-maturing and thermophilic crops. In this area, wheat, soybeans and early or middle-maturing corn, rice, sugar beets, and flax are the dominant crops. The cultivation system has a one-year maturity. South of the Liaohe River Plain, winter wheat, and also inter-cropping or multiple cropping varieties of early maturity crops and pasture after winter wheat are dominant patterns. At present, the most optimal pattern of food crops in the Northeast, is a system of tridimensional inter-cropping of spring wheat and corn, or beans and corn, or corn and potatoes. (Sun 2003).

Construction of the Commodity Grain Base and the Operation of State-Owned Farms The per capita food production of the Northeastern Plain is the highest among China's agricultural areas. The region provides a large quantity of cereal grain and soybeans to the country every year. The total arable area of the region accounts for 17.6% of the country's total arable land. The annual grain output accounts about 17% of the country's total output, soybean production accounts for 40% of the country's output, the food commodity rate is 40%, nearly 5% higher than the national average. After satisfying local demand, the food surplus always has been sent to outside regions. The Northeast Plain is an important national strategic reserve grain base, providing stabilization of prices.

Large state-owned farms have played an important role in the process of building the Northeastern agricultural base. The Heilongjiang Reclamation Area, for example, has the largest group of state-owned farms (114 farms under State ownership). In the past 60 years of development and construction, the Heilongjiang Reclamation Area had produced 243.4 billion kg of grain, of that total about 183.65 billion kg entered the commodity markets. Currently, the Heilongjiang reclamation area produces one third of the total food production for the province, using only about one-fifth of the arable land in the province, of which the commodity grain producing area accounts for about half of the province. About 25% of the grain production is sent out of the province into inter-provincial trade, although most is used for

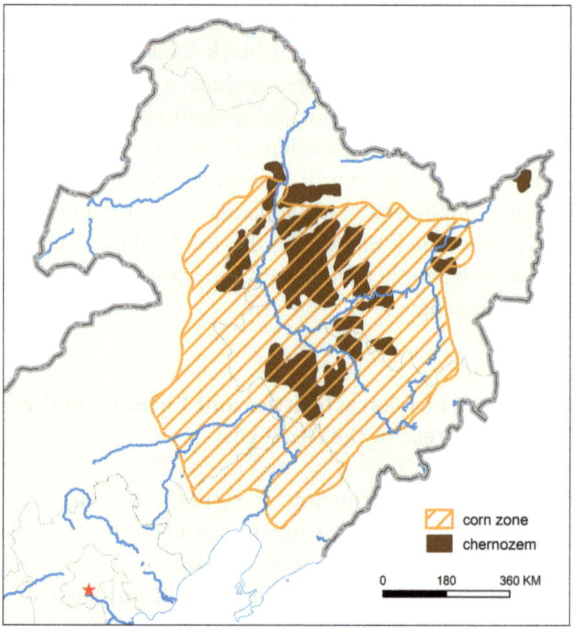

Fig. 10.2 The Northeastern Plain Corn area

domestic consumption in China. In 2010, total grain output of the Heilongjiang Reclamation Area reached 18.2 billion kg, about 17 billion kg of which was grain sold as a commodity. This means that 93 % of production went into commodity markets. The level of agricultural mechanization has reached or exceeded the level common in developed countries. Average annual grain output reached 35.4 metric tons per year per person, the comprehensive mechanization rate is above 93.3 %, of which use of mechanization in dry land farming was about 95 %, and use of mechanization in paddy field (irrigated) agriculture was about 90 %. The Reclamation area also provides services such as tractors and combines to outside areas. (Li 1988).

The Production Centers of Corn and Soybeans Agronomic conditions for corn planting in the Northeastern region are similar to those in the U.S. Corn Belt (described below). Suitable climate and soil conditions all promote corn production. As a result corn yields made up to 6 metric t per hectare or more.

Corn is produced in the northeast of China, in the north of China and in the southwestern portion of China. Of these areas the northeast has the most concentrated production. Jilin Province is one of the world's three golden corn belts (Fig. 10.2). During the past two or three decades, the proportion of grain used for feed and for industrial purposes such as for high fructose corn syrup production has increased, and corn production acreage has gradually expanded. It has increased the fastest in Northeastern China, and it has substituted for sorghum, millet and other low-yielding crops. In 2009, corn acreage accounted for 28.6 % of the agricultural acreage in the country and 42.2 % of the Northeastern region's total crop planting area; production was 46.933 million metric tons in that year, this accounted for 28.6 % of

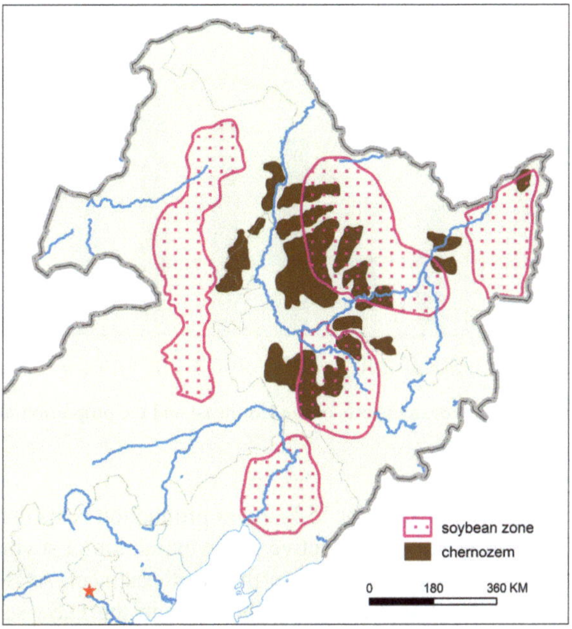

Fig. 10.3 Soybean production & blackland soils of the North-East Plain

the country's total production and 56 % of the region's total grain output. (National Bureau of Statistics of China 2010a, b).

The Northeast has a long history of soybean cultivation. Soybeans are an integral part of Chinese culture and diet with soy sauce and soybean curd finding their way into many typical Chinese foods. The Northeast has intensively cultivated soybean acreage (see Fig.10.3), and soybeans are also an ideal rotational crop. There is great demand in export markets like Japan and in industrial applications such as soy sauce and soybean oil production. Soybean production in the region accounts for about 40 % of the country's total. Harbin, Liaoning and Changchun city are known as the "three soybean warehouses" of China. Soybean commodity rates (the proportion of production in an area going to markets as opposed to being used on the farm itself) are generally 75–80 %, the export rate can be up to 50 % or more. With such high production and a high proportion going into export markets, the high quality soybeans from this region are renowned internationally. The total soybean exports of the Northeast accounts for about 37 % of the country's total (Fig. 10.4). This graph shows growing soybean exports through 2004 and then a leveling off of exports as China's internal demand has grown. It also shows that other areas of China are starting to account for more soybean exports, but that the Northeast is still important. Soybeans from the Northeast are exported to Korea, Japan, other Asian nations and countries of the European Union. (Liao 2007) (National Bureau of Statistics of China 2010a).

Comparing soybean production in Northeastern China with that in the U.S., the U.S. has advantages in soybean production scale (extent of farms), has higher labor

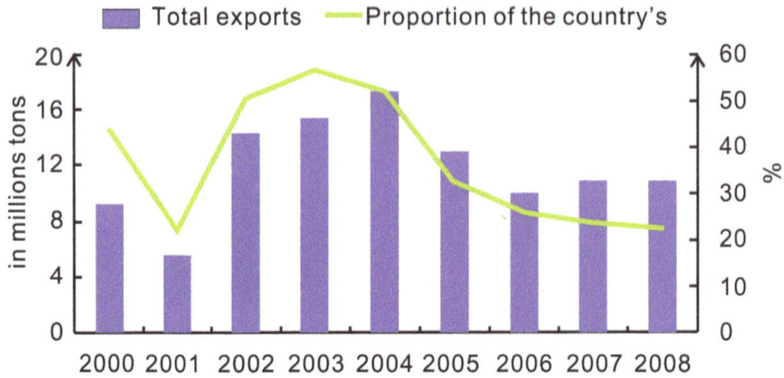

Fig. 10.4 Soybean exports of the Northeast and the proportion of the country's total

productivity, and about 15 % lower production costs than in Heilongjiang Province. The Northeast is not as active in the extraction of soybean oil for purposes of commodity markets. However, in the Northeast such as Heilongjiang, the soy protein content of the soybeans is higher than in the U.S., so soy protein extraction and processing is a very promising field. Industrialization of soybean derived products is a huge global market with great profit potential.

Special Agriculture and Animal Husbandry Products Northeast China is the main wheat producing region in China, accounting for about 50 % of the country's total spring wheat planting acreage and nearly 40 % of the total wheat output. The Sanjiang Plain, the Song-Nen Plain and the Heihe area are the most concentrated wheat growing areas. The Northeast has also developed some economic crops that are specially adapted to local conditions, such as hemp, flax, sugar beets, and sunflowers, which also occupy an important position in the total production picture of the country.

Heilongjiang Province and the western part of Inner Mongolia is one of the top ten pastoral (grazing) regions of China. It contains vast areas of grassland. The region also produces large quantities of other agricultural products such as bran, soybeans, etc. All these factors are conducive to the development of animal husbandry here. This northern portion of this area also has advantages in raising dairy cows, producing milk and milk products (Fig. 10.5).

10.2.2 The North China Plain

10.2.2.1 Geographical Environment and Agricultural Development

The North China Plain is located between 32 and 42°North latitudes. As an alluvial plain of the Yellow River, Huaihe River and Haihe River, it is also known as Huang-

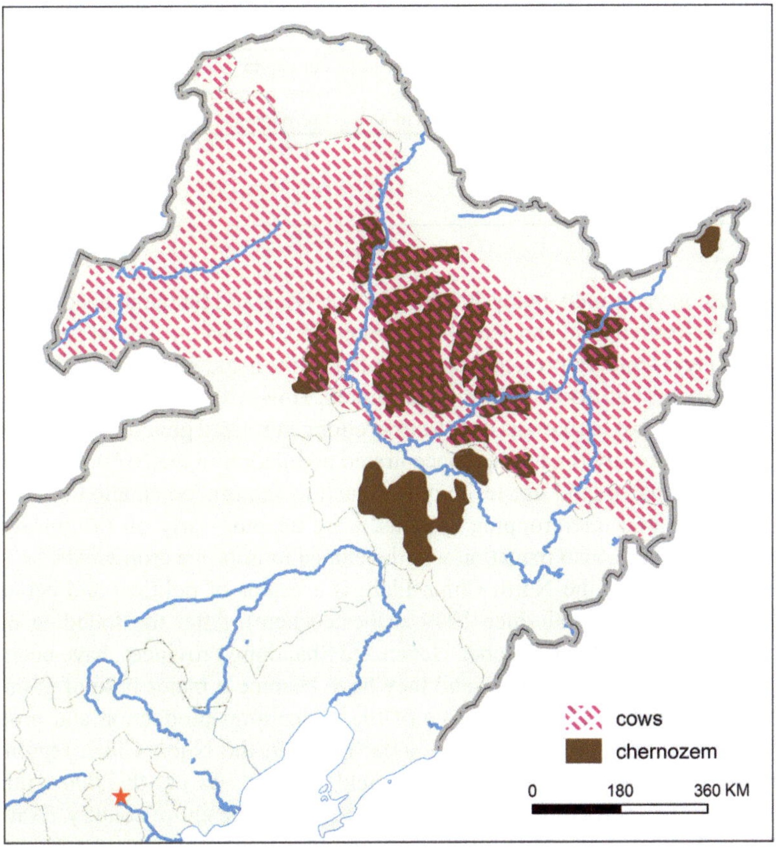

Fig. 10.5 The cattle production area of the Northeast China Plain

Huai-Hai Plain, including the Haihe Plain, Yellow River Flood Plain, Huaibei Plain, etc. It is China's second largest plain with an area of about 310,000 km^2. The region crosses the boundaries of Hebei, Shandong, Henan, Northern Jiangsu and Anhui provinces, and includes Beijing, Tianjin and major cities. The North China Plain is arguably the most ancient agricultural area in the history of China, which has a history of organized agriculture dating back over 3,000 years. China's three "legendary" dynasties the—Xia, Shang and Western Zhou Dynasties, all were founded in different sections of the Yellow River Plain. The North China Plain became the economic and cultural center of ancient China before any other region. This was largely due to the more humid and warmer climate, the friable easily tilled loess (aeolian or wind deposited) soil and good water permeability all of which made for easier cultivation of the alluvial plains. This was coupled with relatively simple dry land farming techniques including the early adoption of simple iron and copper hand tools such as the hoe and spade which predated the use of plows and animal powered implements. (Cheng 1984; Zhou 1993). In the process of development, the North

Table 10.1 Conditions of water resources and its utilization in the North China Plain (National Bureau of Statistics of China 2010)

Region	Total amount of water resources (100 million m^3)	Per capita water resources (m^3/person)	Per hectare water resources (m^3/hectare)
North China Plain	1,990.8	425.3	6,046.3
Total national resources	27,434.3	2,071.1	22,496.9
Proportion of the total (%)	7	21	27

China Plain gave farmers many advantages, low lying and flat land, deep friable alluvial soils, warm humid and semi-humid monsoonal climate—a good combination of sunlight and temperatures, and concentrated warmth and rainfall. The population carrying capacity of dryland farming is limited. However, some historical reasons such as the threat of barbarians and the desire for centralized government control resulted in high-density towns and concentrated population in the North China Plain. Therefore, intensive land use techniques—such as the implementation of crop rotation systems and intercropping systems, were adopted early on to improve the output per unit area, also irrigation was developed to improve crop yields.

In modern times, the North China Plain is a center of political and economic activities (China's capital since 1949 is located here). After the founding of the People's Republic of China, Hebei, Henan and Shandong Provinces, have been designated "agricultural provinces" and they have become a major base of grain and cotton production in China. The role of the agricultural production and processing industry is very important in the area. In 2009, the North China region had 32,925,700 ha under cultivation, and the yield of grain was 191,965,000 t; this accounted for 27 % and 36 % of the respective totals for the entire country. (National Bureau of Statistics of China 2010). The region is centrally located with respect to transportation infrastructure with railways and highways that all strengthen the North China Plain's position in the distribution and processing of agricultural products. The grain wholesale market in the Zhengzhou area of Henan Province, which is the transport hub of the region, is the "leading grain market" of China. It is also the first market approved by the State Central Food Market, the first trading market to introduce market-based mechanisms of access to both present day sales and the use of commodity futures options. Compared with other parts of the country, the balance of land and water resources is problematic in the North China Plain. There is more land and less water available than in most other agricultural regions. The Huang-Huai-Hai Plain accounts for one quarter of the national arable land and is home to one third of national population, but has access to only 7 % of the national total water resources. The total amount of surface water and groundwater are about 2,000 billion m^3. Water resources per unit of cultivated land and per capita are less than 1/3 and 1/4 of the national average respectively (Table 10.1). Coupled with the uneven spatial and temporal distribution of water resources, a serious shortage of water resources exists in the North China Plain. The region shows the effects of global climate change and warming and has experienced a drying trend during the

Table 10.2 Contrasting farmland investment and infrastructure of three agricultural regions (National Bureau of Statistics of China 2010)

Region	Cultivated land (1,000 ha)	Number of units of agricultural machinery per 1,000 ha	Consumption of chemical fertilizer per hectare (kilograms/hectare)	Irrigated area (1000 ha)	Irrigated area as the proportion of the total area (%)
Northeast Plain	21,450	45	236	6,600	31
North China Plain	32,926	31	642	22,347	68
The Middle-Lower Yangtze Plain	14,776	15	781	8,560	58
Country as a whole	121,716	29	444	59,261	49

past 50 years. Water availability and quality has become a major agricultural production constraint, particularly serious during the spring planting season (National Bureau of Statistics of China 2010).

Therefore, subjected to drought and salinization (the development of saline-alkali fields), the North China Plain has developed more low-yielding fields, and this has engendered the need to improve irrigation conditions and the reclaim the saline-alkali fields. The active construction of water conservancy facilities such as irrigation canals and reservoirs makes the area standout nationally. This can be seen in (Table 10.2) which shows that about 68 % of available farm land is irrigated in the North China Plain compared with 49 % nationwide. Regional drought-tolerance is very strong. However, the irrigation water comes mainly from groundwater. The development of irrigation based on over-draft of groundwater aquifers may cause a series of negative eco-environmental impacts, such as a decline in the water table or the peizometric head in confined aquifers, which in turn increases pumping costs and lowers water quality usually due to an increase in salinity as deeper more mineralized water is tapped. Other issues related to groundwater overdraft include the creation of cones of depression, surface subsidence and faulting, seawater intrusion in coastal areas and the loss of flow in springs with both ecological and cultural significance.

10.2.2.2 The Characteristics and Geographic Setting of the North China Plain

In the North China Plain, the main food crops are wheat, corn, and soybeans. The main cash crops are cotton, peanuts, tobacco and sesame. Most areas can meet the temperature and seasonal requirements for a successful harvest three times every 2 years. From north to south, the cropping system changes from harvest once to twice every year. The multiple crop index is high, about 150 % of the national average. (Zhou 2000).

The region that can only sustain harvests three times every 2 years is mainly in the areas with general agricultural conditions and low production levels, such as the Haihe Plain, Shandong and Henan Mountains regions. The crop rotation system is mainly: wheat—cotton—wheat, wheat—soybeans—wheat and wheat—corn—wheat. The region that can sustain harvests two times every year is mainly in areas with better moisture and temperature conditions, such as the Huang-Huai Plain, and the Shandong hills area. The crop rotation system in the cereal crop dominated areas is: wheat—corn, wheat—millet, wheat—soybeans; in cotton dominated areas, it is: cotton—wheat, cotton—potatos; in oilseed dominated areas it is: peanuts—sweet potatoes. In recent years, with the improvement of agricultural production conditions, the double cropping area has expanded each year. The region that can sustain only one harvest per year is mainly in northern and mountain regions with poor agricultural conditions. In these areas the main crops are wheat, cotton, and hemp.

Small Farm Management and Scale Issues in the North China Plain

In the North China Plain and in all other regions of China, a system of household contracts and responsibility for farming was adopted in 1978. Therefore, the main question the development of agricultural resources involves solving is the contradiction between decentralized management at the individual farming family unit level and the benefits that can be derived from adoption of larger scale production methods which require more centralized management.

The North China Plain first achieved large scale production through operations that crossed the region. From the mid-1980s onwards, a group of farm workers that specialized in "plowing, planting, harvesting, and agricultural machinery operation" came into being. Based on the earlier or later maturation of wheat in different regions, these workers moved their operations from south to north. This technique can prolong the season of operations for up to two months or even more. The level of integrated mechanization in the North China Plain exceeds 81 %. The existence of these services combines the benefits of decentralized farm level management and intensive large-scale mechanized production. The wheat reaping technique of using mobile combine harvester teams has transformed thousands of households planting wheat on a small plot into one big farm unit effectively, this makes the scale of agricultural operations and intensivity of wheat production meet or exceed the level of some developed countries, such as France, Germany and the U.K.

In addition to the agricultural product processing enterprises, intermediary companies, and professional associations all play an important role by guiding the farmers in planting activities according to market demand (orders), selection of certain optimal varieties, use of modern technologies and methods and adoption of uniform and appropriate standards. These intermediaries buy the farmers' grain by list (or orders). All these behaviors turn the scattered small-scale production of individual farming families and small cooperatives into a pattern of "small-scale, large-scale

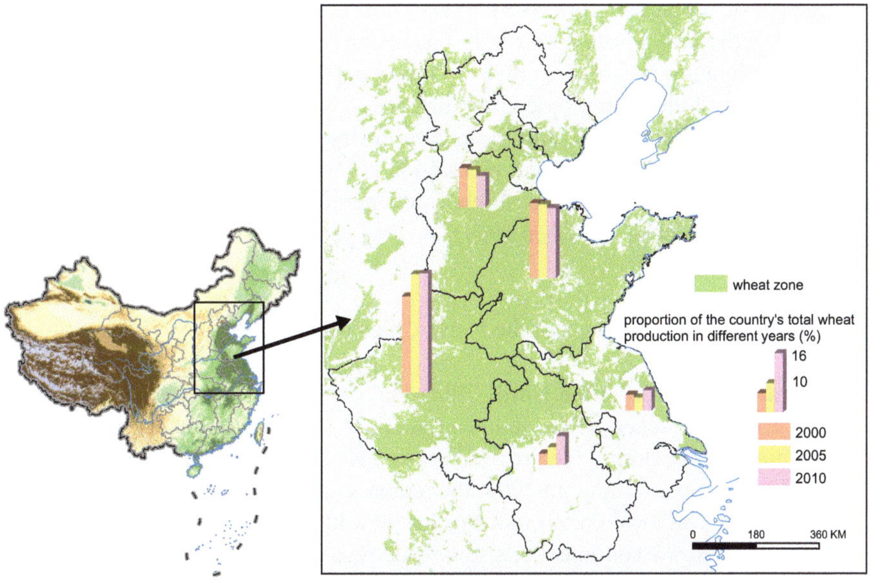

Fig. 10.6 The wheat production areas of North China plain and the proportion by province of China's total wheat production. (National Bureau of Statistics of China 2010)

groups" successfully. The regional government promotes the construction of agricultural industrialization in the form of "enterprise + base + farmers" and "association + base + farmers". In addition to the scattered planting technique, there are various other systems such as some large-scale operations, specialized animal breeding operations, concentrated animal feeding operations, etc. Modernization and intensification are the main development trends in the region.

The Center of Winter Wheat Production

The North China Plain is the most suitable region for winter wheat production in China. Winter wheat is a variety that can overwinter safely under snow in this region and can turn green early in the rapidly warming spring. Moreover, winter wheat is planted in late autumn in fallow fields, which decreases the competition of land and labor with other crops, and also increases the multiple cropping index, and improves the land use rate at the same time. Beside these advantages, the planting time avoids the early autumn floods, which has a positive effect in the stable production of food by avoiding soil erosion due to tillage during wet conditions when rills are likely to form and top soil erodes (Zhou 2000). Therefore, the North China Plain is China's main wheat producing region (Fig. 10.6). With the expansion of the area of irrigated land, soybean, sorghum, millet and other drought-resistant crops have been substituted by wheat gradually. With the promotion and application of advanced cultivation methods, fertilizers and seeds, especially the rotation of wheat and cotton or wheat and corn, the harvest times of wheat have changed from the situation in the past where it was only once per year or three times every 2 years to the situation now where two crops are raised per year or five crops are produced every 3 years. With the multiple cropping index increasing, grain and cotton both get harvested in this region, and the region has become an important production base of grain. The region's wheat production accounts for more than 70 % of the total output of wheat for China as a whole and per unit yield is also higher than the national average (Zhao and Chen 1999).

Special Agricultural Products of the North China Plain

The primary cash crop in the North China Plain is cotton. The North China Plain is ideal for cotton since it has abundant sunlight, adequate moisture and a good temperature regime during the growing season. Hence it is the most important cotton producing area in China. In 2009, the region's production exceeded 40 % of the Nation's total. The local cotton industry has developed rapidly and accounts for a major portion of raw cotton exports to outside regions for use by the textile industry which is in turn the World's largest. Shandong, Henan and Hebei provinces are China's major cotton producing provinces. In addition, peanuts, and sesame seed are the main oil crops. As an important producing region of peanuts and sesame, the North China Plain accounts for about 40 % of the country's planting area. The main peanut producing regions are Shandong, Shanxi and Henan provinces, and Shandong Province ranks first in peanut production. Sesame production is concentrated in the eastern region of Henan.

The Bohai Sea vegetable farm belt of the North China Plain is one of the four major vegetable production areas in China. The vegetables that are raised in green houses here include melons, beans, celery and leaf vegetables. Shandong Province of the North China Plain is China's main vegetable production province. Large scale, large volume and high technology are its main features. The total vegetable output and its scale in Shandong Province rank the area first among the provinces for 11 years successively. Famous vegetable distribution centers include the cities of Shouguang, Changshan and Jinxiang. The Shouguang area of Shandong Province holds the International Vegetable Sci-tech Fair annually to showcase its use of technology, such as hydroponics. Shouguang city occupies an important position in the national vegetable production. In 2009, the growing area of vegetables in Shandong Province exceeded 1.73 million ha, accounting for 10 % of the total vegetable planting area. Annual vegetable production reached 89.372 million t in 2009, accounting for about 14.4 % of annual vegetables output of the whole country. (National Bureau of Statistics of China 2010a, b). In order to achieve optimal planned production of crops during the years from 2008 to 2015, the North China Plain has been designated as the National Superior Production Region for beef and sheep. The area can produce forage crops and hay of over 3,860 t per year, providing a strong resource base for the development of livestock feeding. Such obvious advantages, as the existence of convenient transportation facilities, and proximity to the Beijing-Tianjin-Hebei Megaregion, the "Yangtze River Delta" and the "Bohai Sea" economic circle, have contributed to build a strong marketing foundation and a close connection between production and marketing in this region.

10.2.3 The Middle-Lower Yangtze Plain

10.2.3.1 Geographic Setting and Agricultural Development

The Middle-Lower Yangtze Plain, located between 28 to 33°North latitudes, mainly consists of the middle and lower parts of the Yangtze River flood plain. The Middle Yangtze Plain includes the Jianghan Plain, the Dongting Lake Plain and the Poyang Lake Plain. The Jianghan Plain is an alluvial plain of the Yangtze and Han Rivers, the Dongting Lake Plain is an alluvial plain of the Yangtze River and the Dongting Lake basin; the Poyang Lake Plain is an alluvial plain of the Yangtze River and Poyang Lake basin. The Lower Yangtze Plain includes the coastal plain of the Yangtze River, the Chaohu Plain and the Yangtze River Delta. Compared with the Northeast Plain and the North China Plain, the Middle-Lower Yangtze Plain is fragmented. The provincial boundaries that cross the Middle-Lower Yangtze Plain include those of Hubei, Hunan, Jiangxi, Anhui, Jiangsu, Zhejiang, Shanghai and small parts of other provinces. (Zhou 2000).

The Middle-Lower Yangtze Plain has a long history of human occupation and agriculture spanning about 2000 years. During the late Western Han Dynasty, a large number of people migrated to the south from the Central Plains of China, and promoted the development of the area. The area is humid with a subtropical monsoon dominated climate: hot in the summer and warm in the winter, with abundant rainfall. Sunlight, temperature and moisture conditions are good; rain and warmth come at the same season. All the conditions for agricultural productivity are ideal for multiple cropping and also good for the formation of abundant surface runoff, the development of rich soils and the growth of thermophilic crops. As population pressures have increased, the region has adopted intensive precision farming as its main farming method. Due to numerous and well-designed water conservancy facilities such as irrigation canals and reservoirs built during its long history of agricultural development, the region has becomes the "main paddy field" of China. Here, a wide distribution of rich soils and other agriculturally desirable physical conditions have produced the most concentrated area of paddy soil in China. During long-term flooded conditions, except for an extremely thin oxidized layer at the surface, the whole non-plowed soil profile below the paddy soil is in a reduced condition. High organic matter content will accumulate in such conditions, the soil has excellent water retention and cation exchange capacity characteristics, hence added nutrients are well retained and the solid "plow pan" below the ploughed layer becomes an important basis for superior grain yield. Abundant precipitation and low-lying terrain make for dense river networks and numerous lakes in this area. It is also an important place for freshwater fish farming and fishing. The "five great

Fig. 10.7 Landscape of Yangtze River Plain. Intersecting rivers, widespread lakes and well-developed food production makes it known as the "land of plenty". Inhabitants and agri-cultural production is distributed along the rivers and lakes. An oblique aerial-view of the south bank of the Yangtze River in Jiangxi Province

lakes" inter-connect with the Yangtze River, and also provide a wealth of irrigation water and hydro-electric energy which ensure the stable development of agriculture and supports towns that are an integral part of the base. The region benefited from the opening of the Beijing-Hangzhou Grand Canal, which began to be built in the Spring and Autumn period about 700 BCE and was continued during the Sui, Song and Tang dynasties, and basically completed 1,800 years later in the Yuan dynasty. The well-developed water transport facilities (for barge and sea-going ship traffic), play a very important role in regional trade, exports and food transport. The proximity to excellent transport facilities makes the transportation of agricultural products much easier and less costly in this region than in many other areas of China. (Sun 2003; Tang 1998).

Therefore, once it was initiated, agriculture developed rapidly and it built a solid economic base in the region, this in turn contributed to the development of urbanization and industrialization. So the region has occupied an important position as a national economic center for a long time. Therefore the level of agricultural intensification and comprehensive development of the Yangtze River is the highest in China. The region's rural economy is more highly developed than other rural areas in China. A solid industrial base also provides support for manufacture of fertilizer, pesticides, and to the technology and machinery sectors. Dense urban population and intensive agricultural areas also provide a broad-based market. However, during the past 3 decades, with integrated economic development, especially the growing level of urbanization, prices for potential industrial land have risen dramatically. Farmers have chosen to sell their land and it has been converted to industrial and residential uses. Thus the acreage of agricultural land in the Middle and Lower Yangtze Plain has seen a sharp decrease. The impacts of these changes have caused some of the counties in the region to change from net grain exporters to grain importers. The conflict between use of land for construction of industrial plants and housing and its use for farming has limited the expansion of agriculture in the area.

The agricultural status of the area has declined; however the agriculture of this region still plays an important role on a nationwide basis. (Xu 2002).

The main problems the area is facing are the prevention and control of flooding and drought. In early summer, cold air coming from north can still invade the area and meet the warm air of the south. The month-long period of "plum rains" occur where these air masses meet. This is an important climatic feature of the region. In the summer, typhoons also bring heavy rains. "Plum" rains and typhoons both are an important supplement to the summer rainfall in this area. However, their large variability from year to year may lead to droughts and floods. Areas covered and affected by flood disasters in the past make up about 30 % of the country. Therefore, the Yangtze River region needs to focus resources on improving the ecological environment, managing rivers and lakes, strengthening flood control and water conservancy facilities.

10.2.3.2 The Characteristics and Location of the Region

Temperature conditions in the Yangtze River Plain make it possible to achieve two harvests per year in the north and three harvests per year in the south. The southern area is suitable for the growth of a variety of subtropical fruits such as plums, lychees, citrus and navel oranges. The cropping index of the region is very high, it is over 230 % in Tai Lake Plain for example.

The Grain Base with the Highest Comprehensive Level

More developed agriculture, higher production levels, and high crop yield per unit area all indicate a high degree of agricultural intensity in the Yangtze River Plain region. As China's most important production area for rice, rapeseed, silk and fish, it is known as the "land of plenty". It is also an important base for the production of commodity grain, cotton and oil (derived from seeds). Rice for human consumption is the main agricultural product and the region also produces other economic crops. The area has outstanding natural conditions for agricultural production in both the northern and southern parts. The cultivation of cash crops has a comprehensive regional character. The region is also China's main producing area for rapeseed and sesame seed, much of which is used for oil production. Most cash crops and food crops are grown in rotation, such as a cotton-wheat rotation, or rice-rapeseed rotation. (Liu 1996).

The Center of Rice Production

The main type of cultivated land in this area are paddy fields used to grow rice, dry lands are also used to grow rice and paddy fields that are converted to and from dry lands depending on the season or other factors are used for rice production as well. The main food crops in the region are rice and wheat, the proportion of rice is larger than wheat. Therefore, rice production plays the leading role in the food production

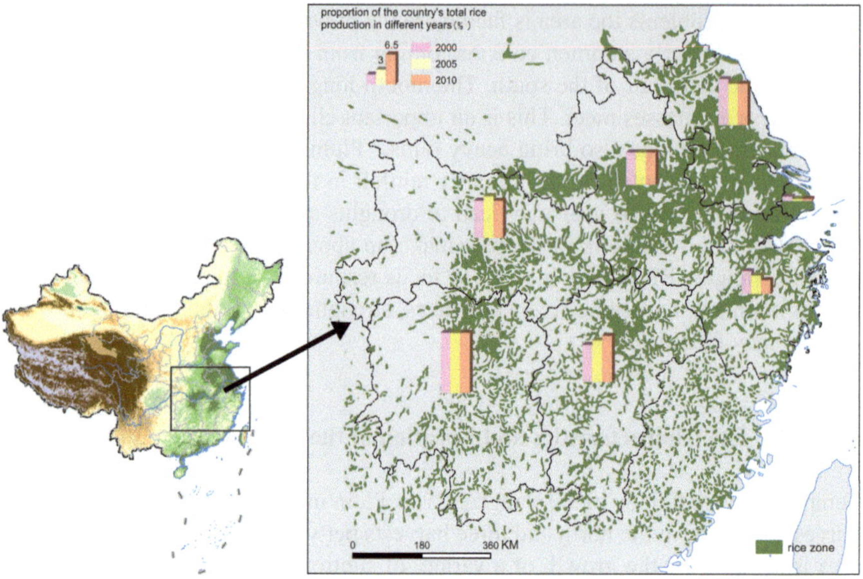

Fig. 10.8 The rice distribution and production in The Middle-Lower Yangtze Plain and its relationship to China's total rice production. (National Bureau of Statistics of China 2010)

of the Yangtze River basin. The cropping system in the region is mainly based on double cropping of rice, the area devoted to double cropping of rice accounts for about 70 % or more of the land in the region that grows rice. The total rice planting area of the region accounts for 58 % or more of the region's total sown area.

Rice production in the region makes up approximately 42 % of the country's total rice production. It is the largest rice producing area and commodity grain production base in China (Fig. 10.8). Hunan, Hubei, Jiangxi and Anhui are all important provinces that contribute to grain price stability in China. The Tai Lake Plain, Jianghan Plain and Dongting Lake Plain, are all areas focused on producing rice. So, they are called the "warehouse of rice". The Yangtze River Delta, with the Tai Lake Plain and the Lixia River plain at the core, is also a famous high yield region for rice production.

Other Crops: Cotton, Silk and Tea

In 2009, the cotton planting area and total production of the region account for 26.5 and 20 % of the country's respectively, occupying an important position in China's overall cotton production. China in turn is the number 1 cotton producer on the planet in most years. Hubei and Jiangsu are important cotton-producing provinces, respectively, ranking No. 5, and 7, and are China's highest-yield cotton growing regions. As China's largest hemp producing region, the area's hemp output accounts for nearly 1/3 of the total for the whole country. The main fiber species grown are hemp, ramie, jute and kenaf. The district's rapeseed and also peanut acreage is very large. It is therefore an important oil production base in China. The northern portion of the Yangtze River Plain is where sesame is more commonly grown. This area accounts for about half of the country's total output.

In addition, the Taihu Lake Basin is one of the three famous silkworm rearing areas, the cocoon production accounts for 56 % of the country's total and China is the world's largest producer of raw silk as well as silk-based textiles. Zhejiang province has throughout history been the main producing area of mulberry bushes used in raising silk worms. The silk industry in the area is long famous and it is known as the "house of silk". Silk has become an important export, with China accounting for 80 % of the world silk trade. The U.S. does not produce silk, and other producers like Japan and Thailand do not have as much land that is suitable for growing mulberry bushes, therefore starting in the last century, the region has become the most important raw silk export as well as silk textile manufacturing center in the world. (National Bureau of Statistics of China 2010b).

This region is also China's most important tea producer. It has large areas of tea plantations. Both the high yield and the extensive area of cultivation help account for half of the county's production. Zhejiang is the most important tea producing province (Hunan is second). In these provinces are the climatic conditions and hill slopes that produce micro-climates and soil conditions that yield many famous varieties of tea, such as Fujian's Yancha tea, Jiangxi's Lushan Mist tea, Zhejiang's Longjing Tea and Jiangxi's Biluochun Tea. China is one of the world's largest tea producers. In 2008, the planting area of tea was 1.258 million ha, accounting for about 43 % of the world total; the tea yield reached 1.275 million t, accounting for about 32.75 % of world tea production. In China, there are about 20 provinces, cities and autonomous regions that produce tea; about 80 million workers are employed in the industry. The tea industry is a major pillar of the rural economy and tea is a symbol of Chinese culture and history. (National Bureau of Statistics of China 2010b).

10.3 Two Major Agricultural Regions of America: The High Plains and The Corn Belt

10.3.1 Introduction

The Corn Belt and High Plains agricultural regions are two of the most important areas in America (the United States of America or U.S.) devoted primarily to agricultural production. Both these huge areas are extensive multi-state regions with rich soils that produce vast surpluses of cereal crops that in turn support sophisticated systems of animal husbandry in modern concentrated animal feeding operations (CAFO's). The main difference between the two regions is a climatic one. The Corn Belt receives considerably more rainfall which in turn makes corn production more feasible on a rain-fed basis. In contrast, the High Plains are considerably

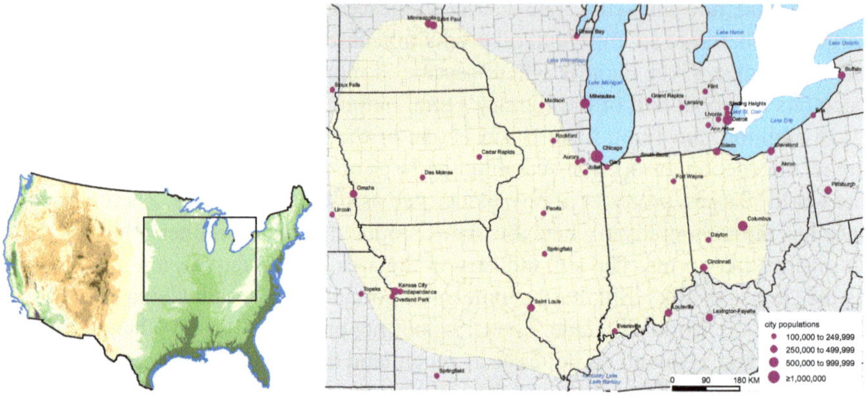

Fig. 10.9 The Corn Belt region with State and county boundaries and major cities

dryer and without irrigation, corn is usually not viable and wheat production is dominant in this region, particularly in those areas without irrigated farmland. With irrigation, other crops including corn but also including cotton are important in the High Plains. While there are many important population centers in the Corn Belt as shown in Fig. 10.9, the density of cities decreases to the west. Chicago, Minneapolis and Cincinnati are all major cities on the edge of the Corn Belt region and serve to move agricultural commodities from the hinterland of the region to global markets. Note that the Corn Belt region has a generally far higher density of farms than surrounding areas (Fig. 10.11). These densities were estimated using a per square mile basis for each county by the U.S. Department of Agriculture Census of Agriculture in 2007. In general, areas with rich soils can support more farms and smaller farms can survive financially. Some mountainous counties have many small farms, so some areas with poor farming do have a high farm density by this measure. For example, in 2009 the average farm size in the State of Iowa (the core of the Corn Belt region) was 331 acres (or about 150 ha). The total number of farms in Iowa was 92,600, these farms produced about $ 24 billion in agricultural products and about 60 % of farms were single family farms with farmers living on them. About 12 % of Iowa's population was engaged in agriculture in some way, but only about 7 % were full time farmers. In the U.S. as a whole, about 2 % of the population lives on farms and 1.9 % of the population was primarily engaged in agriculture in 2007. In contrast in Oklahoma, a State primarily in the High Plains region, the average farm size was 405 acres and there were a correspondingly lower number of farms, and the value of farm production was slightly less than $ 6 billion. (Economic Research Service USDA 2007). The High Plains region, while larger in size than the Corn Belt, has a much smaller population and most of its major cities are at the edge of the region as shown in Fig. 10.10. There is a scarcity of large cities in the High Plains region with the exception of Denver, Colorado which is on the edge of the region adjacent of the Rocky Mountains that define the western edge of the region.

The average size of farms in the Corn Belt region is smaller than the more extensive nature of farms (and ranches) that are prevalent in the High Plains region. This

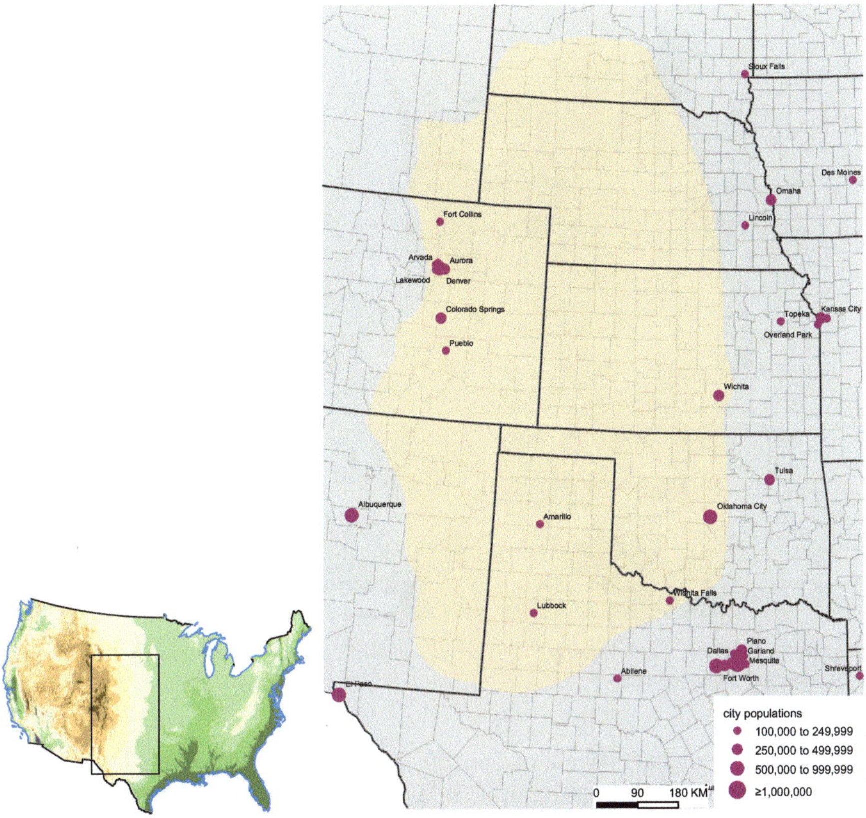

Fig. 10.10 The High Plains region, with State and county boundaries and major cities

difference in the density of farms in the two regions is illustrated in Figs. 10.11 and 10.12. The farm density in the western U.S. is far lower than in the eastern portion of the country, indicating most areas of the High Plains have much larger farms. The drier conditions there require more extensive farming practices, particularly in areas with center pivot irrigation. The exceptions are areas around Lubbock, Texas, and Scotts Bluff, Nebraska where agriculture is more specialized, as well in the Boulder, Colorado area where organic farming is a factor (the largest herbal tea company in the U.S. is headquartered in this county, which has many small herb farms).

10.3.2 The Corn Belt Region

10.3.2.1 Geographical Environment and Agricultural Development

Corn is the most valuable agricultural commodity produced in the U.S. with an estimated value in 2010 of $ 66 billion. In the U.S., corn was raised on 79 million acres

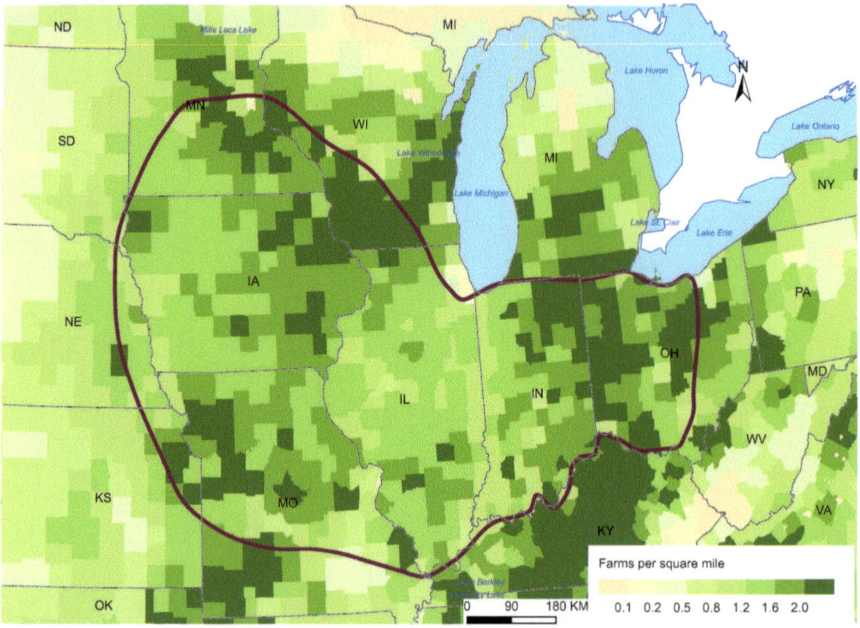

Fig. 10.11 Corn Belt Farm Density. A density of two farms per square mile implies that a farm is 320 acres or 129 ha in extent

(32 million ha) in 2009 (World of Corn 2010). The U.S. is the world's leading corn producer and exporter. In 2010, of 92 million metric t of worldwide corn exports, the U.S. accounted for 50 million metric t. The next largest exporter, Argentina, supplied 15 million metric t. China imported a net 1 million metric t of corn in 2010. Japan imported 16 million metric t and was the largest net importer. (USDA 2010). Corn is used in many ways. These include corn production for human consumption, corn for animal feed (one of the largest uses), and corn for the production of high fructose corn syrup, which is mostly used as a sweetener particularly in carbonated soft drinks like Coca Cola. Corn is also used in the manufacture of ethanol and for many other industrial purposes such as adhesives and plastics. Ethanol from corn includes both the traditional corn whiskey (bourbon) that for hundreds of years was an important way to conserve and trade corn. More recently corn is used for the production of ethanol for motor fuel added to gasoline in mixtures as high as 15 % ethanol. The percentage of corn going into ethanol production has been increasing, in 2009 this use accounted for 32 % of corn output. However, the price of oil and subsidies for production of fuel grade ethanol has an impact on production levels. While corn is produced in many parts of the United States, including the southeastern coastal plains, the valleys of Appalachia, the High Plains and in the west coast, the region of most concentrated corn production is an area generally referred to as the Corn Belt. This region is centered on the state of Iowa but also includes most of Illinois, and Indiana, the western two thirds of Ohio, southern Minnesota,

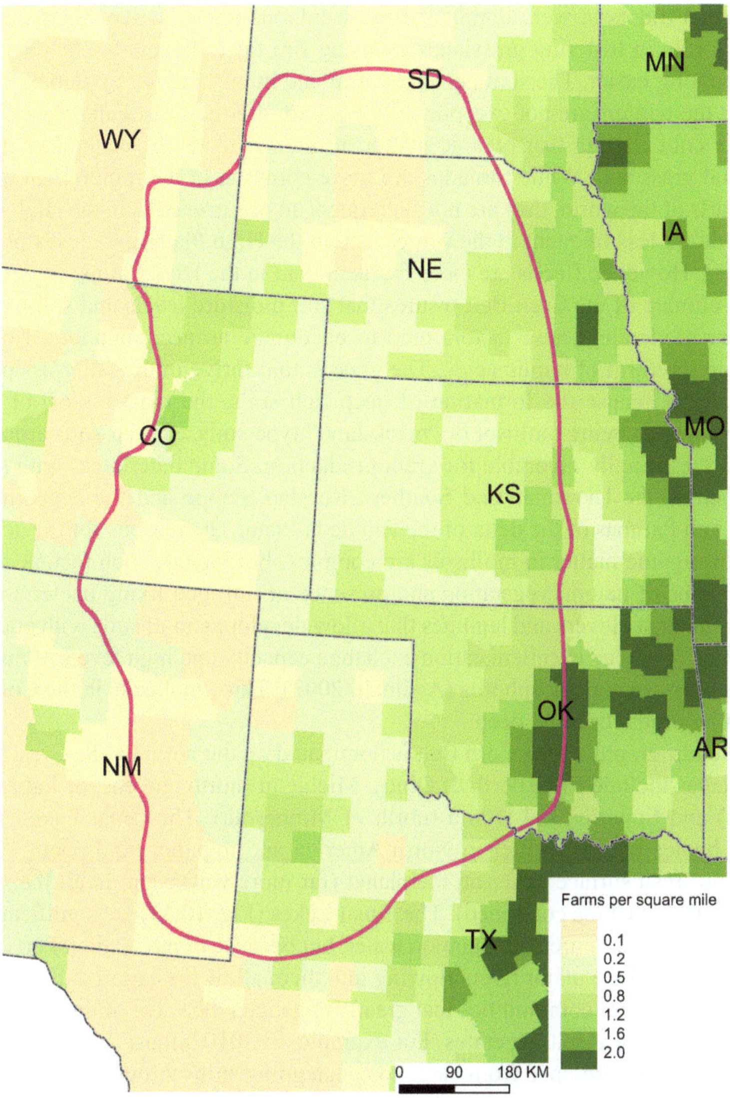

Fig. 10.12 High Plains Farm Density per county. Density is much lower in the High Plains. A Density of 0.1 Farms per square mile means a farm is 6,400 acres (2,591 ha)

most of Missouri and parts of eastern Kansas and Nebraska. This region is defined by climatic factors and geomorphology that supported mostly grassland vegetation and small patches of forest that in turn contributed to the formation of soils that are ideal for corn production.

The Corn Belt has frequent but generally gentle spring, summer and fall rains and winter snows which once supported lush grasslands. But precipitation was generally insufficient to support heavy forests, except along rivers and streams. The forests that did exist were dominated by deciduous trees. Native Americans also played a role in fostering grasslands by using fire for millennia to clear forests to make hunting easier. The summers are warm and humid but not extremely hot nor dry and the winters are not exceptionally cold nor are they particularly windy. But they are cold enough that organic matter in the soil tends to accumulate under a perennial grass sod. While tornados are more common in the region than in most other parts of the world, they are not as frequent an occurrence as in the High Plains. Blizzards are less frequent in the winter than in the High Plains and dust storms are unknown. However, floods are more frequent than in the High Plains.

The climate of the Corn Belt insures that soil moisture levels and soil temperatures are moderate. These factors tend to encourage an accumulation of organic matter in the form of humic acids. The grasses that thrive in the tall grass prairies of this region cause the formation of deep rich soils, the most typical of which are classified as being mollisol or "black-land" type soils. Similar soils around the world are especially favorable for grain production. Some other areas with similar soils include the Ukrainian and Southern Russian Steppe and the Argentine and Uruguayan Pampas of the delta of the Rio de la Plata. The reasons for the accumulation of organic matter in mollisols are complex, but include such factors as deep root systems of natural vegetation and presence of nitrogen fixing bacteria in root nodules in most clovers and legumes that allow development of soils with good tilth and structure and an excellent cation exchange capacity and high levels of nutrients such as nitrogen and phosphorus (Aydinalp 2003). Thus, mollisols in the Corn Belt are fertile, productive and deep.

The hydrography of the Corn Belt is dominated on the north by the Great Lakes. These lakes include Erie (north of Ohio), Michigan (north and east of Indiana and Illinois) and Superior and Huron (north of Minnesota). The Great Lakes are the largest bodies of fresh water in North America and together hold about 17% of the liquid fresh surface water on the planet (far more water than in all the surface water bodies in China combined). The Great Lakes (Fig. 10.13) are significant economically not as a source of irrigation water, but because of navigation. Ports on the Great Lakes and the major rivers flowing into them allow the export and movement of bulk agricultural commodities that greatly enhanced the value of the surpluses of grain that the Corn Belt generates. For example, in 2010 almost 2 million metric t of grain was trans-shipped from barge to ocean going ships through the ports of the Great Lakes Seaway. Shipment by barges and ocean going freighter is far more economical than any alternative means of transport. This is not to say that rivers are not important also, the Mississippi River is a tremendous highway for barge transport. An extensive series of locks as well as some canals allows grain from the Corn Belt to easily reach the mouth of the river south of New Orleans. Also, canals interconnect the Great lakes and Mississippi River systems and these interconnections are in the heart of the Corn Belt, mostly in Illinois, and much grain transits through the greatest metropolis in the Corn Belt; Chicago. Historically, Chicago's wealth was based on the trade in grain and livestock. Chicago remains the site of the commodity exchanges that set prices on a global basis (Great Lakes Seaway 2010).

Fig. 10.13 Major hydrographic features of Central North America

Most of the water used in farming in the Corn Belt falls as rain or snow, but ground water is abundant and economically important, particularly that water which is extracted from buried valley aquifers. These are alluvial deposits in valleys carved out by glaciers and filled with coarse grained glacial outwash deposits that hold large quantities of high quality water that are essential for large scale agriculture such as swine CAFO's. The buried valley aquifers are numerous in Ohio, Indiana, Iowa and Minnesota and yield purer water than the many rivers that bisect the region flowing towards the Mississippi River or the Great Lakes (Fig. 10.13).

Transportation infrastructure is of greater density in the Corn Belt than the High Plains because population density is far higher. The population center of the United States (lower 48 states) is in the Corn Belt in Missouri, although it has been moving south and west through the Corn Belt for many decades. The State of Ohio has as its motto that it is the "Heart of it all", thus proclaiming the centrality of its location, while Chicago is home of one of the busiest airports in the U.S. and now the headquarters of many airlines and the Boeing Aircraft Company. While the High Plains has been losing population (except in a few energy production-focused areas), the Corn Belt continues to grow in population. The primary transportation infrastructure skirts the Great Lakes passing through hubs like Saint Louis, Chicago, Minneapolis, and west toward Omaha and Dallas.

Land tenure and partitioning patterns are systematic in the Corn Belt due to the prevalence of the Public Land Survey System (PLSS) in the region. As with the High Plains (except for Texas) and other parts of the Western regions of the U.S.,

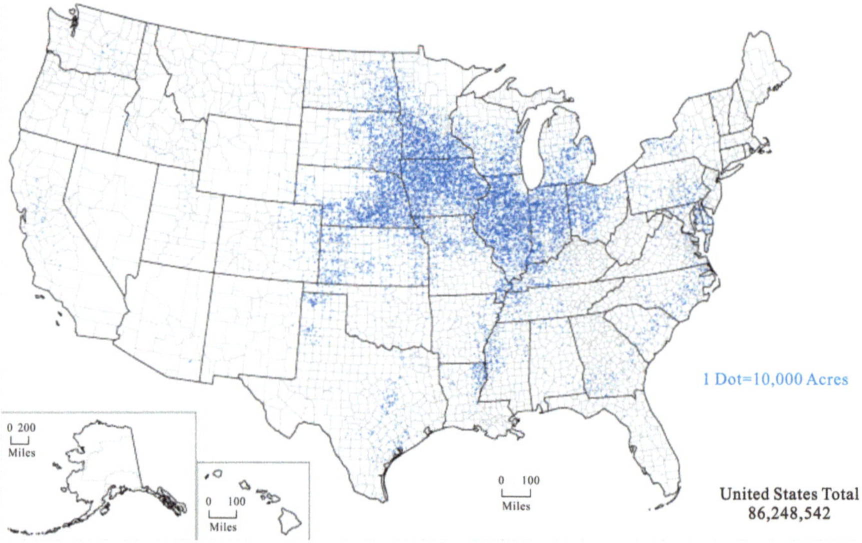

Fig. 10.14 Corn Production for grain based on the 2007 Census of Agriculture, shows the dominance of the "Corn Belt" in production of the most valuable U.S. crop. (Map courtesy of U.S. Department of Agriculture National Agricultural Statistics Service 2007. http://www.agcensus.usda.gov/Publications/2007/Online_Highlights/Fact_Sheets/Geographic/ag_atlas.pdf)

the division of land by the PLSS into a one square mile (640 acre or 214 ha) grid is the dominant land partitioning system present in the region. The great richness of the soil and the reliable rainfall meant that farming was more common than ranching (grazing animals) and that the efficient size of a farm could be 160 acres (53 ha) or less in the Corn Belt. Thus, the numbers of farms of over 1,000 ha are few in the Corn Belt, but farms and ranches of this size or larger are common in the High Plains and many other areas farther west. Also, the towns in the Corn Belt were located not along rail lines, but along rivers, like the cities of Cedar Rapids, Indianapolis and Minneapolis/Saint Paul and where rivers entered the Great Lakes, as in the case of Chicago and Cleveland. In Iowa and surrounding Corn Belt states, small diversified family farms of 160 acres (53 ha) or less became prosperous. The farms in the nineteenth and early twentieth century were diversified. They typically grew corn, had fruit orchards and grew vegetables (mostly for consumption by the farmer or limited barter) and raised cows, swine, chickens, geese and had draft horses. The advent of tractors and farm trucks in the 1920s did not change this pattern much except that draft horses became redundant. Many farmers also added small scale industry such as cheese manufacturing or furniture production, some began to specialize in dairy cattle or swine or just corn. Farm size remained small with adjacent farms being acquired by wealthy farmers, but remaining as family farms. The surplus in wealth generated by moderate sized family-owned farms (See Fig. 10.11) is demonstrated by the fact that Iowa is not only one of the wealthiest U.S. states, but it also has the fairest distribution of income. The Corn Belt region boasts the highest proportion of population with college degrees in the U.S. (Chronicle of Higher Education 2012).

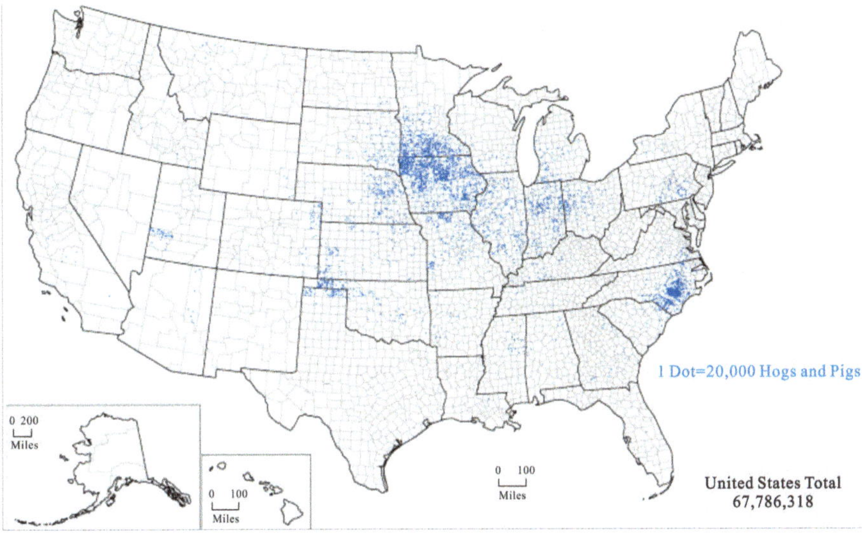

Fig. 10.15 Swine production; each dot represents 20,000 swine. The dominance of the Corn Belt is clear. It shows two other importance areas for swine, one centered in North Carolina and another in the panhandles of Texas and Oklahoma. (Map courtesy of U.S. Department of Agriculture National Agricultural Statistics Service 2007. http://www.agcensus.usda.gov/Publications/2007/Online_Highlights/Fact_Sheets/Geographic/ag_atlas.pdf)

The fact that small family farms can survive and even thrive in a place like Iowa is demonstrated by the success of the Amish. The Amish left Germany in the nineteenth century. They are mostly family farmers and craftsmen making products like furniture. They hold to nineteenth century modes of dress. They also continue to use horses to plow and to pull the wagons and buggies in which they travel. They communally share responsibility for projects like building barns. They are emblematic of the values of hard work and extended family and inter-family cooperation (Hostetler 1993). More recently, large scale production of swine has been attracted to the region by the abundance of corn. Historically, and still today, among the Amish, the farms that raised swine had an enclosed space (sty) where swine could be kept in cold weather and where sows could farrow safely, but most swine spent time outdoors in a pasture. Today most swine are raised in concentrated animal feeding operations (CAFO's).

10.3.2.2 Agricultural Resources of the Corn Belt

Corn

Corn is "King" in the Corn Belt and is primarily raised on family owned farms that, while primarily growing corn, may rotate to other crops as a means of soil conservation and pest control and often have other production of cattle, swine, wheat, or soy beans. In the past, dairy cattle were common in the northern Corn Belt and

Wisconsin cheese is famous. However, since the 1960s, the dairy industry has been shifting away from the northern Corn Belt to less densely populated States of the far west like Oregon, Arizona and Idaho. Thus corn is even more a dominant product in the Corn Belt than in past years. Yields from the best varieties of corn under ideal conditions can exceed 300 bushels per acre (18 metric t per hectare), but in most cases yields of corn are about half that. The average grain yield in Iowa is about 150 bushels of corn per acre (9 metric t per hectare). Note that the bushel is a customary measure of weight of corn in the U.S. and is equivalent to about 40 kg of corn while there are 2.47 acres in a hectare. Iowa is the center of the corn belt (See Fig. 10.11) had over 17 million acres planted in corn in 2010 and the value of the crop was over $ 11 billion in 2010, out of a total value for corn produced in the U.S. of about $ 66 billion nationally in 2010 (ERS-USDA 2010). These production totals for corn far exceed those of any other crop in the U.S., especially in the Corn Belt. Corn is thus by far the most valuable crop and soybeans and wheat have less than half the value of corn production.

Swine

Swine can be viewed as "value added corn". Swine can adjust to being raised in a confined space. This has led to the development over several decades of the Concentrated Animal Feeding Operation (CAFO's). In swine CAFO's, a large barn may contain as many as 2,000 sows. Economies of scale in swine processing (slaughter) mean however that the processing facilities are large centralized facilities and while located in regions containing numerous swine CAFO's they may be more than 100 km distant from the CAFO where the swine was raised. A few firms dominate the swine CAFO business. These include Tyson, Con-Agra, and Seaboard Farms. Usually the swine CAFO is operated on land owned by a farmer and leased from a farmer and the farmer is hired to be manager of the barn. (Pig Process 2009). The system is increasingly export oriented, with large quantities of pork shipped from the U.S. to Japan, for example. The same firms which have developed the swine CAFO industry are building similar farms in many other places, including Mexico, Russia and China.

There are several negative externalities associated with the large scale swine CAFO. One is the potential for water pollution and odor from the liquid waste that swine generate. The waste from swine in the typical CAFO falls through the steel gratings of the floors of each pen into a collection system. Sometimes the waste is stored underground in a concentrated form, more frequently it is piped to an open "lagoon" where it is mixed with more water and rain. This water/animal waste mixture is then typically pumped to center pivot irrigation systems that spray it on crops such as corn that will in turn be feed to the same or other swine. Not all swine in the U.S. reside in the Corn Belt, but in 2009, 29 % were in Iowa, another 11 % were in Minnesota and another 13 % were in Illinois and Indiana together, adding those in other parts of the Corn Belt, the majority of swine in the U.S. were raised in the Corn Belt (Fig. 10.15). Where these animals ended up ultimately was in places often far from the Corn Belt, about 2 million live swine were exported to Mexico in 2008.

10.3.3 *The High Plains Agricultural Region*

10.3.3.1 Geographical Environment and Agricultural Development

The "High Plains" while something of a term of art, generally refers to a region of the Western United States characterized by relatively high altitude and very flat prairies (steppe type) lands. These are located east of the Rocky Mountains but mostly to the west of the center of the United States (which is located in Kansas). The region includes the "panhandle" of Texas, the western half or more of Oklahoma, and western Kansas, also eastern Colorado, and much of Nebraska; the region also includes parts of South Dakota, north-eastern New Mexico and eastern Wyoming. Altitudes in the region range from 500 to 2,500 m above sea level, higher than average for the U.S. and that along with the flat character of the land accounts for the name High Plains.

The region is characterized by a relatively harsh and extreme weather with hot, dry and windy spring and summer weather and temperatures as high as 40 °C with long dry spells and infrequent but intense thunder storms. The winters are cold and windy with occasional blizzards, but generally have cold and clear days and limited snowfall. Temperatures as low as 40° below zero Celsius are seen, particularly in Wyoming and the far western edge of the region which has higher elevations. The outstanding weather phenomenon of the High Plains is the tornado. More tornadoes occur here than anywhere else on the planet. These intense and often destructive wind storms derive from thunderstorm super cells that are created by the convective mixing of warm moist air that comes up from the Gulf of Mexico to the south that then comes in contact with cold dry air masses funneled down from the arctic along the Rocky Mountains. Tornadoes can have winds in excess of 500 km per hour, the highest on the planet. These storms can be part of outbreaks and can cause millions of dollars in damage. Much of the damage is to agricultural crops and infrastructure, although scores of lives are lost in typical years in tornadoes in the High Plains.

Another feature of the high plains that is significant is the propensity for drought. The region can have prolonged droughts, the most infamous being the "Dust Bowl" of the years 1931–1937. This drought was centered on the southern portion of the region and caused millions of refugees to flee the dust storms and crop failures that occurred. Another drought starting in 2010 caused over $ 10 billion in agricultural (and forestry) losses in Texas alone. Since the 1950s groundwater resources, particularly of the Ogallala aquifer (see Fig. 10.16) have been exploited and this has lessened reliance on rain fed (dry land) farming in the region and made farmers more impervious to droughts. The long term sustainability of the overdraft of this and other aquifers in the region is in question and this in turn imperils the long term economic outlook for the region which in general is seeing a decline in population. Recent development of wind energy in the region is an outgrowth of some of the same factors that made the dust storms of the Dust Bowl period so severe (Worster 2004). Dust storms are a feature of spring weather in many of the southern and western parts of the region. This is something that the region has in common with Northern China. Even when there is not a drought, the region is generally dry with

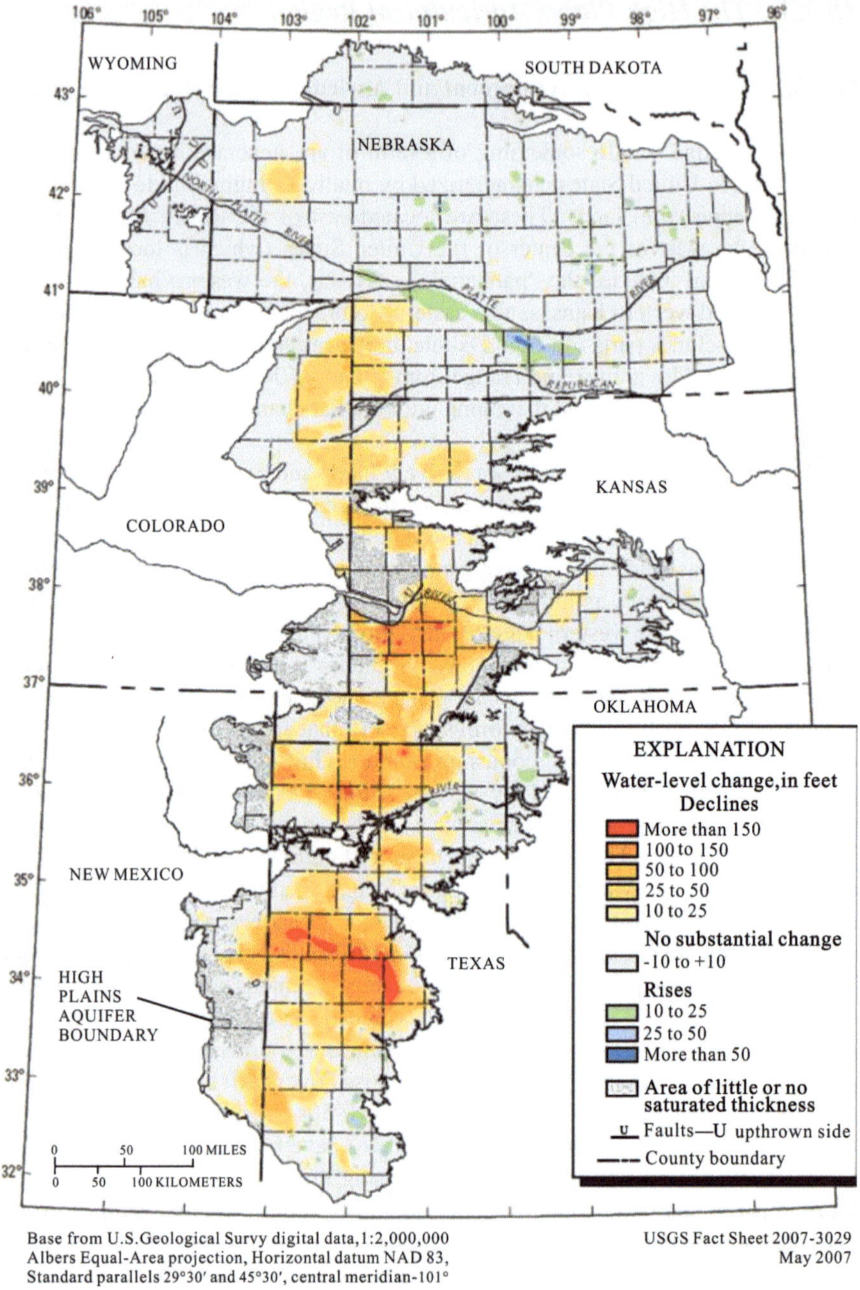

Fig. 10.16 Map of the Ogallala (*High Plains*) Aquifer, the most important regional water source. The aquifer is being over-drafted with draw-downs that exceed 50 m. In Nebraska water availability is large while in the panhandle of Texas, the aquifer is in danger of depletion. (Courtesy: U.S. Geological Survey 2007)

total precipitation ranging from 250 to 500 mm with potential evapotranspiration being significantly higher due to low humidity and frequent wind. Wind erosion control measures such as wind breaks are a feature of the region, although they are less common than in the period immediately after the Dust Bowl. Precipitation lessens from East to West across the High Plains, with snow being more common in the north than south of the region, but low humidity causes a lot of the snow cover to sublimate.

The High Plains are a large area with a diverse range of soils. The most common soils are mollisols developed under the grasslands that dominated the region. These mollisols are however thinner than those common in the Corn Belt. The wind and thinner soils make them prone to erosion and soil conservation is an important issue in the High Plains. Also aeolian soils are present in the region. As with many areas of the North China plain, the High Plains have a considerable region dominated by wind-blown soils. These soils cover many areas of the panhandle of Texas and Eastern New Mexico.

The most important hydrographic feature of the region is invisible except in its effects. It is the Ogallala (and other) aquifers that underlie the region. The water in the aquifer can be considered "fossil" water as it was deposited in deep alluvial sediments fed by ice age glaciers that covered the Rocky Mountains to the West. These glaciers gradually receded over the last 15,000 years and the current recharge of the aquifer from rainfall is negligible. There are however rivers that cross the region. These rivers mostly originate in the Rocky Mountains and the water in them is a valuable commodity and the source of many contentious struggles between various states and private parties. Rivers such as the South Platte, North Platte, Arkansas, Pecos and Canadian cut across portions of the region but do not have adequate flow to satisfy the total demand for water. Since these rivers can be flood prone, flood control reservoirs that also have a water supply function have been constructed on some of the rivers, but the extreme flatness of the high plains means that reservoirs are less than full and have net evaporation, thus they do not conserve water, although they may make water available at times and in places that would not otherwise be the case. (Kromm and White 1992).

Numerous railroads and Interstate highways crisscross the flat High Plains. The High Plains had early access to railroads with the advent of the transcontinental railroad (Union Pacific-Southern Pacific) in the 1860s as well as the Atchison, Topeka and Santa Fe, the Missouri Pacific and other rail lines. The romantic era of the cowboy driving vast herds of cattle across the unfenced High Plains to riotous rail heads like Dodge City, Kansas lived on in memory and movies long after the invention of barbed wire began to close off the open range in the 1870s. The invention of refrigerated rail cars and the growth of a slaughter house industry in cities such as Chicago, Illinois, Omaha, Nebraska and Fort Worth, Texas also were part of a growing and integrated infrastructure. The scientific efficiency (if not the cleanliness) of these early slaughterhouses was famous, they had, it was said, "a use for every part of the pig, but the squeal". This rail and meat processing infrastructure was added to by highways connecting far-flung cities during the twentieth century. These include the Lincoln Highway in 1919 which was the first transcontinental paved highway and which cut through the High Plains and the more famous and later Route 66 which connected Chicago with Los Angeles cutting through the heart of the High Plains in the 1930s. The Interstate Highway System has superseded these earlier roads and connects all the major economic centers of the High Plains with ports and processing centers.

The Interstate Highway System and rail is the circulatory system of the High Plains and determines the siting of facilities, particularly for grain processing and storage and large-scale cattle feed lots and swine concentrated animal feeding operations (CAFO's). These facilities are a critical component of export-oriented agricultural production (Federal Highway Administration 2010).

In earlier years, many grain elevators and milling and processing facilities were owned by farmer cooperatives. Although many such cooperatives still exist, increasingly, corporate agricultural giants like Cargill, Archer Daniel Midlands (ADM) and ConAgra, JBS and Swift control the processing facilities, particularly loading facilities where grain is shipped by rail, barge and truck to demand centers and for export. One demand center is the large-scale feedlot operations and CAFO's. Cattle feedlots are situated near transportation hubs and calves born and raised on distant ranches throughout the western U.S. are transported to them, typically by truck on State, U.S. and Interstate highways to centralized feedlot operations. At cattle feedlots they are fed cereal grains and gain as much as 500 kg in weight in a few months prior to slaughter. Since shipping full grown steers is harder than transport of calves, most slaughter houses are in close proximity to the concentration of feedlots which in turn are located centrally in the High Plains, although the dryer conditions of the western portion of the High Plains are more favorable for these extensive outdoor operations than the climate further east. The largest of these located in Eastern Colorado has in excess of 100,000 cattle being feed simultaneously.

The history of land tenure in the High Plains is one tied to the issue of land partitioning systems. In most of the U.S., the public land survey system (PLSS) was used to convert lands roamed by nomadic (as well as settled) Native American (Indian) tribes into a regular pattern of farms, and towns and cities. The Federal Government provided land to railroads and to settlers. The land was divided into numerous 1 square mile (640 acre or 300 ha) square sections. Most land was given away to pioneer settlers as part of the Homestead Act which started in 1862 and continued until most land of any agricultural value had been given away, typically before 1890. The Homestead Act provided each pioneer farmer (or rancher) with 160 acres of land at no cost with the requirement that he live on and improve the land before it became his private property. The Federal Government also allocated large areas to the state governments to fund higher education. An exception to this system was Texas which from 1836 to 1845 was an independent republic. When it joined the Union it retained ownership of all (not yet private) lands which at that time was over 100 million acres in extent. These lands included the vast acreage of the Texas panhandle which is an important part of the High Plains. These lands were sold and/or given away by the State of Texas (Nebraska Studies 2009). Over time, the original pattern of a single family farm on each 160 acres began to break down. Unlike in many other parts of the world where in time subdivision of farms in ever smaller plots took place, in the High Plains the reverse process has taken place. The efficient size of farms in the days prior to the advent of tractors in the 1920 may have been 160 acres, but after the arrival of mechanization, the efficient size of a farm or ranch in the High Plains increased to several times that, usually several thousand acres. Many ranches particularly in Texas and Wyoming far exceeded this size. An extreme case was the XIT Ranch in the Texas High Plains which covered over 3.1 million acres (1.3 million ha) and was the largest private land holding in the U.S. The result of the growing size of ranches and farms was movement of some population to towns in the region, but overall a drop in population in the region as a whole. The growth of CAFOs, cattle feedlot operations and meat processing facilities has also caused a shift in the character of the population, with many Hispanic migrant and seasonal workers taking jobs in these facilities. In the areas of the High Plains outside such centers of intensive meat processing as Guymon, Oklahoma, or Amarillo, Texas, the declining population is both aging and still very much Anglo-Saxon in origin. Some Asian immigrants took up farming in the High Plains, most of these farmers were Japanese immigrant wheat farmers, only a very few Chinese, who had worked originally on the railroads, settled down to farming in the High Plains.

Farming patterns have adjusted both to the larger size of land holdings and the advent of deep wells capable of taping the underlying aquifers. Thus what were exclusively wheat farms supplied with water by the infrequent but intense summer rains and winter blizzards have become more diverse with crops like corn, and cotton also being produced there. Despite growth of these other crops, wheat continues to be the predominant crop in terms of acreage in the High Plains. It has the advantage that it lends itself to production on a vast scale and does not require irrigation. As the aquifers underlying the High Plains continue to be over-drafted, the costs are ever increasing amount of energy needed to pump water from aquifers with a declining pressure head. Many High Plains operations are "farm-ranches". This region has its more fertile areas devoted to crops like corn in rotation, some farms may have orchards in limited areas, some may grow oats or barley, but most of the land is devoted to either wheat or pasture for cattle. These diversified operations of a farm-ranch give its owners the flexibility to raise crops and livestock and cereals to meet various market conditions. This is not so much a matter of being self-sufficient, as in mixed farms in the past, as all of the operations are on a scale that far exceeds the needs of the few workers involved. The diversification is more a matter of taking advantage of the character of the land on an extensive holding to produce the crop that is most suited to that portion of the farm-ranch and to be able to take advantage of different market conditions such as an increase in corn prices in one year and a drop in wheat demand at another time. Of course, there are other patterns with some areas with poorer soil, colder weather or dryer conditions (or all three, as in Wyoming) being devoted solely to grazing with areas with richer soils, more moisture and warmer winter temperatures such as Deaf Smith County, Texas or central Kansas growing corn more frequently.

A contrast between the Corn Belt and High Plains is provided by the farming statistics for Wyoming. In 2007 Wyoming had only 11,000 farms (mostly cattle ranches or mixed wheat and cattle "farm ranches"). The average size of these ranches was 1,470 ha. Many of the ranches were even larger, with 27% being over 2,000 ha in size. In spite of their vast size, many of these ranches are owned by a single family and with seasonal migratory labor and help of cowboys they often are operated by an individual man and woman and their immediate family. The 11,000 farms and ranches in Wyoming held 732,883 cattle in 2007. (Economic Research Service, USDA 2007). This great extent is in sharp contrast to the situation in a State like Iowa, where the average farm is about 134 ha in extent. The average farm in China is about 1 ha (or slightly less) compared to about 175 ha in the U.S. in 2007, but in areas with climates similar to the High Plains, particularly the higher elevation portions of the region like Wyoming, the size of farms and ranches (grazing lands) is correspondingly larger and areas like Inner Mongolia or Tibet might approximate the average farm (ranch) size in cold arid regions of the High Plains. Wyoming is not a top cattle producing State, but the five leading beef cattle producing states in 2007 were Texas, Kansas, California, Nebraska, and Oklahoma in that order. California is not in the High Plains region but the other 4 states are. All of these four States extend outside the High Plains, but some of the best cattle raising areas in these states are in the High Plains region.

10.3.3.2 Agricultural Resources of the High Plains Region

Wheat

Wheat is gown in many parts of the U.S., but the High Plains are particularly important (Fig. 10.17). Wheat is the crop that the High Plains have been most known for since the later part of the nineteenth century. The rich loamy soils formed over millennium by grassland vegetation and the flat expanse of the High Plains lent itself to wheat farming once plows that could turn the thick heavy sun hardened sod and reapers that could harvest wheat on a large scale were invented by John Deere and Cyrus McCormick respectively. Tractors and combines further mechanized the process of wheat planting and harvesting. The most common wheat grown in the High Plains were varieties of wheat like "Turkey Red" and hard red winter wheat. Wheat yields per acre are actually lower in the High Plains than for the U.S. as a whole, but the extent of farms and the ease of harvesting vast flat acreages more than compensates for lower than average yields. For example, in the Texas portion of the High Plains, average yields are about 35 bushels per acre (about 400 kg per hectare), compared to a national average of 46 bushels per acre in 2010. Texas wheat production, which is mostly from the High Plains, was estimated at 122.5 million bushels in 2010, up from 61.3 million bushels in 2009. Harvested acreage for wheat in Texas, at 3.5 million acres in 2010, was up 43% from 2009. Other States in the High Plains region had winter wheat production that totaled 1.48 billion bushels in 2010. Irrigation can significantly increase wheat production in the High Plains. Typically, irrigation doubles wheat yields in the High Plains. As a result over 1 million ha of irrigated wheat was raised in the High Plains of Texas alone in 2009 (Fig. 10.18). (National Agricultural Statistics Service 2010).

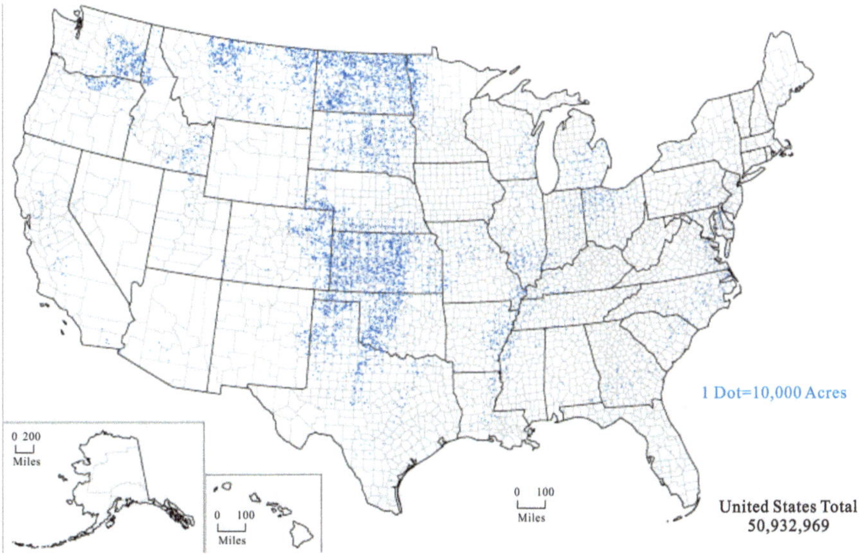

Fig. 10.17 Map of wheat production in 2007. Each dot represents 10,000 acres of wheat. Note that the High Plains, particularly Kansas, North Dakota and Eastern Washington have the highest concentration of wheat farms. However, the yield per acre is higher in some of the southern areas, particularly when irrigated. (Map courtesy of U.S. Department of Agriculture National Agricultural Statistics Service 2007. http://www.agcensus.usda.gov/Publications/2007/Online_Highlights/Fact_Sheets/Geographic/ag_atlas.pdf)

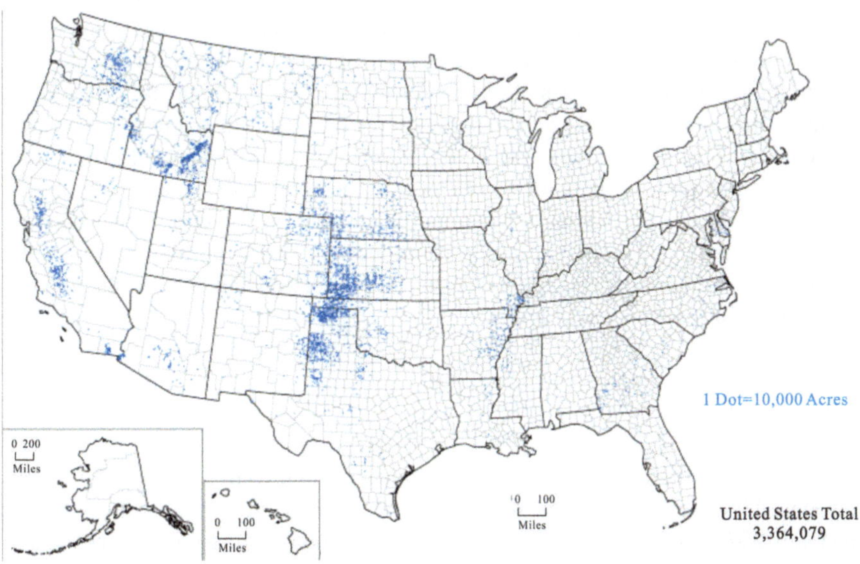

Fig. 10.18 Map of irrigated wheat farms by county in 2007. Note that each dot corresponds to 10,000 acres of irrigated wheat. Irrigating wheat is a common practice in parts of the High Plains, & Idaho including some of the most productive counties of the panhandle of Texas. (Map courtesy of U.S. Department of Agriculture National Agricultural Statistics Service 2007. http://www.agcensus.usda.gov/Publications/2007/Online_Highlights/Fact_Sheets/Geographic/ag_atlas.pdf)

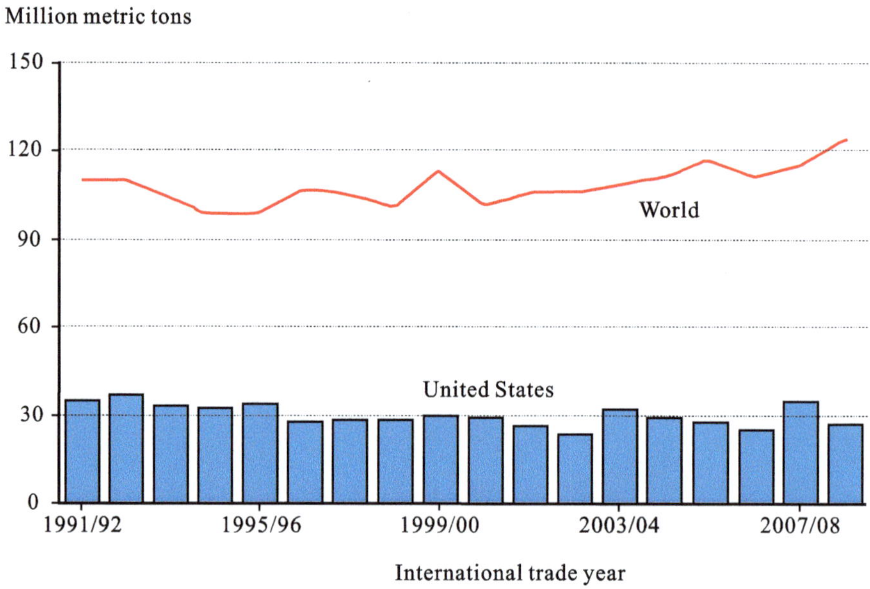

Fig. 10.19 Graph of U.S. wheat production versus total world wheat production. U.S. wheat production has been close to 30 million metric t per year. The U.S. is the world's leading wheat exporter, China produced more wheat than the U.S. recently, and this meets internal demand. (Courtesy U. S. Department of Agriculture Economic Research Service Foreign Agricultural Service 2010)

Harvesting is often carried out by roving teams of contractors that move fleets of combine harvesters from farm to farm. The system depends on the date when wheat is ready to harvest, being earlier in the southern part of the High Plains and then later in the season farther north. A similar system is used in China and is described earlier in this chapter. The U.S. is a major source of wheat for world food security. The chart in Fig. 10.19 shows the U.S. contribution to world wheat production. The U.S. top export markets for wheat include the nations of Nigeria, Egypt, and Mexico, but at present China is not a major destination for U.S. wheat exports, because China is self-sufficient in wheat production. However, many areas of the developing world are dependent on U.S. wheat production and U.S. production has an important impact on world prices.

Cattle

Raising cattle in this area predates wheat production in the High Plains. Cattle were introduced into Texas by the Spanish in the seventeenth century and by the period of the Civil War (1861–1865), vast herds of more than one million half-domesticated longhorn cattle ranged throughout Texas. These were driven by the legendary cowboys to rail heads in places like Dodge City, Kansas and Cheyenne, Wyoming to be loaded onto rail cars, typically for shipment to slaughterhouses in places like

Omaha, Nebraska and Chicago, Illinois. Today there are many vast cattle ranches in the Western U.S. including throughout the High Plains, but the typical pattern does not involve raising cattle from birth to slaughter in the High Plains, but involves bringing calves raised in family-owned "cow-calf" operations throughout the Western U.S. to nearby auction barns throughout the region. At these auctions buyers bid on the calves and the calves are then shipped to feedlot operations, frequently located in the High Plains, where they are fed intensively with grain brought from closer at hand than the cattle themselves. The system is driven by logistical factors. The weight of the feed a steer eats at a feedlot is perhaps 100 times the weight that the steer gains. It costs much more to ship the grain used than the calf. Thus the feedlot should optimally be located closer to the areas of surplus grain production and along a rail line (or better yet near the junction of several rail lines). (McLachlan, 2008). An example of such a centralized and optimally located facility is the "Five Rivers" cattle feedlot in Greely, Colorado in 2012 owned by Brazilian agricultural conglomerate JBS. It holds approximately 120,000 cattle at any given time. It is located in the northwestern periphery of the High Plains, so it can draw grain from the areas to the south and east and cattle from areas that are less productive for grain but can still support cattle ranching to the north and west. Calves are shipped to this facility from as far away as Nevada and Oregon, adding to many that come from the mountains and valleys of the Rocky Mountains region as well as the High Plains itself. Raising a calf requires considerable space and grazing land, but not necessarily the lands with the richest soils. It is better to raise the calves in other areas such as Montana, New Mexico or East Texas not otherwise as suitable for crop production or large scale grain production as the heart of the High Plains and ship the calves hundreds or even thousands of kilometers in a large truck over modern highways and/or by rail to the feed lot operation (Fig. 10.20).

Of course not all cattle are raised on feed lots, there are numerous dairy cattle operations in the region, many of these farms are cooperatives which share milk processing facilities. There are also an increasing number of organic dairies with organic milk fetching a premium price (California Farmer 2010). Dairy cattle are likely to be in areas with richer grasslands and are not moved as far as the vast distances that meat cattle travel in the modern integrated operations found in the High Plains. Traditionally, the Hereford breed has been dominant in meat cattle production, but the Black Angus and various cross breeds are becoming more popular, perhaps because of higher quality meat. The preferences of consumers in a global market for beef have had some effect on cattle production choices, with some producers starting to produce Wagu and other premium oriental varieties of beef for export to Japan and South Korea. While corn-fed dry aged beef from such areas as the Omaha, Nebraska region has long been considered a premium product, grass fed oriental breeds like the Wagu command a much higher price and are becoming more commonly raised. Also beef from this breed is starting to be demanded by gourmets in the U.S., not only in Japan and Korea.

Cattle are used to produce both dairy products (milk and cheese) as well as meat, but dairy cattle are raised on different types of farms and using different systems than beef cattle. Wisconsin, a state partly in the Corn Belt and California are the

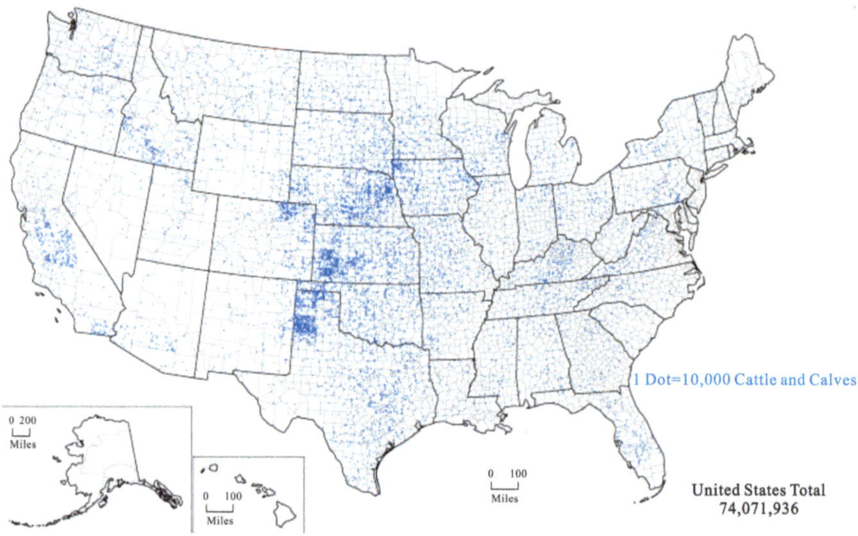

Fig. 10.20 Where cattle and calves were sold at auction. 1 *dot* is equal to 10,000 sold. The High Plains, (Texas panhandle, western Kansas and central Nebraska and north eastern Colorado) are the most important sources for cattle. (Map courtesy of U.S. Department of Agriculture National Agricultural Statistics Service 2007. http://www.agcensus.usda.gov/Publications/2007/Online_Highlights/Fact_Sheets/Geographic/ag_atlas.pdf)

top dairy cattle states with Texas being the third leading dairy cattle State. Texas is the leading beef cattle raising State, however. Inside of a large State like Texas, areas with many dairy cattle are different and separated in most cases from those areas where many beef cattle are raised. Dairy cattle are concentrated in areas with better grazing in the central portion of the State of Texas near Waco. Beef cattle production is more widespread in Texas, but is concentrated in the portion of the "panhandle" of Texas that is within the High Plains. Frequently beef cattle are born and spend their early months with their mother cows and at a sufficient age (and weight) the calves are sold at local auctions. The cow-calf operations are often small family farms with a few hundred hectares of land, the feed lots are often vast mechanized facilities that could be in another State. Many of the cow-calf operators only supplement their income by farming and depend on other jobs for their livelihood. The feedlots are usually closer to sources of grain, while the cow-calf ranches are in areas with good pastures.

Beef is processed in yet other even more centralized facilities often in states like Oklahoma and Nebraska. Most of the beef is consumed inside the U.S. However, in 2011 501 million pounds (238 million kg) of beef was exported from the U.S. The value of the beef exported in 2011 was slightly over $ 1 billion. In 2011 the top 4 countries importing U.S. beef were Japan, Mexico, South Korea and Canada. In 2011, no U.S. beef was exported to China. This is because in 2003 China instituted a ban on imports due to bovine spongiform encepalopathy (mad cow disease) in the U.S. herd. In 2002 the value of U.S. beef exports to China was $ 150 million, since

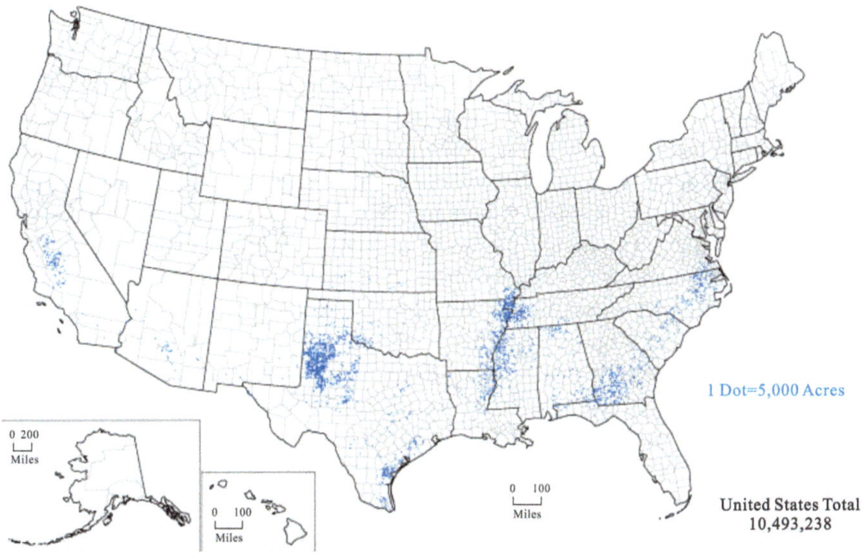

Fig. 10.21 Cotton Production in 2007. Note that the High Plains, particularly the panhandle of Texas is very important in cotton production. In 2010 Texas had the highest production of cotton in the U.S. (Map courtesy of U.S. Department of Agriculture, National Agricultural Statistics Service 2007. http://www.agcensus.usda.gov/Publications/2007/Online_Highlights/Fact_Sheets/Geographic/ag_atlas.pdf)

then the trade in beef in either direction is non-existent. Although only two cattle with BSE have been found in the U.S., as of 2012, the ban remains in effect.

Other Crops

While wheat and cattle are the mainstays of agriculture in the High Plains, some other crops and animal products (poultry in particular) are also raised there. The High Plains are not a very diverse area. The vast extent of lands, the harsh climate and the large scale logistically driven farming practices do not lend themselves to multiple specialty crops. Prices for commodities are now an important part of the decision to plant a particular crop or rear a particular animal. Corn prices in the U.S. are driven by a subsidy on ethanol for fuel that has prompted a rise in corn production and conversion of what was usually wheat land in the High Plains into corn production. Recent high prices for cotton have spurred more cotton production in the High Plains (Fig. 10.21). Texas is a center of cotton production, growing over 4 million bales (about 1,000 metric t) worth over $ 1 billion. The cotton is raised on 6 million acres, mostly in the panhandle. Texas is first among U.S. states in cotton production and the U.S (See Chap. 4) is the world's second largest cotton producer. China is the largest producer of cotton raising 25 million bales compared to 17 million bales raised in the U.S. in 2010 (Nationmaster 2010).

There are crops in the High Plains besides wheat and cattle and to a lesser extent cotton and corn that are important. Swine are increasingly being produced, not least because other more densely populated areas are complaining about the odor of large scale swine CAFO's. The large swine farms of in the Texas panhandle (some of which are Japanese owned) and in the Guymon, Oklahoma area are good examples. There are some other types of livestock raised in the region including chickens and turkeys, particularly in the south eastern area that does not get as cold. There is also one uniquely American animal: the bison (American buffalo) which has made a comeback. Today there are over 500,000 bison in commercial ranches, mostly in the High Plains. Bison are naturally well suited to the climate of the High Plains. Their meat has less fat and more protein than beef and their pelts and leather fetch premium prices. One major difficulty in bison production is that ordinary fences cannot contain the animals, so high wire mesh or electrified fences with large wooden or steel posts set in concrete are often required to contain bison herds. The bulls, which can weigh 1,000 kg, are particularly difficult to manage. Ted Turner (founder of the CNN News Network) is the owner of the most private land in the U.S. In 2010 he owned over 2 million acres of ranch lands, much in the High Plains and he has over 50,000 bison on his ranches in New Mexico, Oklahoma, Kansas, Nebraska, South Dakota and Montana. The scale of his operations allows him to profit from diversification, including wind and solar energy production and tourism and recreational hunting on his lands. (Turner 2010).

Sorghum, barley, oats, and sugar beets are all produced in the High Plains. Other root crops like potatoes and onions are grown; there is limited production of fruit. One of the most important recent economic diversifications in the High Plains has been another sort of farm: the wind farm. One factor that has driven the population down historically has been the high and seemingly unceasing winds in the High Plains. A source of misery during the dust bowl and during summer dust storms and winter blizzards is now a source of prosperity. Wind turbines do not significantly interfere with extensive wheat or cattle operations. Many of the ranches with wind turbines now in operation also have oil or gas wells that are now often depleted, but the new energy bonanza will not run dry in the foreseeable future. China is presently one of the largest manufactures of wind turbines, with most of the huge turbines arriving through the Ports of Houston and Corpus Christi and moving in several pieces on trucks by Interstate highways up to the panhandle of Texas and other areas that are transverse by power lines taking the energy generated in the High Plains to cities as far away as Dallas, Chicago and Denver. Given the reliance of the High Plains on rapidly depleted groundwater, the existence of a low cost perpetual energy source will play a role is sustaining irrigated agriculture and in fact makes the economy of the region less reliant on farming of any sort (High Plains Wind 2010). Whether wind energy will help to reverse the decline in population remains to be seen.

10.4 Conclusions

The world has several great "Bread Baskets" and "Rice Bowls". These are areas where due to rich soils, adequate water resources and available arable lands and infrastructure, cereal production can far exceed local demand. The surplus then is available for export or for livestock production. These areas may meet national demand as in India and the Nile Valley or they may produce such a surplus that domestic demand is greatly exceeded as in the Ukraine and Kazakh steppe, the Argentine pampas, the rice paddies of Thailand, the grain belt of Australia and the Corn belt of the Midwestern U.S. and the High Plains of America and the Canadian prairies. China also has a number of food production centers that exceed regional demand and provide for some international exports. Those areas producing the greatest regional surpluses are the Northeast China Plain, the North China Plain and the Middle-Lower Yangtze Plain.

In these three key regions in China natural and socio-economic factors have produced differences in cropping systems and the types of crops raised in the three main agricultural production regions (bases). The Northeast China Plain mainly grows wheat, corn and soybeans. The North China Plain mainly grows wheat, corn, and cotton, and the Middle Lower Yangtze Plain mainly grows rice, rape seed, mulberry (for silk production) and tea; fishing is also highly developed in this region. In recent years, the center of gravity of agriculture in China has shifted from the Middle-Lower Yangtze Plain to the North China Plain and the Northeast Plain.

The production of the three main agriculture bases constitute about three-quarters of the total food production of the whole country. This in turn guarantees China's food security, and also, to a great extent, eases the world food security situation. China's major grain producing areas are also its main consumer areas—as the eastern plains are densely populated. With the development of an industrial oriented economy, the agricultural land area is shrinking. Therefore, the most important agricultural development imperative must be to sustain and enhance grain production lands by use of modern production technology and improve the intensity and efficiency of agricultural production in these vital regions.

The Corn Belt and High Plains are two of the most important agricultural regions in the U.S. The economy of each is based on production of cereals, primarily corn in the Corn Belt and wheat in the High Plains. These crops in turn support many other industries, in particular livestock production and ethanol manufacture. Swine production is centered on the Corn Belt and beef cattle ranching and feedlots in the High Plains. Climate and soils related factors account for the productive differences between these regions. Historical and physical factors account for their settlement and ownership patterns, with the Corn Belt supporting more and smaller, mostly family owned farms and the High Plains supporting larger often corporate owned ranches, feed lots and irrigated farms. Geographical factors such as transportation networks, availability of large tracts of land and central location all play a role in the economic significance and high production efficiency of these regions. The global importance of these areas has grown recently for a variety of reasons including the development of large scale concentrated animal feeding operations, many geared toward export and the use of corn for production of ethanol. In any case, with a growing world population, these two regions will play a vital and growing role in the world economy.

References

Aydinalp, C. (2003). Some important properties and classification of mollisols. *Journal of Central European Agriculture, 4*(3), 225–236.

California Farmer. (2010). Advantages of organic dairy farms. http://www.californiafarmer.com/story.aspx/study/shows/organic/dairy/advantages/9/44001. Accessed 21 Dec 2012.

Cheng, L. (1984). *Agrogeography of China*. Beijing: Agriculture Press.

Chinese Academy of Sciences, Institute of Soil and Water Conservation. (1961). *Soil of the North China Plain*. Beijing: Science Press.

Chronicle of Higher Education. (2012). Interactive map: Proportion of adults, 25 to 34, with college degrees. http://chronicle.com/article/Interactive-Map-Proportion-of/65009/. Accessed 21 Dec 2012.

Economic Research Service. (2007). U.S. Department of Agriculture, state fact sheets for Iowa and Oklahoma, 2007.

Economic Research Service. (2010). U.S. Department of Agriculture, corn data for 2010.

Federal Highway Administration. (2010). The Eisenhower interstate highway system. http://www.fhwa.dot.gov/interstate/homepage.cfm.

Great Lakes Seaway. (2010). History of the Great Lakes Seaway. http://www.greatlakes-seaway.com/en/pdf/slsdc_newsletter_fall_2010.pdf.

Hao, T. Y. (2010). Research progress and prospects of reducing chilling damage in Northeast China.

High Plains Wind. (2010). Wind energy resources of the High Plains. http://www.highplainswindandsolar.com/resources/high-plains-wind-energy.html. Accessed 21 Dec 2012.

Hostetler, J. (1993). *Amish society* (4th ed.). John Hopkins University Press. Baltimore, Maryland, USA.

Kromm, D., & White, S. (1992). *Groundwater exploitation in the High Plains* (Development of Western Resources). Lawrence: University Press of Kansas.

Liao, Z. F. (2007). *Status of soybean export trade of China*. Economic Review, (12):25–26.

Li, Z. (1993). *Natural geography of Northeast of China*. Beijing: Higher Education Press.

Li, Z. Q. (1988). *Economic geography of the Northeast economic zone*. Jinlin: Northeast Normal University Press.

Liu, S. J. (1996). *Chinese geography*. Beijing: Higher Education Press.

National Agricultural Statistics Service. (2007). Overview of Hog and Pig production. USDA Washington.

National Agricultural Statistics Service. (2010). Texas wheat production http://www.nass.usda.gov/Statistics_by_State/Texas/Publications/pr13510.pdf. Accessed 20 Dec 2012.

Nationmaster. (2010). World cotton production. http://www.nationmaster.com/graph/agr_cot_pro-agriculture-cotton-production. Accessed 20 Dec 2012.

National Bureau of Statistics of China. (2010a). *China statistical yearbook 2010*. Beijing: China Statistics Press.

National Bureau of Statistics of China. (2010b). *Rural China statistical yearbook 2010*. Beijing: China Statistics Press.

Nebraska Studies. (2009). The homestead act and the settlement of Nebraska. http://www.nebraskastudies.org/0500/frameset_reset.html? http://www.nebraskastudies.org/0500/stories. Accessed 20 Dec 2012.

Pig Process. (2009). New rules for CAFO's. http://www.pigprogress.net/news/us-epa-cafo-rule-to-have-big-impact-id2256.html. Accessed 20 Dec 2012.

Chapter 11
Research Cooperation between Chinese and American Geographers and its Significance in the Quest for Sustainable Development

Peijun Shi, Clifton W. Pannell and Tao Ye

11.1 Introduction

When comparing the geographies of China and the United States we note both countries are large, middle-latitude territories roughly the same size in area with a major ocean on the east and southeast flank. Yet there are key differences as China is enclosed on the southwest, west, and northwest by great mountains and basins. By contrast the U.S. has an ocean on its west which links it to Asia. These geographic realities have helped shaped the historical evolution of the two great countries. China is an ancient civilization and culture and today is on a rapid trajectory of economic growth and social change as it modernizes and urbanizes. It contains the world's largest population of 1.35 billion. The U.S. is a relatively young country with a population of 314 million and has an advanced economic and technical system, although today the country faces serious financial challenges owing to a recent period of excessive spending and inappropriate fiscal management.

China and the United States are the two largest economies in the world. China has the largest market in the world, while the United States has the most advanced technologies, which should provide very good conditions for Sino-US cooperation in science, technology and economy. The United States is among the countries with the largest per-capita energy consumption and exerts significant impact on global energy, environment and climate. China, perhaps the fastest growing economy, with

P. Shi (✉) · T. Ye
State Key Laboratory of Earth Surface Processes and Resource Ecology,
Beijing Normal University, No. 19, XinJieKouWai St., HaiDian District,
100875 Beijing, China
e-mail: spj@bnu.edu.cn

T. Ye
e-mail: yetao@bnu.edu.cn

C. W. Pannell
University of Georgia, 520 Cloverhurst Ave.
Athens, GA 30606 USA
e-mail: cpan@uga.edu

R. Hartmann, J. Wang (eds.), *A Comparative Geography of China and the U.S.*,
GeoJournal Library 109, DOI 10.1007/978-94-017-8792-5_11,

its huge population base, makes its total resource consumption, especially energy, rank top in the world. People are constantly concerned that China's rapid development will exhaust world resources and destroy the global environment. In different ways, China and the United States both exert substantial influence on global energy, environment and climate, and both are of critical importance to global sustainable development. How do geographic realities affect the outcomes of economic development and technologic change in these two countries that impact the environment, climate change and energy usage, and what are the contributions of geographers in each country to the study and resolution of on-going and future problems and challenges?

Since the formal establishment of China-US diplomatic relations in 1979, we have seen significant efforts developed to realize solid cooperation between China and the US for sustainability. At the government-to-government level, cooperation on environment, energy and climate issues are the most valuable in terms of sustainability, and also a significant part of the relationship between China and the US. Moreover, geographers from China and the US have been working on sustainability of environment and resources and other topics (Fan et al. 2003). However, as we shall see researchers and scholars in each country have employed varying methodologies and approaches derived from different academic and political traditions, although we now begin to see some evidence of the beginning of convergence especially among Chinese geographers who have completed their graduate education outside of China. In this chapter, the details of the cooperation at both government and academic levels shall be briefly summarized to discuss future efforts required.

11.2 Government-to-Government Cooperation

11.2.1 Cooperation in the Past

Since 1979 when the Sino-US diplomatic tie was formally established, some periodic characteristics emerged in Sino-US cooperation in fields like environment, energy and climate, which were closely related to the international political environment and the diplomatic background of the two countries. In general, the cooperation in the above fields can be divided into four main periods: the starting period of cooperation in the 1980s, the temporary suspension period in the 1990s, and the re-development period after "9 · 11" and the accelerated development in the Obama period.

11.2.1.1 1980s: the Starting of Cooperation

From the establishment of the Sino-US diplomatic tie to the end of the 1980s, the Sino-US relationship evolved successfully in various fields. After the *Third Plenary*

Session of the Eleventh Central Committee (1978), China determined to focus on economic construction, implement the policy of reform and opening-up and actively introduce foreign capital and advanced foreign technologies and equipment. In doing this, it was willing to cooperate with developed countries including the United States. During this period, the Sino-US relationship was normalized. On the other hand, the US government supported China's reform and opening-up, as the US realized the huge market potential in China and its economic and political benefits. The US invested heavily in China, provided advanced technologies and developed bilateral trade and an economic relationship.

During this period, the cooperation was mainly inter-governmental, and political interest surpassed economic interest. Both governments were very much enthused and actively created policy conditions for cooperation, reflected in the US government's relaxation of restriction on export of technologies to China and the Chinese government's active implementation of reform and opening-up to create conditions for Sino-US cooperation. During the warming-up period, cooperation projects were limited, at a smaller scale and less deep level. China was at the starting period of reform and opening-up; its environmental awareness and energy demands were limited and the extent of opening-up was not large. The US interest in China was more from strategic need, and cooperation in environmental protection was not a key issue in the Sino-US relationship.

Specific fields of cooperation between China and the US in terms of environmental protection in this period mainly included monitoring and control of air pollution, ozone protection and acid rain. There were several important pacts and agreements signed on environmental protection. In 1980, the first cooperation agreement on environmental protection technology was signed between China and the US. On February 5, 1980, the leading group of the environmental protection Office of the State Council of China and the US Environmental Protection Agency signed the Protocol on Technological Cooperation between China and the US in Environmental Protection, under which four appendixes were formed afterwards, involving environment and health, pollutant control, environmental change and impact and global environmental problem. Other agreements followed.

During this period, Sino-US energy cooperation was mainly concentrated in petrochemical energy such as coal, petroleum and natural gas, as well as hydropower generation and water resource utilization. Besides, China and the US also explored cooperation in peaceful utilization of nuclear technology. On August 28, 1979, China and the US signed the cooperation agreement on hydropower generation and water resource utilization. On March 15, 1980, the appendixes of cooperation agreement were signed in Beijing. On September 20, 1982, the Chinese government and the US government signed the cooperation agreement on hydropower generation and water resource utilization from 1982 to 1984. Several famous water conservancy and hydropower projects known to us now were included such as the *Three Gorges Dam* project. In 1985, the two countries signed an agreement on cooperation in peaceful utilization of nuclear energy. In May 1986, after the Sixth Meeting of the Sino-US Joint Economic Committee, the two parties signed two agreements to fund the feasibility study of two projects of disposal hazardous wastes in China and power station turbine reconstruction. The signing of these agreements indicated the support of the two governments for promoting Sino-US energy cooperation.

11.2.1.2 1990s: Cooling Period

In early 1990s, due to dramatic changes in the world political situation, western countries like the US started to reassess their relationship with China, and the Sino-US relationship temporarily entered a low-ebb period. As a result, Sino-US cooperation in environmental protection was diminished, and many agreements signed earlier could not be implemented. The US projects and funds for environmental protection in China were gradually reduced, and for a time the US Environmental Protection Agency did not have a special fund for cooperation with China. Compared with corporations in other western countries, corporations in the US lost many commercial opportunities in the competition for the Chinese environmental protection market. However, by the late 1990s, China had successfully withstood the pressure and sanctions imposed by western countries and safeguarded its fundamental national interests. China and the US started to restore their cooperation in environmental protection. After all, Sino-US cooperation met both parties' interests; therefore, from 1994 to 1995, Sino-US cooperation was reinvigorated after a short period of suspension.

During this period, the cooperation areas were expanded based on the ones of the previous period. New cooperation highlights emerged besides the strengthening of cooperation in air pollution and river pollution, and new environmental policies. For example, legislation of emission trade in China gradually became an important part in Sino-US environmental cooperation. In terms of energy, besides strengthening of cooperation in petroleum, coal and natural gas, the utilization of clean energy, development of new energy and environmental protection became important elements in Sino-US energy cooperation.

The US transnational corporations and organizations started to play a role in Sino-US cooperation in environmental protection and energy. Some large US investment companies started to invest in China and export advanced equipment to China. China's huge demands also became the focus of attention for US entrepreneurs, who saw a promising prospect in the extensive Chinese market. Economic globalization also had a complementary impact on Sino-US cooperation. China needed funds, technologies, advanced equipment and management abilities of the US, while the US needed China's market.

During this period, the Sino-US agreements on cooperation for environmental protection followed. Specifically, in March 1997, Premier *Li Peng* and the visiting US Vice President *Al Gore* co-presided over the "Sino-US Environment and Development Symposium" in Beijing. After the meeting, the Chinese National Planning Commission and the US Department of Energy signed the *Sino-US Proposal on Energy and Environmental Cooperation*. The main areas of this agreement were clean energy, urban air pollution control and electricity application in rural areas. Afterwards, the US Environmental Protection Association and the Chinese government and research department cooperated to carry out the research on Chinese policy and legislation in emission trade. In April 1999, State Administration of Environmental Protection of China (SAEP) and the US Environmental Protection Agency (EPA) signed the intention for cooperation on "Feasibility Study on Reducing SO_2 Emission by Using the Market Mechanism in China". In September 1999, EPA and the SEAP signed an agreement to carry out research and pilot work on total emission control and emission trade under the Sino-US cooperation framework. *Benxi*City and *Nantong* City in China were appointed as the first two pilot cities. Numerous other agreements including some related to the peaceful use of nuclear technology soon followed.

11.2.1.3 Post "9.11"Period: Strengthened Dialogue and Cooperation

In early 2000s, some changes took place in the Sino-US relationship. The Sino-US spy plane incident and the unilateral policy promoted by *George Bush Jr.* in the beginning of his administration led to another low ebb of Sino-US relations. However, the "9.11" incident in New York City occurred soon afterwards. China showed its understanding and support for the US anti-terrorist actions, and the US government finally came to realize that the largest threat to the US in the new millennium was terrorism instead of China. A closer Sino-US relationship was resumed. On the basis of bilateral cooperation, cooperation within a multi-lateral framework also started to be an important component in Sino-US cooperation, especially in terms of environmental cooperation. Meanwhile, the serious threats to international energy resources increased potential dangers of Sino-US competition for energy, and the intensified competition were not favorable for either country's economic development. Therefore, China and the US both called for strengthening of dialogue and cooperation. During this period, China's accession to the World Trade Organization further increased China's opening-up, and its energy demand growth made China a great attraction for US enterprises. However, in terms of climate issues, former President *Bush* thought that the requirements of the *Kyoto Protocol* did harm to economic development in the US and thus the US withdrew from *the Kyoto Protocol.* No substantial progress was made in Sino-US cooperation in climate change.

The breadth and depth of Sino-US cooperation in environmental protection and energy were increased in the early twenty-first century. Changes also occurred in forms of cooperation, and the influence of non-governmental communication was gradually increasing. Sino-US cooperation in environmental protection and energy was developing rapidly and in good shape, with less and less influence from political factors. In April 2000, US Former Vice President *Al Gore* signed the *Sino-US Joint Communiqué on Cooperation in Environment and Development*, further promoting Sino-US communication and cooperation in energy, environment and development, including protecting global climate and bio-diversity. A number of specific agreements and communiqués followed which elaborated cooperation on energy usage and efficiency, hazardous wastes, and climate issues.

In terms of climate change, in June 2007, the Chinese government released China's National Plan for Coping with Climate Change for the first time, which gave a comprehensive description of China's position and policy in climate change, drawing extensive attention. In the "*Eleventh Five-Year Plan*", the Chinese government put forward the constraining indexes, i.e., the energy consumption per unit GDP would be reduced by about 20 % during the period of the "Eleventh Five-Year Plan", and the total emission of main pollutants would be reduced by 10 %. Meanwhile, the US call for changing passive climate policy and adopting more powerful measures to reduce greenhouse gas emission was increasing, and the concern with the issue of climate change in US society was also substantially rising.

The rapid warming of the issue of climate change made it more urgent for China to carry out international cooperation in climate change. This strong intention of international cooperation was reflected in the detailed list of technologies needed for coping with climate change in China's National Plan for Coping with Climate Change. The huge changes of the climate politics in the US, to some extent, also increased the US's intention to strengthen Sino-US cooperation in coping with climate. In terms of the US government, in the Sino-US strategic and economic dialogue started in 2006, the US would list energy and environment as one of the six key cooperation fields. In October 2007, the US Assistant Secretary of State Negroponte delivered a speech at the banquet of the US-Sino Relationship National Committee and listed climate change as one of the five global challenges with which the US should intensify cooperation with China.

In considering how Chinese and American geographers/climatologists and atmospheric scientists might cooperate successfully in research ventures, it would be useful to get the key national scientific organizations in each country to contact major departments and research centers to determine the nature and extent of current research on climate change. Further, surveys should be conducted to determine which units and researchers would like to do cooperative and comparative research on specific issues related to climate change both from an historical and contemporary perspectives in the U.S. and China. Such collaborative research could be encouraged and stimulated through research funding that would be specifically designated for the most significant research topics by funding agencies such as the National Science Foundation. An example of a research agenda and topic that would be of critical importance would be the ongoing study of carbon emissions in each country and how they affect other countries through the global atmospheric circulation system. Analysis could be undertaken at a variety of geographic scales ranging from the local to the regional to the global.

11.2.1.4 The Obama Era: Cooperation towards the Future

On January 20, 2009, *Obama* was elected the 44th US President, ushering the US into the Obama Age. Since the start of his administration, *Obama* put forward many guidelines in his election campaign and adopted many corresponding measures in his administration. The Obama government attached great importance to climate change, new energy and green economy and vigorously advocated "*New Energy for America*" centering on developing clean energy, reducing energy consumption and improving energy efficiency. Several experts in energy and climate such as *Steven Chu* were appointed as high-ranking officials of the government. In the American Recovery and Reinvestment Act signed by *Obama* on February 17, 2009, 10% of the budget for economic reinvigoration was to be used in green energy projects. On June 26, 2009, the American Clean Energy and Security Act were released, aimed at reducing greenhouse gas emission in the US and reducing the US's reliance on foreign oil.

In the Obama era, the Sino-US cooperation in environmental protection, energy and climate change was carried out in a comprehensive manner, and Sino-US cooperation entered a new historical period. In November 2009, during President Obama's visit to China, the two parties formally signed the "*Memorandum of Understanding between the Chinese and American Governments on Strengthening Cooperation in Climate Change, Energy and Environment*". The two parties reiterated that they would continue to promote the 10-year cooperation and announced agreement on an energy efficiency action plan under the 10-year cooperation framework. In Mid-May of 2010, the Sixth Meeting of the 10-Year Cooperation Joint Working Group was held, incorporating the relevant results of 10-Year Cooperation into the second round of Sino-US strategic and economic dialogue. During this period, besides energy cooperation in clean coal technology and coal bed gas development, there is potential for cooperation in fields such as energy conservation, energy efficiency improvement, renewable energy development, nuclear energy and potential technology as well as relevant standards, policies and economic incentive mechanisms.

11.2.2 Problems and Challenges

Although China and the US achieved significant progress in government-to-government cooperation on environment, energy and climate change, problems and challenges have always been an important issue. There are still differences, obstacles and challenges for cooperation.

China and the US are two independent countries. Although we share a unique planet earth, the countries are different. Any government-to-government cooperation has to follow the national interest. If the cooperation is believed to be positive to national interest, it will be carried out. Otherwise, it will not, as China and the US are such different countries. The most critical issue is the divergence between socialism and capitalism, and the difference between a developing country and a developed country. For a worthwhile explanation, see the 1984 prefatory essay of Professor *Huang Bingwei*. The differences undermine mutual trust and consequently hamper the support from the US to China.

The disparity in ideology has always created difficulty in China-US cooperation, although the "Cold War" ended more than two decades ago. It is one of the fundamental approaches of North-to-South cooperation that the developed countries provide international aid to developing countries. However, socialist countries were excluded from the list of benefit countries. Official aid provided to China by the US government is very modest in sharp contrast to the support from Japan and the EU. The influence of difference in ideology can also be observed by the change in the attitude of the US to China's need of cooperation induced by political events. In the 1980s, China started reforms and an opening policy and urgently needed funding and technology. As we have seen in an earlier section, China and the US started cooperating on environment and energy issues in this period. However, since 1989, the US enacted serious economic sanctions on China. The sanctions stopped the export from the US to China, particularly of high-tech products. The US opposed any proposal to provide China with loans by international financial institutions. Also, the peaceful utilization of nuclear energy in China was affected.

China's environmental problems have not always been used as a reason for cooperating with China by the US government. Frequently these environmental concerns have been used to convince outsiders that China is a serious threat to the global environment. Most developed countries finished the stage of rapid progress with severe environmental pollution and high resource consumption. Therefore, many people are concerned that the rapid development of China will produce considerable environmental pollution and consume a substantial amount of global resources causing serious challenge to the world. Some propose that the development of China should be restricted rather than helped with funds and technology to avoid environmental and resource problems. The environment diplomatic report of the US in 1997 claimed that in 1995 the consumption of coal, grain and meat in China had surpassed the US, and the green gas emissions from China were the second in the world. When President *Clinton* met Chairman *Jiang Zemin* in the mid-1990s, he claimed that the largest threat of China to the US was not military, but resulted from environmental problems. China was taking the place of the US and becoming the largest environmental polluter in the world. Americans believed that the carbon dioxide and sulfur dioxide emitted from Chinese factories could travel as far as the eastern Pacific and damage weather, atmosphere and climate of the US. Therefore, the US did have a sense about China's environmental problems, but the strategy was to blame, not to help.

In the energy field, China and the US have more conflicts. Due to the limited total stock of fossil fuels, China and the US as the two largest consumers in the world will inevitably meet in the energy market. The US is now the largest crude oil consumer, taking up 25 % of the worlds' total consumption, with annual import of 500 million t. In past years, the US has been making great efforts to explore new and clean energy sources such as alcohol and bio-fuels to reduce its reliance on crude oil and imports. In year 2010, its import rate of crude oil dropped below 50 %, which is the first time in 13 years. Of greater potential significance is the recent discovery and recovery of large reserves of shale gas in the US. This is natural gas trapped in shale formations that can now be released through a form of mining called hydraulic fracturing or "fracking." While the current amount of shale gas mined now accounts for about 35 % of natural gas consumed in the US, the total production of natural gas is projected to rise by 44 % by the year 2040, and the US is committed to a policy of energy independence in part based on its large reserves of natural gas and associated oil deposits (US Energy Information Administration 2012).

In contrast, China's dependency rate on imported oil increased from 47.2 % in 2007 to 56.5 % in 2011. Nevertheless, China's technology on clean and renewable energy sources is much less developed. *Kissinger* has his famous saying: "control oil and you control nations". In the twenty-first century, it is not the oil but the technology of clean and renewable energy that may determine the power of a country. Therefore, it would be extremely difficult for China to get the advanced technology from the US, due to the lack of trust and the fear of threat, although the action would benefit the planet earth.

There is also significant divergence between the perspectives to climate change of China and the US. As the top developed country, the US is leading its developed friends claiming that all countries should be treated equally and contribute to emission reduction according to the current emission level. The US quit the *Kyoto Protocol* with the excuse that large developing countries such as China and India should not be exempted from emission reduction. However, the developing countries believe that climate change is the result of cumulated carbon dioxide in the atmosphere mainly attributed to carbon emission by developed countries since the industrial revolution. In this sense, the developed group should be more responsible. Also, they are expected to provide developing countries with funds and technology to reduce emission in the development process. From Copenhagen in 2009 to Cancun in 2010, and to Durban in 2011, there has been no actual progress in climate consultation before the Rio + 10 conferences in 2012.

11.2.3 Future Perspectives

11.2.3.1 Cooperation: A Win-Win Game for China and the U.S.

With notable rapid economic development, China is suffering from the huge cost of eco-environment degradation, which also brings concern from other countries. In June 2005, the journal *Nature* published a famous paper "*China's environment in a globalizing world: How China and the rest of the world affect each other*" (Liu and Diamond 2005), indicating that if the consumption level of China reaches the standard of the developed countries, global resources consumption and environmental impact will be two times greater than the current (year 2005). At the closing part of the paper, the authors said

> … In the past two decades, China has created an economic miracle. We hope that, over the next two decades, China can also create an environmental miracle and set a good example for other nations to achieve both socioeconomic and environmental sustainability. The outcome will affect not just China, but the entire world. (p. 1186)

Another famous article is from the *Guardian* (Watts 2005), saying that

> In the nineteenth century, Britain and Europe taught the world how to produce. In the 20th, the US taught us how to consume. If China is to lead the world in the twenty-first century, it must teach us how to sustain. (Website http://www.guardian.co.uk/world/2005/sep/20/china.jonathanwatts)

As a rising and developing country facing serious issues related to global environmental change, China assumes great responsibility. Nevertheless, as a developing country with the largest population in the world, China faces the problems of limited resources and technology availability. Consequently, international cooperation with technical and financial support in environment and energy problems for China is needed, perhaps even essential. The demand for advanced technology makes China a huge market. Most of the environmental protection products and energy technology are below the international standard, and therefore considerable amount of products rely on import. In this sense, international cooperation is urgent for China.

Many of the major environmental policies, institutions and laws in China are from the developed countries, especially the US, based on their experience for success and lessons learned. Much of the terminology and underlying activities are also introduced from the US to China, e.g. *Environmental Impact Assessment*, *Emission Trading*, *Clean Production*, *Eco-compensation*, *Environmental-taxation*, etc.

China's huge demand in environment and energy brings the US opportunities. As the country with the most advanced science and technology in the world, the US has invested substantially in developing environment and energy products. These products have been sold to China and brought back profits. For instance, in China the cost of constructing a nuclear power plant is estimated to be 2 billion USD, of which half will be left to foreign investors because China does not possess full technology to finish the project. It is estimated that a 1 billion USD contract will bring 5,000–7,000 employment opportunities, most of which are in high-tech industries due to the complexity and intelligence-intensive nature of the project (Tremayne and Waal 1998).

As we see above, on the one hand, China benefited from the international cooperation on environment and energy. Technology and funds have brought China significant progress in environmental protection and sustainability. On the other hand, the support to China is positive to the national interest of the US, because high tech exports win considerable revenue and employment opportunity. The cooperation on environment and energy is a win-win game for China and the US.

The continuing economic growth and development of China offers good opportunities to the U.S. and a fertile ground for collaborative research between American and Chinese geographers in examining the expanding trade relations between the two countries. This trade is based on the idea of comparative advantage in the factors of production and regional specialization of production. Both countries benefit from such trade at least theoretically. Economic geographers in both China and the U.S. are ideally suited to examine and analyze the various issues in this bilateral trade relationship to help determine the nature and mutual benefits of such trade. Moreover, the nature of global trade is a key factor in the continuing economic growth of both countries, and such analysis extends far beyond the bilateral trade between the two. Examples of the collaborative work of Chinese and American geographers on trade matters that extend to the global economy are available (Liu et al. 2009).

11.2.3.2 Cooperation for Planet Earth

The benefit of China-US cooperation on environment, energy and climate issues is far beyond the bilateral context. The process of global environment change and globalization make humans increasingly aware about the globalized environment problems, e.g. climate change, ozone destruction, land degradation, water pollution, desertification, and reduction in biodiversity. The world now is highly interconnected. Countries are closely linked to each other, not only in socioeconomic activities but also in environment and energy issues. Acid rain and dust storms may easily cross country borders. International rivers can easily induce drought in downstream countries if the upper stream is not well managed. Deforestation, desertification and industrialization can accelerate carbon emission and increase the uncertainty in future global climate. Also we see the impact of natural disasters is global. Examples are the impact of Iceland's volcano eruption and the 2011 eastern Japan great earthquake and tsunami.

The extent and damage such natural events cause indicate another area of potential cooperation for Chinese and American geographers. Geographers, with their interest and abilities to study and understand both natural and human-associated phenomena are in an excellent position to conduct research on the issue of disaster risk reduction. For many years Professor Gilbert White and his students have examined natural hazards and the behavior of humans in ignoring or diminishing the risks or dangers of such natural hazards. None of the problems can be solved by a single country, and international cooperation is needed. Sharing research findings and methodologies for analysis on hazards can begin the important goal of cross-national cooperation in disaster risk reduction. Sustainability issues are beyond races, nations, and ideology. China-US cooperation brings benefit to both China and the US. It also makes a great contribution to global sustainability and sets up a good example for other countries to follow.

11.2.3.3 Future Perspectives

There are problems and challenges, but there are also a number of achievements obtained in the past decades in China-US collaboration. In the current international environment, every country looks beyond its own borders for the sake of its national interest. Conflicts may result, but opportunities for collaboration always exist as long as there is common interest. Also, the divergence can always be resolved through bilateral and multilateral communication. With the increasing concern about the sustainability of our planet earth, we should be optimistic about the China-US cooperation on environment, energy and climate in the coming decade. China's development strategy, environmental protection policy and energy policy requires considerable input in related fields and industries. The environmental protection and energy technique market is booming, and it will continue to bring American companies lots of opportunities. The international community welcomes deepened China-US cooperation on environment and energy.

It is expected that more framework agreements will be signed between China and the US on environment, energy and climate issues, which will further strengthen the government-to-government cooperation. As indicated by China's Vice Premier *Wang Qishan*, China and the US have many points of common interest. China is the largest developing country in the world. It is at the stage of rapid economic development and structural change. Energy-efficient and environment-friendly technologies have huge market potential in China. The US is the largest developed country in the world. It has advanced technology and plenty of experience in energy efficiency, clean energy, and eco-environment protection. The relationship between China and the US on this matter is complementary. A series of agreements will be signed between the two largest countries in the future to further discuss how specific problems can be handled, e.g. greenhouse gases, water pollution, and sulfur-dioxide emission.

Cooperation is expected to have a broader context. Besides the traditional cooperation on greenhouse gases and pollution, several new fields are going to be incorporated in the collaboration framework. For instance, since 2004, China and the US began to discuss how to use an emission trade quota system to control emissions from the power industry and reach the national goal of emission-reduction.

In summary, as the two largest countries in the world, and the countries with the most energy consumption, China and the US are mutually dependent on environmental protection, energy and climate issues. The further cooperation between the two countries will definitely bring benefit to the people of both countries, and the entire globe.

11.3 Collaboration between China and U.S. Geographers

11.3.1 Historical Perspective

11.3.1.1 1949~1979: Early Collaborations

Although there had been some contact and communication between Chinese and American geographers prior to 1949, it was limited owing to the Second World War and the civil war in China which followed. While several geography departments were established in some of China's major universities during the 1920s, China's Institute of Geography was first organized in 1940 under the Academia Sinica (Hsieh 1959). The rupture in diplomatic relations between the People's Republic of China and the United States thereafter cut off scholarly and academic linkages, and the Cold War which followed the Korean War precluded virtually all contact between China and the United States for a number of years.

The reorganized Academia Sinica which included the Institute of Geography was moved to Beijing in 1953. This became the Chinese Academy of Sciences (CAS) designated as China's premier research agency and center. One of its Divisions, "*Earth Sciences*" included Geography. As Hsieh (1959) noted following this reorganization,

> China has taken the Soviet Union as its model in the transformation of an agricultural country into an industrialized one. It recognizes that geography can play an important role in national construction in surveying the physical environment and natural resources and in the planning of development projects concerned with irrigation, land use, soil erosion, transportation and hydrography.

It is clear the focus of this Soviet-inspired "socialist" geography was mainly on various fields of physical geography as well as applied research that would serve the fundamental interests of the state and its people. It is also clear that this focus on physical geography and natural resources development allowed geography to take its place within the Chinese scientific establishment because it copied the Soviet model of geographic science. (Wu 1958; Huang 1984).

Yet human geography managed to survive through its metamorphosis and linkage to the natural environment and resource analysis. *Wu Chuanjun*(1958), for example, one of China's leading economic geographers, indicated that the Soviet approach to geography was based on fieldwork, sound technical knowledge of the physical environment, and the interrelation of physical and economic geography (Wu 1958). In 1963 at the 3rd National Assembly of China's Geographical Association, 343 papers were presented and focused on topics such as agricultural and physical regionalization, land use, water conservation, soil and land resources, agroclimate regions, agricultural mapping and climate cycles (The Third National Assembly, Annual Meeting of the Geographical Society of China 1964; Pannell 1980). In this period, other topics for Chinese geographers included water resources, river management, flood control, conservation, reclamation, and navigation (Chang 1975). On the other hand, studies on economic geography emphasized most heavily transportation, coal, petroleum, iron and steel, machines and cotton textiles, besides the farming of rice, wheat and cotton. The pattern of focus was appropriate since it was the period of China's recovery and reconstruction after the war.

During this period, China was not open to western countries, and we can hardly find any study by Chinese geographers on foreign countries or even the global issues. On the other hand, the obstacle of diplomatic relationship also impeded the work of American geographers who were interested in mainland China. American geographer *Chang* (1975) made a survey of the studies of western geographers, including 100 books (monographs, or atlases) and 50 articles in American geographical journals. He found that about one third of these books were exclusively on China. The rest of the books paid attention to border issues, heavy industry, communes, minorities and local developments with respect to China. The journal articles were concentrated on the reappraisal of China's resources foundation, progress made in transportation, redistribution of industries, urban growth, water resources and desert studies.

Jack Williams (1969) surveyed American geographers to assess their willingness to conduct research on China. According to Williams, among the approximately 4,000 geographers listed in the 1967 directory of the Association of American Geographers, only 65 expressed interest in mainland China. Furthermore, among the 26 geographers who gave first choice to China, only 6 of them published important studies on China (Table 11.1). As Williams (1969) indicated, the result

Table 11.1 Survey statistics (Williams 1969)

	Choice of China	Some knowledge of Chinese	Native Speakers
First	26	17	8
Second	27	11	6
Third	12	4	1
Total	65	32	15

> … suggests a depressing lack of interest … in a country that is inevitably destined to play a major role in world affairs despite the turmoil and disunity presently displayed, a country with one-quarter of the human race within its borders, with resources sufficient to make it a major economic force and certainly a country with one of the most enduring cultures ever created by man.

The reasons for the depressing lack, as Williams summarized, were the inaccessibility of China and the difficulty in learning Chinese language. Williams (2003) later provided an updated picture of the work of American geographers on China with an emphasis on those who are members of the China Geography Specialty Group (CGSG) of the Association of American Geographers. As he indicated the progress in the work being done in the last three decades is impressive, although the research agenda remains limited.

The period from 1966 to 1976 witnessed the Cultural Revolution in China. In this period, Chinese geographers were much less productive due to the domestic disorder. Yet some good scientific work in geography continued. Perhaps the best scholarly work in geography was seen in physical geography and its subfields of geomorphology and hydrology. While much geographic research had previously been done in regional classification of soils, mineral resources, and climate, in geomorphology and hydrology increasing emphasis was placed on process and analytical understanding of the mechanisms of geomorphic and hydrologic forces(Kikolski 1964; Stoddart 1978). Meanwhile work in cartography and related fields such as remote sensing lagged (Pannell 1980).

11.3.1.2 Post-1979 Era: Getting to Know and Work with Each Other

Knowing Each Other by Conferences Even before China and the US established formal diplomatic relations in 1979, mutual visits and cooperation between Chinese and American geographers had begun. In 1977 the first official visit of American geographers under the auspices of CAS took place (CAS, Institute of Geography 1977; Ma and Noble 1979; Ma 2007). A group of 10 American geographers spent 3 weeks in China and visited several university geography departments as well as research institutes doing geographic research. The following year, the Ohio Academy of Sciences sponsored a return visit of 10 of China's most senior and promising young geography scholars. CAS and university scholars in China presented briefings on the status of geographic research and study in China, and complementary field excursions and briefings on the latest geographic research in the US were presented in return during the visit of the Chinese geographers to the United States in 1978. The highlight of the October, 1978 visit of the Chinese geographers was a meeting held at the Wingspread Conference Center of the Johnson Foundation in Racine Wisconsin. A number of leading American geographers convened for an exchange of ideas and information on the state of geographic research and scholarship in both countries. The importance of these initial meetings and exchanges cannot be overemphasized, as they soon led to a flow of graduate students from China as well as joint research activities and exchanges among a number of scholars and institutions (Ma and Noble 1979; Ma and Noble 1981; Ma 2007).

Soon thereafter, Chinese geographers started to participate in geography academic conferences held in the US. The most popular events are the Annual Meetings of American Association of Geography (AAG) and the International Geographical Union (IGU). The first formal report written by Chinese geographers regarding geography annual conferences was in 1981, by *Zhang Piyuan* (1982). It is the first introduction to Chinese Geographers about the AAG Annual conference, in which the conference structure, value of geography, and the latest trends in geographic scholarship were summarized. In the report, the author emphasized the trend of quantitative approaches and metrics in geographical research. Quantitative analysis in geographic scholarship emerged in the US in 1950s and accelerated in the 1960s, while Chinese geographers started such research in the 1980s and reached its climax in the 1990s. Later, the full text of the speech of Prof. *Ronald F. Abler* on AAG Annual Conference, 1987 was translated into Chinese and published in the Chinese journal *Human Geography* (Alber and Wang 1991). AAG events on 1994, 2010 and 2011 were also formally reported to Chinese geographers in journals (Zhou 1998; Zhao and Zhang 2010; Zhang and Chai 2011).

The settings of conference topics were always reported to introduce the latest trends in geographical research. Yet, participants were sensitive about the interest and scope of research presented at the AAG conference which made them feel anxious. For instance, in the report of the 1994 AAG event, Prof. *Yixing Zhou* (Zhou 1998) pointed out that the innovations in methodology (for example, the introduction and application of computers for geographical information systems (GIS) and digital cartography) were advancing geographical research in the US. Also, he found that topics and working groups such as cultural studies, social justice, geographical perspective of women, and human rights were completely strange to him as a geographer. These topics illustrate the divergence between the geographical scholarly emphases of China and the US which have not converged till now (e.g. social justice or human rights), or the frontiers of geography which were later introduced to and well developed in China (e.g. studies with GIS, and women's perspective).

Cooperation Platforms There are important platforms for Chinese and American geographers, which have played a significant role in promoting cooperation. The working group of AAG, CGSG aims to promote the study of the geography of China, including Taiwan, and to serve as a clearinghouse of information for persons interested in Chinese geography. It also seeks to increase contacts with Chinese geographers and encourage professional activities, including the development of collaborative research projects. A report on some of the recent research activities of the CGSG and its members may be reviewed in the essay on the geography of China compiled for a major compendium of recent scholarship in American geography (Fan et al. 2003).

The working group began research on China as early as 1953. In 1973, the Committee on Chinese Geography (CCG) was formally established within the AAG structure, and *Rhoads Murphey* of the University of Michigan was elected to serve as the committee chair. In 1979, CGSG was established in Philadelphia and the CCG was formally terminated and merged into the CGSG.

The major activity of CGSG includes academic study and publication on China issues, journal publication, organizing panels for the annual AAG meetings and maintaining a newsletter to keep its members informed on their activities as well new publications and research projects. It also presents annual awards for the best student papers to encourage the younger generation in the geographical study on China.

Examples of Specific Projects Chinese and American geographers also carried out a substantial number of joint research projects. From 1981 to 1983, Chinese experts went to the United States to cooperate with US experts in doing research, with beneficial progress made in numerical value forecast and numerical value simulation experiments. In 1982 and 1983, experts of China and the US exchanged mutual visits several times and prepared cooperation plans in the climate area. Chinese scientists and US scientists held the Sino-US Climate Academic Symposia in Beijing, to discuss issues like climate research cooperation and put forward feasible measures. In 1987, the scope of cooperation of this project started to expand to research on annual tree rings, climate change and climate simulation. The Sino-US monsoon cooperation research carried out from 1983 to 1993 was highly effective. This project for atmospheric science cooperation was carried out with the US government with funding from the China Meteorological Administration and the National Natural Science Fund. China and the US carried out cooperation in terms of the formation mechanism of East Asian monsoon, the relationship between East Asian monsoon and Indian monsoon, monsoon and Chinese weather and climate, monsoon's low frequency vibration and inter-annual vibration characteristics, putting forward many new views, providing some basis for long-term forecast and cultivating a group of excellent climate study experts. In 1988, the two countries carried out Sino-US mesoscale meteorological cooperation and held three mesoscale meteorological academic symposia in 1988, 1990 and 1992. For more information, see the websites of the branches of the Institute for Atmospheric Physics (IAP, http://cmsr.iap.ac.cn) of the Chinese Academy of Sciences such as the Center for Monsoon Research.

11.3.2 Problems, Challenges and Opportunities

There are obstacles and challenges for geographer-to-geographer cooperation between China and the US beyond the diplomatic and political issues. The largest discrepancy is in the field, focus and scope of study. As known to us, although geographical research covers a wide range of topics in both China and the US, their main foci are different. American geographers are more concerned about human issues in their study and concentrate on the human-environment relationship. They have both advanced physical geography and human geography, but the latter has always been dominant. In contrast, most Chinese geographers concentrate on physical geographical research. They prefer to discuss human-environment relationship from the angle of the natural environment, but less from the perspective of human agency. Human geography in China lagged in the early socialist years. In the US, geography is often regarded as a social or human science, although it has frequently been described as an earth science (National Research Council 1965). This varies by departments, although it is true that the number of human geographers far outnumbers the physical geographers. In China, geography is viewed as a natural science (Wu et al. 1984).

The discrepancy has its historical cause. During the period of 1949–1979, geographical study in China was deeply influenced by that of the Soviet Union. At the beginning, Chinese geographers had limited knowledge of human geography, from theory to methodology. Human geography in China started its rapid development only after the reform and opening. The cooperation between Chinese geographers and foreign scholars also contributed to the process.

Second, geography in China is a discipline with specific goals. Geographical research is expected to provide useful knowledge and analysis for policy-making and to produce social value. During the past six decades, Chinese geographers have been making contributions to the development of China, whose projects and findings were directed at national needs (Lu and Cai 2001; Wu 1981). Therefore, its focus was on a national scope, with concentration on natural resources assessment and agricultural development in the 1950s and 1960s (Zhao, 1981). Since the 1980s, many geographers were involved in large national projects, mainly working on physical geographical study, e.g. the *Three-Gorge Dam* construction, dust storms and grain-to-green policy. Consequently, many publications on land resources, land utilization, environmental protection, regional planning as well as natural disaster reduction were seen in journals (mainly Chinese).

The scope of geographical study clearly indicated the bias of Chinese geographers for physical geography. China is a country with highly centralized power. In this sense, Chinese geographers prefer macroscopic and top-down approaches, in which physical processes and patterns are essentially important, while human dimensions are less essential. When providing consultancy to the government, the perspectives of Chinese geographers are typically focused more on the natural aspects, and the scale of analysis is generally at the national or regional level.

While these varying perspectives have limited research collaboration in the past, as Chinese geographers have expanded their studies in recent years to include a greater focus on various subfields of economic and urban geography, there appears to be increasing opportunity for collaborative study. Examples would be work in transportation, tourism, trade, urbanization and urban social and economic studies (Liu et al. 2009; Cai and Chan 2009). China's dynamic economic growth has led to rapid migration of rural people to urban areas which in turn has created an urgent need for new policies and solutions to emerging new social and economic problems and concerns (Han and Pannell 1999). Chinese and western geographers have frequently interacted and collaborated to study these problems and issues and to provide additional expertise that can be useful for policy planners in China (Chan and Zhang 1999; Fan and Sun 2008; Zheng et al. 2008; Li et al. 2009).

One of the most positive trends since the 1980s is the large number of Chinese students and scholars who have come out of China for graduate education in Hong Kong, North America, Europe, and Australia. Many of these young and now middle-aged scholars have completed their graduate education and assumed important faculty or research positions in major universities or research institutes in China or remained in their host countries. See for example the work of George Lin at the University of Hong Kong (Lin 1997) and Kam Wing Chan at the University of Washington (Chan 2009). With their expanded perspectives on the nature of contemporary geography, these scholars have assumed an increasingly important role in defining the academic and intellectual trajectory of geography in China today. Those Chinese scholars that have elected to pursue academic careers outside China often maintain close contact with their friends and former classmates in China and increasingly share in collaborative research projects (Li and Wei 2010). In this way, it seems clear that there has been a growing focus on the importance of human agency in geographic research among some scholars in China. It is likely that this trend will grow stronger as the power of geographic methodologies such as GIS, digital cartography and various spatial statistics is recognized for analysis of economic and social problems in the arena of applied scholarship and knowledge (Veeck et al. 1995; Yu and Wei 2003; Tang and Pannell 2009).

Recent trends in geographic scholarship indicate the persistence of some divergence of scholarly work and perspectives between China and the US, although there is growing evidence of limited convergence as better understanding and appreciation of the value of human geography research grows in China. This is clearly seen in the examples cited above of extensive recent collaborative work among geographers in China and those in the U.S.

11.3.3 Future Perspective

11.3.3.1 Common Interest, Obligation, and Duty in Sustainable Development

As stated before, geography is a discipline that seeks to facilitate the activities and practices of human beings in their quest for dealing harmoniously and productively with the challenges and opportunities of natural environments. This is true for Chinese and American geographers (Wu et al. 2008; "Rediscovering Geography:" Patricia Gober's speech on 1994 AAG Annual Conference). Since the 1970s, international geographical research has been switching from the nature-dominated environmental change to human-dominated environmental change (Messerli et al. 2000; National Research Council 1997). Geographers have been increasingly concerned about participating in programs regarding global issues and sustainability of natural environments for the proper and long-term use of future generations. Consequently they have rarely been so motivated to support national and government policy-making (Lu and Cai 2001). When sustainability comes as a global challenge, Chinese and American geographers share a common outlook. Right after the 1992 Rio conference on sustainable development, seminal articles were published in *Annals of theAssociation of American Geographers* (Wilbanks 1994) and *Acta Geographic Sinica* (Zhang et al. 1994; Huang 1996). In the twenty-first century, we believe Chinese and American geographers will cooperate even more closely on this topic.

11.3.3.2 Strategic Directions for Geographical Research in the Twenty-First Century

In 2010, the US National Research Council (NRC) released the report—*Understanding the Changing Planet: Strategic Directions for the Geographical Sciences* (NRC 2010), providing geographers a stepping stone for launching discussions about the strategic issues facing the planet and geography as a discipline (Sui 2011). In the report, 11 strategic directions in 4 groups were listed (Table 11.2).

The 11 directions listed above in Table 11.2 cover a wide range of research fields and topics. Directions in group A are mainly about human-environment relationship, emphasizing human impact to the natural environment and the vulnerability of human-environment system. Although the title of group B uses the word sustainability, it is mainly about demographic issues, population and heath. While geographers in China have not engaged in these matters, it is clear there are great opportunities for geographers in China to expand their research agendas once they seek competency in geographic training for doing research on demographic, social, and medical/epidemiological issues. A good example of where such geographic research would assist in solving a serious threat is the potential use of geographic tools such as GIS and analysis in tracking an exploding epidemic such as occurred in China and Hong Kong during the SARS outbreak. See, for example, the recent collaborative studies of *Li* and *Wei* (2010) and *Zheng* et al. (2008). These are very important areas of geographic research in the United States and have significant roles for applied policy making as well as achieving a sustainable society and future for all citizens. Group C is mainly about social, economic and political issues, in geographic perspective, and their impacts on sustainability. Group D is about technological issues.

At the beginning of 2000s, Chinese geographers also had wide discussions on the future development of geographical research, particularly research in China. Here we list the strategic directions summarized by representative Chinese geographers (Zheng and Chen 2001), Table 11.3.

As we see from Table 11.2 and 11.3, there are some fields and interests in common between Chinese and American geographers. Human-environment relationship is still the core of geographical research and the key to sustainable development. The focus on global environmental change, carrying capacity of key natural resources and ecological systems, and technological advance in geographical information and its impact all serve the goal of discussing human-environment relationship.

Table 11.2 Strategic directions for the geographical sciences according to the NRC report (National Research Council 2010)

A. How to understand and respond to environmental change
1. How are we changing the physical environment of Earth's surface?
2. How can we best preserve biological diversity and protect endangered ecosystems?
3. How are climate and other environmental changes affecting the vulnerabilities of coupled human–environment systems?
B. How to promote sustainability
4. Where and how will 10 billion people live?
5. How will we sustainably feed everyone in the coming decade and beyond?
6. How does where we live affect our health?
C. How to recognize and cope with the rapid spatial reorganization of economy and society
7. How is the movement of people, goods, and ideas changing the world?
8. How is economic globalization affecting inequality?
9. How are geopolitical shifts influencing peace and stability?
D. How to leverage technological change for the benefit of society and environment:
10. How might we better observe, analyze, and visualize a changing world?
11. What are the societal implications of citizen mapping and mapping citizens?

Nevertheless, the difference between geographical study in China and in the US can still be observed from the directions and ideas listed above. Chinese geographers typically approach research from the perspective of physical geography. Among the topics listed in Table 11.3 for China, 70% were selected from the perspective of the natural environment. This has changed little since the statement of then president of the Geographical Society of China, Professor *Huang Bingwei*, in 1984 when he averred "*It is estimated that no less than 70% of the work belongs to physical geography and its various branches*." In contrast, 10 out of the 11 strategic directions in Table 11.2 for the US were drawn from the perspective of human agency. The difference could be attributed to the historical development of geographical study in China in the socialist period with its derivation from the Soviet model of geographical scholarship with its strong focus on physical geography and natural resources. American geographers have mainly focused on human dimensions. As stated before, the difference is a challenge for collaboration and cooperation of Chinese and American geographers going forward.

11.4 Conclusions

11.4.1 Global Perspective

China and the US have so many similarities in geographical conditions. Due to historical reasons, the development of China and the US is not balanced. The US is the most developed country with the most advanced technology. China is the developing country with the largest population and potential human capital. They are

Table 11.3 Disciplinary frontiers of geographical research in China (Zheng and Chen 2001)

Comprehensive study on the processes and spatial patterns of the terrestrial surface
Geophysical processes in water resource cycling
Land evolution (soil generating and land degradation)
Bio-geo-chemical processes and its health impact
Holistic research of natural regional systems
Global environmental change and regional responses
Palaeogeographical environment change of Holocene, particularly the last 2,000 years
Integrated study on polar region, alpine region and cryosphere
Land use/land cover change and its driving forces
Global environmental change and its impact on environmentally-vulnerable regions
Adaptation strategies to global environmental change
Natural resources and ecosystem reconstruction
Allocation and sustainable use of key natural resources (water and land)
Ecological and environmental service evaluation
Integrated study on eco-environmentally vulnerable regions
Sustainable regional development, Mechanism of human-environment relationship
Human settlement and urbanization
The impact of human factors on natural environment
Coupled human-environment system dynamics
Geo-informatics and strategies of digital Earth research

also different in ideology, the difference between capitalism and socialism, which sometimes brings the two countries into conflict.

Yet the conflict may have less to do with ideology than with the rapidly growing economic and military power of China during its so-called "rise to power" and the resulting idea of a "power transition" with the United States (Lai 2011). Such a power transition raises the spectra of a serious challenge to American security interests both in the western Pacific region and globally. Nevertheless, both China and the US are facing similar environmental protection, energy shortage and climate change problems. Those problems force both countries to come and discuss the solutions. The pursuit of a sustainable planet goes far beyond country boundaries, racial differences, and ideological divergence, making long term cooperation between the two countries desirable and possible.

11.4.2 Joint Responsibility for the Future of Planet Earth

There are significant historical episodes of cooperation between China and the US geographers, which can be traced back to early twentieth century. China and the US cooperation on soil-water conservation started as early as 1920s. American geographer Walter Clay Lowdermilk once helped Chinese people handle matters of desertification and environmental governance. Advanced soil conservation technology was introduced to China from the US, while experiences accumulated by Chinese people were brought back to the US. The cooperation among geographers from the two countries significantly promoted modern and contemporary development of geography in China. Nevertheless, due to the deep impact of the Soviet Union, there was limited human geography research in China before the opening and reform (Wu 1958; Huang 1984). Consequently, physical geography has been leading China's geographical study for decades. In contrast, geographical research in the US pays much more attention to the human dimension, and therefore the cooperation was muted.

In the last two decades, cooperation between Chinese and US geographers has increased. Global sustainable challenges such as environmental problems, energy problems and climate change issues bring geographers together. In addition, as China's economy continues to grow rapidly and as the country shifts from an agricultural society to an urban-based economy and society, many new problems emerge. These present a number of new opportunists that are appropriate for geographical analysis as China reorganizes its spatial system and economy, modernizes and builds its cities, towns and urban infrastructure, and creates new transport systems and means of communication. It is here that geography as both a physical and human science can expand its horizons, as a new generation of geographers in China pursues new avenues of research in the service of creating a more just, sustainable, and livable environment and society for all of China's citizens. Geographers in the United States will no doubt be delighted to assist our Chinese colleagues in pursuing these new research agendas through collaboration and cooperation as well as in assisting in the continuing education of China's geographers who seek access to and understanding of the most recent trends and methodologies in American geography. As contemporary geographers on both sides, we are obligated and responsible to promote further cooperation in geographical research, for the well-being of the people of both China and the US, and the rest of the world.

References

Alber, R. F., & Wang, Z. H. (1991). What shall we say? To whom shall we speak? *Human Geography, 2,* 16–23, 66–72.

Cai, F., & Chan, K. W.(2009). The global economic crisis and unemployment in China. *Eurasian Geography & Economics, 50*(5), 513–532.

CAS, Institute of Geography. (1977, Aug 12). *Oral presentations to the Ohio academy of science geography Delegation at the Institute of Geography*. Beijing.

Chan, K. W., & Zhang, L. (1999). The Hukou system and rural-urban migration in China: Processes and changes. *China Quarterly, 160,* 818–855.

Chan, K. W. (2009). The Chinese Hukou system at 50. *Eurasian geography and economics, 50*(2), 197–221.

Chang, K. S. (1975). The geography of contemporary China: Inventory and prospect. *The Professional Geographer, 27*(1), 2–6.

Fan, C. C., Ma, L. J. C., Pannell, C. W., & Tan, K. C. (2003). Geography of China. In G. Gaile & C. Willmott (Eds.), *Geography in America at the dawn of the 21st century* (pp. 668–678). Oxford: Oxford University Press.

Fan, C., & Sun, M. J. (2008). Regional inequality in China, 1978–2006. *Eurasian Geography and Economics, 49*(1), 1–20.

Han, S. S., & Pannell, C. W. (1999). The Geography of privatization in China. *Economic Geography, 75*(3), 272–296.

Hsieh, C. M. (1959). The status of geography in communist China. *Geographical Review, 49,* 535–551.

Huang, B. W. (1984). Preface. In *Geography in China* (p. 2). Beijing: Science Press.

Huang, B. W. (1996). On earth system science and sustainable development strategy (1). *Acta Geographica Sinica, 51*(4), 350–354.

Kikolski, B. (1964). Contemporary research in physical geography in the Chinese people's republic. *Annals of the Association of American Geographers, 54,* 139–154.

Lai, D. (2011). *The United States and China in power transition. Strategic studies Institute.* Carlisle: U.S. Army War College.

Li, Y. R., & Wei, Y. H. D. (2010). A spatial temporal analysis of health care and mortality inequalities in China. *Eurasian Geography & Economics, 51*(6), 767–787.

Li, Z. H., Ma, L. J. C., & Xue, D. S. (2009) An African enclave in China: The making of a transnational Urban space. *Eurasian Geography & Economics, 50*(6), 699–719.

Lin, G. C. S. (1997). *Red capitalism in South China.* Vancouver: University of British Columbia Press.

Liu, J., & Diamond, J. (2005). How China and the rest of the world affect each other. *Nature, 435*(7046), 1179–1186. doi:10.1038/4351179a.

Liu, W. D., Pannell, C. W., & Liu, H. G. (2009). The global economic crisis and China's foreign trade. *Eurasian Geography & Economics, 50*(5), 497–512.

Lu, D., & Cai, Y. L. (2001). Geography in China: As sciences change direction. *Advance in Earth Sciences, 16*(5), 467–472.

Ma, L. J. C. (2007). Building the first bridge of contact: A personal account of the 1977–1978 exchange program between American and Chinese Geographers. In *The 30th year of Sino-US Geography exchange.* Nanjing: Institute of Geography, CAS, Nanjing Institute of Limnology and Natural Resources.

Ma, L. J. C., & Noble, A. (1979). Recent developments in Chinese geographic research. *Geographical Review, 69*(1), 63–78.

Ma, L. J. C., & Noble, A. (Eds.). (1981). *The environment: Chinese and American views.* New York: Published for the Ohio Academy of Sciences by Metheun and Co.

Messerli, B., Grosjean, M., Hofer, T., et al. (2000). From nature- dominated to human-dominated environmental changes. *IGU Bulletin, 50*(1), 23–38.

National Research Council (NRC). (1965). *The science of Geography.* Washington, DC: National Academy Press.

National Research Council (NRC). (1997). *Rediscovering geography: New relevance for science and society.* Washington, DC: National Academy Press.

National Research Council (NRC). (2010). *Understanding the changing planet: Strategic directions for the geographical sciences.* Washington, DC: National Academies Press.

Pannell, C. W. (1980). Geography. In Leo Orleans (ed.), *Science in contemporary China* (pp. 167–187). Stanford: Stanford University Press.

Stoddart, D. R. (1978). Geomorphology in China. *Progress in Physical Geography, 2,* 187–236.

Sui, D. Z. (2011). Introduction: Strategic directions for the geographical sciences in the next decade. *The Professional Geographer, 63*(3), 305–309.

Tang, Y. B., & Pannell, C. W. (2009). A hybrid approach to land use and cover classification. *GI Science and Remote Sensing, 46*(4), 365–387.

The Third National Assembly and Aid to Agriculture, Comprehensive Scientific Annual Meeting of the Geographical Society of China. (1964). *Acta Geographica Sinica, 30*(1), 78–84.

Tremayne, B., & Waal, P. (1998). Business opportunities for foreign firms related to China's environment. *The China Quarterly, 156,* 1029.

U.S. Energy Information Administration. (2012). Energy in brief, what is shale gas and why is it important? http://eia.gov/energy-in-brief/article/about-shale-gas.cfm.

Veeck, G., Li, Z., & Gao, L. (1995). Terrace construction and productivity on loessal soils in Zhongyang county, Shanxi province, PRC. *Annals of the Association of American Geographers, 85*(3), 450–467.

Watts, J. (2005). The railway across the roof of the world. The Guardian, Tuesday 20 September 2005. http://www.guardian.co.uk/world/2005/sep/20/china.jonathanwatts. Accessed 18 July 2012.

Wilbanks, T. J. (1994). Sustainable development in geographic perspective. *Annals of the Association of American Geographers, 84*(4), 541–556.

Williams, J. F. (1969). American geographers and China. *The Professional Geographer, 21*(5), 354–357.

Williams, J. F. (2003). Geographers and China. *Issues and Studies, 38*(4), *39*(1), 217–247.

Wu, C. J. (1958). The geographical organization and new trends of development in geography in the USSR. *Acta Geographica Sinica, 24*(4), 438–456.

Wu, C. J. (1981). Delineation of China's agricultural regions. In C. W. Pannell & C. L. Salter (Eds.), *The China geographer, No. 11* (pp. 27–40). Boulder: Westview Press.

Wu, C. J., Wang, N. L., Lin, C., & Zhao, S. Q. (1984). Geography in China. In *The geographical association of China*. Beijing: Science Press.

Wu, D. T., Wu, Q. X., Liu, R. W., & Song, J. P. (2008). Comparative study on the development venation of geography in China and America during the past hundred years—based on the statistical analysis of Acta Geographica Sinica and Annals of the Association of American Geographers. *Advances in Earth Science, 23*(6), 553–561.

Yu, D. L., & Wei, Y. H. D. (2003). Analyzing regional inequality in post-mao china in a GIS environment. *Eurasian Geography and Economics, 44*(8), 514–534.

Zhang, G. Y., & Chai, Y. W. (2011). The 2011 AAG annual conference was held in seattle. *Geographical Research, 30*(5), 963–964.

Zhang, P. Y. (1982). On the annual conference of association of American geographers. *Geographical Research, 1*(3), 89–90.

Zhang, S., Zhang, P. Y., & Wang, E. Y. (1994). Sustainable development and geography. *Acta Geographica Sinica, 49*(2), 97–106.

Zhao, S. Q. (1981). Transforming wilderness into farmland; An evaluation of natural conditions for agricultural development in Heilongjiang Province. In C. W. Pannell & C. L. Salter (Eds.), *The China Geographer, No. 11* (pp. 41–57). Boulder: Westview Press.

Zhao, X., & Zhang, G. Y. (2010). Chinese scholars participating the 2010 AAG Annual Conference. *Acta Geographica Sinica, 65*(7), 891–893.

Zheng, D., & Chen, S. P. (2001). Progress and disciplinary frontiers of geographical research. *Advance in Earth Sciences, 16*(5), 599–606.

Zheng, S. Q., Long, F. J., Fan, C. C., & Gu, Y. Z. (2008). Urban villages in China: A survey of migrant settlements in Beijing. *Eurasian Geography & Economics, 50*(4), 425–446.

Zhou, Y. X. (1998). On the new directions of geographical research in the US from observations on the 1994 AAG Annual Conference. *Economic Geography, 18*(4), 16–18.

Chapter 12
Epilogue

Rudi Hartmann and Jing'ai Wang

Now that the manuscript to *A Comparative Geography of China and the U.S.* is coming to its completion we cannot help but recall the 8 years of work on this book together with our geographer colleagues from both China and the U.S., from the original idea, outline, organization, project funding, review and revisions to completion. We highly appreciate all the institutions, teachers and students who have been part of the formulation and realization of the book with their input.

The idea of launching a book project on the topic of A Comparative Geography of China and the U.S. developed over a number of years and the beginnings of this intellectual and organizational venture can be retraced on parallel tracks which eventually came together after several meetings Professors Jing'ai Wang and Rudi Hartmann had in Beijing in 2007.

Jing'ai Wang's idea of a book was sparked in 2005, the year the undergraduate course "China Geography" at Beijing Normal University (BNU) was identified as a model course with national significance. In the process of developing the course appropriate textbooks and other reading materials had to be selected. The idea of compiling a book on the comparative geography of China and the U.S. transpired and was subsequently pursued. In 2007, the Regional Geographical Teaching Team at Beijing Normal University qualified as the National Team. Jing'ai Wang as the leader of a BNU faculty team published a "China Geography Tutorial" textbook which included comparative topics of China and the U.S.

Rudi Hartmann's interest in teaching a Geography of China in a comparative context was steered by two events. A first visit to the People's Republic of China in 1990 (to Shanghai, Hangzhou and Suzhou) at the occasion of an IGU Geographic Education Symposium held at the University of Hong Kong would eventually result

R. Hartmann (✉)
Department of Geography and Environmental Sciences, University of Colorado Denver, Denver, Colorado, USA
e-mail: rudi.hartmann@ucdenver.edu

J. Wang
School of Geography, Beijing Normal University, Haidian District, Beijing China
e-mail: jwang@bnu.edu.cn

R. Hartmann, J. Wang (eds.), *A Comparative Geography of China and the U.S.*, GeoJournal Library 109, DOI 10.1007/978-94-017-8792-5_12,

in a longer stay in China and more exposure to China's geography. The occasion was teaching two geography classes to Chinese students of the International College Beijing, a joint venture of the University of Colorado Denver with China Agricultural University, during Spring Semester 1996. Here, first contact with Chinese geography colleagues in Beijing including Professor Lansheng Zhang from Beijing Normal University materialized. Back at home in Denver, a new Geography of China course was laid out soon after and became a standard part of the teaching programs of the University of Colorado Denver. An American-Chinese education conference held in Beijing in December 2005 was an opportunity to reconnect with BNU, this time with Professor Xiuqi Fang who served as host for a 2007 visit to BNU and as the crucial link and connection to Jing'ai Wang harboring the idea to a book.

Meetings of Jing'ai Wang and Rudi Hartmann in May/June as well as in October 2007 when Professor Peijun Shi joined in as well led to the first full outline of the book. The agenda of the book project was advanced in meetings that followed in 2008 and the years after with mutual visits in China and the U.S. The 'roadmap' to the book saw the following stops, phases and involvements:

- 2008: Jing Chen's participation (on behalf of Jing'ai Wang) at the panel session "Teaching the Geography of China" at the AAG Meeting 2008 in Boston (Wang et al. 2008)
- 2009: Jing'ai Wang's participation in the Las Vegas AAG Meeting, with a paper presentation on the Wenchuan Earthquake 2008 (Shi et al. 2009) and Rudi Hartmann's visit to Beijing, with teaching 'A Geography of North America' segment for the National Geographical Teaching Team at BNU
- 2010: Two special sessions, "A Comparative Geography of China and the U.S.: Physical and Human Geography Perspectives" and "Comparing China and the U.S." Geographic Agendas, Potentials and Prospects, at the Washington, D.C. AAG Meeting, with participation of a core group of the book team from China and the U.S. and the layout of an agenda, general statements and programmatic assessments of the comparative approach (see, for instance, Wang et al. 2010 and Hartmann 2010), with the signing of a book contract
- 2011: Jing'ai Wang and other BNU geographers' participation in the Seattle AAG Meeting and more meetings advancing the book project, followed by Rudi Hartmann's visit to Beijing with participation in BNU's Regional Geography Salon and later meetings at the occasion of the annual Chinese Geographic Education Conference in Xining, Qinghai Province with fieldtrips to Lhasa, Tibet
- 2012: Jing'ai Wang and other BNU geographers' participation in the New York AAG Meeting, with further meetings of the book editing team and a first full print out version of the book, followed by Rudi Hartmann's visit to Beijing during the Thanksgivings' Week in November
- 2013: Jing'ai Wang and other BNU geographers' participation in the Los Angeles AAG Meeting, with a joint paper session "Comparing China and the U.S.: Geographic Perspectives in an Era of Globalization and Emerging Asia", two book editing meetings and discussion of a third version of the book, followed by

Rudi Hartmann's visit to Beijing in September and the submission of the complete manuscript in October 2013.

All the meetings in China and the United States included field trips and local workshops facilitating a better mutual understanding of the diverse geographies of both countries. Thousands of kilometers/miles were traveled by the editors, contributors and associated staff members of the book as well as a multitude of e-mail messages and progress reports exchanged. And yet, the promise of the comparative approach of a regional geography of China and the U.S. is not yet fulfilled, and items on the agenda for work on this front remain.

During the 8 years of work on the book the socio-economic, cultural and ecological environment in China, in the U.S. and worldwide has undergone slight or significant changes. While this outlook cannot give a complete list of observations of such trends which may have had an impact on the living conditions in China and the United States, a few items should be presented as future tasks of a comparative geography approach.

Mobility, travel and tourism in China and the U.S. have seen tremendous growth and change. After decades of slow but continuous increase in recreational and business travel China has recently become a leading nation in international tourism (outbound travel) as well as in the provision of services, accommodation and transportation for the domestic tourism market at home. Besides the availability of more international, national and regional flights it is the new high speed train infrastructure that has changed the mobility of Chinese citizens and the regional economies of the country because of a greater interconnectedness across the nation. While the U.S. has made only more adaptive changes in the improvement of local and regional mass transit systems, the agenda in the implementation of new transportation technologies has also reached the policy makers in Washington, D.C. and in a number of states including California.

The potential of a changing cultural and social geography in China and the U.S. has not been fully explored yet. One of the more remarkable changes in the uses of languages in both countries are a wider acceptance of English in communication in China, in particular among the younger generation, and the increasing adoption of Mandarin in the U.S. high school system. An increasing exchange of students (by number and time spent) both ways, Chinese students attending U.S. colleges and universities as well as American students finally making the decisive step to learn about Chinese civilization abroad, are encouraging signs for the prospect of a better mutual understanding.

Global climate change as well as a continued globalization in the world economy and their impacts on China and the U.S. should be addressed in any future comparative regional studies. While it is not easy to fully integrate such trends in comparative regional analysis, it is a growing concern for many. Regional expressions of climate change and the greater reach of a globalizing economy can be examined certainly for China and the U.S., and comparative notes on short-term or long-term changes in both countries can highlight such global trends.

There is a need for continued efforts in a comparative geography on many fronts. The potential that comparative studies hold in geographic education can hardly be overstated. A path to a better mutual understanding of the regional geographies of the two countries goes a long way, beyond pre-conceived images and stereotype contrasts. Geographic education in form of a comparative regional analysis can make the difference.

References

Hartmann R. (2010). *Comparing the geographies of China and the United States.* Paper presented at the AAG meeting, Washington, D.C., 18. April 2010.

Shi P., Wang J., Chen J., & Hao L. (2009). *Disaster effect assessment and post-disaster response of Wenchuan Earthquake, China.* Paper presented at the AAG meeting, Las Vegas, 4. March 2009.

Wang J., Shi P., Ge Y.-J., Zhang J.-S., & Zhao D. (2010). *Geographical patterns and driving forces: A China-USA comparison.* Paper presented at the AAG meeting, Washington, D.C., 18. April 2010.

Wang J., Su Y., Chen J., & Yue Y. (2008). *Geographic education in China: Developing a model course China geography.* Paper presented at the AAG meeting, Boston, 18. April 2008.

Index

R. Hartmann, J. Wang (eds.), *A Comparative Geography of China and the U.S.*,
GeoJournal Library 109, DOI 10.1007/978-94-017-8792-5,

MIX
Papier aus verantwortungsvollen Quellen
Paper from responsible sources
FSC® C105338

If you have any concerns about our products,
you can contact us on
ProductSafety@springernature.com

In case Publisher is established outside the EU,
the EU authorized representative is:
Springer Nature Customer Service Center GmbH
Europaplatz 3, 69115 Heidelberg, Germany

Printed by Libri Plureos GmbH
in Hamburg, Germany